UNIFORM FRECHET ALGEBRAS

NORTH-HOLLAND MATHEMATICS STUDIES 162

(Continuation of the Notas de Matemática)

NORTH-HOLLAND – AMSTERDAM • NEW YORK • OXFORD • TOKYO

UNIFORM FRECHET ALGEBRAS

Helmut GOLDMANN

Mathematisches Institut
der Universität Bayreuth
Bayreuth, F.R.G.

1990

NORTH-HOLLAND – AMSTERDAM • NEW YORK • OXFORD • TOKYO

ELSEVIER SCIENCE PUBLISHERS B.V.
Sara Burgerhartstraat 25
P.O. Box 211, 1000 AE Amsterdam, The Netherlands

Distributors for the United States and Canada:

ELSEVIER SCIENCE PUBLISHING COMPANY, INC.
655 Avenue of the Americas
New York, N.Y. 10010, U.S.A.

Library of Congress Cataloging-in-Publication Data

Goldmann, Helmut.
Uniform Fréchet algebras / Helmut Goldmann.
p. cm. -- (North-Holland mathematics studies ; 162)
Includes bibliographical references.
ISBN 0-444-88488-2 (U.S.)
1. Uniform algebras. I. Title. II. Title: Frechet algebras.
III. Series.
QA326.G64 1990
512'.55--dc20

90-6870
CIP

ISBN: 0 444 88488 2

Printed in the Netherlands

To my parents, and to all my friends for just being there.

PREFACE

By a uniform (Banach) algebra we mean a uniformly closed algebra of complex-valued continuous functions on a compact Hausdorff space. These algebras have been studied intensively and there exists a vast literature on this subject.

But there are important examples of algebras of continuous functions on noncompact spaces which are not Banach algebras, we mention for example Hol(X), the algebra of all holomorphic functions on a domain X in $\mathbb{C}^n$. Hence it makes sense to generalize the term uniform algebra in an appropriate way. This leads to the definition of uniform Fréchet algebras. The term refers to an algebra of complex-valued continuous functions on a hemicompact space which is complete with respect to the compact open topology. (By a hemicompact space we mean a Hausdorff space X such that there exists a countable compact exhaustion $(K_n)_n$ of X such that each compact subset K of X is contained in some K_n, e.g. let X be a locally compact and σ-compact Hausdorff space.) Examples are the algebra of all continuous functions on a hemicompact k-space and Hol(X), the algebra of all holomorphic functions on a hemicompact reduced complex space X.

An important problem in the theory of uniform (Banach) algebras is the question of the existence of analytic structure in the spectrum of a uniform algebra A, i.e. the question as whether parts of the spectrum can be endowed with the structure of a complex space so that all elements of A - more precisely all Gelfand transforms of elements of A - become analytic functions with respect to this structure. We adopt this question for uniform Fréchet algebras. So the main part of this book is devoted to the problem when a given uniform Fréchet algebra is topologically and algebraically isomorphic to Hol(X), X a suitable complex space. We obtain a function algebraic characterization of certain classes of algebras of holomorphic functions. In particular we give various characterizations of Hol(X) in the case that X is

- an n-dimensional Stein space,
- an n-dimensional Stein manifold,
- a Riemann domain over $\mathbb{C}^n$,
- a domain of holomorphy in $\mathbb{C}^n$,
- a polynomially convex domain in $\mathbb{C}^n$,
- a logarithmically convex complete Reinhardt domain,
- a domain in $\mathbb{C}$, etc..

These results have been mainly obtained by Arens, Brooks, Carpenter and Kramm.

The material is presented in three sections. In the first chapter we collect definitions and results from the theory of Banach algebras which are constantly used in the book. Since this material has been well reported in many books, cf. for example [GAM], [STO], we omit the proofs of the results in most cases but give references.

In the second chapter we consider the algebra Hol(X) in greater detail. On the one hand it serves us as the first example of a uniform Fréchet algebra, on the other hand, since we want to characterize Hol(X) within the class of uniform Fréchet algebras, we collect properties of it which will be discussed later in the general setting of uniform Fréchet algebras. The necessary background in the theory of several complex variables is well presented in textbooks, e.g. in [G/R], [GR/R], [HÖR]. I have included these preliminary chapters to make this book more accessible for a reader who is not so familiar with these theories.

The second part is devoted to the study of uniform Fréchet algebras resp. more generally to the study of Fréchet algebras. One starting point are surely the papers of Arens [ARE 6] and Michael [MIC] in the early fifties. A basic result is the projective limit representation of each Fréchet algebra. It enables us to develope this theory along the lines of the theory of Banach algebras. Many results which are known for Banach algebras remain true for Fréchet algebras with almost the same proofs. In particular we mention the holomorphic functional calculus and consequences of it like Shilov's idempotent theorem etc..

Problems arise in connection with the question as whether each multiplicative linear functional on a Fréchet algebra is automatically continuous. This question, known as Michael's problem, has been intensively studied (cf. the paper of Dixon and Esterle [D/E]), but only partial answers have been obtained. So to carry over some results for Fréchet algebras, like the uniqueness of norm theorem for semisimple Banach algebras, one has to develope completely new proofs.

Differences between both theories appear in connecion with peak points. There exists no analogue for Fréchet algebras to Rossi's maximum modulus principle.

The third part is devoted to questions of analyticity in one form or another. As mentioned above we characterize various classes of algebras of holomorphic functions. Moreover we study properties of holomorphic functions, like the maximum principle, Liouville's theorem, Montel's theorem etc., in the general setting of uniform Fréchet algebras and ask to what extend they are independent of the existence of analytic structure in the spectrum. We mention an example - due to Wermer - of a uniform Fréchet algebra which satisfies the maximum principle and the identity theorem but whose spectrum does not even contain an analytic disc.

There exist several other books which deal with the theory of more general topological algebras, like locally multiplicatively convex algebras or natural systems, see [BNS], [GUI], [MAL], [MIC], [RIC 2], [ZEL]. Some of them contain also results on Fréchet algebras. Most results of part 3 and some results of part 2 are presented for the first time in a book.

In [KRA 8] my academic teacher Bruno Kramm announced his book "Nuclear function algebras and Stein algebras - holomorphic structure in spectra", which should also appear in this series, but which has not been finished due to his tragic death in 1983. The present book is not the book Bruno had in mind. However parts of the book, in particular the part on the elementary theory of uniform Fréchet algebras, owe much to a lecture of Bruno, held at the university of Munich in 1980. I hope this book reflects also some of the spirit of his work on this subject.

Finally, it is a pleasure for me to thank Peter Pflug from Vechta for his constant encouragement during the preparation of this book. Many valuable suggestions and improvements are due to him. I am also indebted to Sandra Hayes from Munich for stimulating discussions. Robert Braun, Paul Fischer, Stefan Müller-Stach and Roland Weinfurtner did much of the proof reading, Elisabeth Guth typed a first version of the manuscript.

CONTENTS

PART 1

BANACH ALGEBRAS, ALGEBRAS OF HOLOMORPHIC FUNCTIONS
AN INTRODUCTION

This part is devoted to classical definitions and properties of Banach algebras and algebras of holomorphic functions. We put together main definitions and results which are of constant use in the book. Most of the results were given without proof but with references.

CHAPTER 1

AN EXCURS ON BANACH ALGEBRAS

Banach algebras will serve us as an important class of examples of Fréchet algebras. Moreover it will turn out that in some sense they are the local building stones of Fréchet algebras. So, to develope the theory of Fréchet algebras we shall frequently have to use results from the theory of Banach algebras.
In this chapter we put together results which will be needed later. Of course this excurs is by no means complete and cannot replace the detailed study of Banach algebras, which is presented in many textbooks (see the list in the bibliography), but intends to make the book more readable for the reader, who is not familiar with this theory.
In the first part of this chapter we give an introduction to the general theory of Banach algebras. In the second part we state deeper results, like the holomorphic functional calculus, Rossi's maximum modulus theorem etc. . In this part we omit the proofs but give always a detailed reference. In most cases we cite the books of Stout [STO] and Gamelin [GAM].

(1.1) General theory

(1.1.1) Throughout this book we only consider associative and commutative algebras over the field of complex numbers. With the exception of chapter 3 we also assume that our algebras have an identity, i.e. there is an element 1 in the algebra A such that $1 \cdot f = f$ for all $f \in A$.

(1.1.2) Definition. i) A Banach algebra (B-algebra) is an algebra A, which is a Banach space and in which multiplication and the norm are linked together by the inequality

$$\|fg\| \le \|f\| \, \|g\|, \; f,g \in A.$$

Moreover $\|1\| = 1$.

ii) A Banach algebra is called uniform Banach algebra (uB-algebra) if

$$\|f^2\| = \|f\|^2 \text{ for all } f \in A.$$

(1.1.3) Examples. i) Denote by $C^k([0,1])$ the algebra of all k-times continuously differentiable (complex-valued) functions on the unit interval with pointwise defined operations. Then $C^k([0,1])$ is a B-algebra with respect to the norm

$$p_{k+1}(f) = 2^k \sup \{|f^{(i)}(t)| : t \in [0,1], \; i = 0,1,\ldots,k\}$$

where $f^{(i)}$ denotes the i-th derivative of f. $C^k([0,1])$ is not a uniform algebra.

ii) Let K be a nonempty compact Hausdorff space. Let C(K) denote the algebra of all (complex-valued) continuous functions on K with pointwise defined operations. Then C(K) becomes a uB-algebra when endowed with the sup-norm

$$\|f\|_K = \sup\{|f(t)| : t \in K\}.$$

iii) Let $K \subset \mathbb{C}^n$ be a compact set. We define

$$P(K) = \{f \in C(K) : f \text{ can be approximated uniformly on K by polynomials}\},$$

and

$$R(K) = \{f \in C(K) : f \text{ can be approximated uniformly on K by rational functions which are analytic on K}\}.$$

P(K) and R(K) are uB-algebras with respect to the sup-norm and we have $P(K) \subset R(K) \subset C(K)$.

For other examples cf. [STO] and [GAM].

(1.1.4) Let B be a B-algebra. We introduce the spectrum $\sigma(b)$ of an element $b \in B$, $\sigma(b)$ is just the spectrum of the operator

$$T_b : B \longrightarrow B,\ x \longrightarrow bx.$$

We shall show that $\sigma(b)$ is a nonempty compact set. As a consequence we get the theorem of Gelfand-Mazur which states that each B-algebra which is a field is isomorphic to $\mathbb{C}$.

Definition. $\sigma(b) = \{\lambda \in \mathbb{C} : \lambda \cdot 1 - b \text{ is not invertible in } B\}$.

(1.1.5) Proposition. The set of all invertible elements of a B-algebra B

$$B^{-1} = \{b \in B : \text{There is } b^{-1} \in B \text{ so that } b \cdot b^{-1} = 1\}$$

is an open subset of B.

Proof. Let $b \in B^{-1}$. Let $a \in B$ such that

$$(*) \qquad \|b-a\| < 1/\|b^{-1}\|.$$

For $n \in \mathbb{N}$ set

$$s_n = b^{-1} \sum_{j=0}^{n} (b^{-1}(b-a))^j,$$

then for $n>m$

$$\|s_n - s_m\| \leq \|b^{-1}\| \sum_{j=m+1}^{n} (\|b^{-1}\|\ \|b-a\|)^j.$$

It follows that $(s_n)_n$ is a Cauchy sequence in B, since $\|b^{-1}\|\|b-a\|<1$ by $(*)$. Hence the series

$$b^{-1} \sum_{j=0}^{\infty} (b^{-1}(b-a))^j$$

converges in B and one can prove in the same way as for the geometric series that it is just the inverse of a, i.e.

$$(**) \qquad b^{-1} \sum_{j=0}^{\infty} (b^{-1}(b-a))^j = a^{-1}.$$

Thus

(***) $\{ a : \|b-a\| < 1/\|b^{-1}\| \} \subset B^{-1}$, for each $b \in B^{-1}$.

□

<u>(1.1.6)</u> Let $G \subset \mathbb{C}$ be a domain, and let $f : G \longrightarrow B$ be a mapping. We say that f is analytic if for each $z_o \in G$ there is a neighbourhood U of z_o in G such that

$$f(z) = \sum_{k=0}^{\infty} b_k(z-z_o)^k, \text{ for all } z \in U,$$

with coefficients $b_k \in B$. The theory of analytic functions with values in Banach spaces can be developed in an analogous way to the usual theory of analytic functions, in particular one has the theorem of Liouville, i.e. if $f : \mathbb{C} \longrightarrow B$ is a bounded analytic function, then f must be constant.

<u>Theorem.</u> $\sigma(b)$ is a nonempty compact set.

<u>Proof.</u> i) $\sigma(b)$ is closed.
Let $\lambda \in \mathbb{C}\backslash\sigma(b)$, i.e. $\lambda \cdot 1-b$ is invertible. Let μ be an arbitrary element of the disc

$$|\lambda-\mu| < 1/\|(\lambda\cdot 1-b)^{-1}\|.$$

Then

$$\|(\lambda\cdot 1-b)-(\mu\cdot 1-b)\| = |\lambda-\mu| < 1/\|(\lambda\cdot 1-b)^{-1}\|,$$

hence $\mu \in \mathbb{C}\backslash\sigma(b)$ by (***) of the proof of proposition (1.1.5).
ii) $\sigma(b)$ is bounded.
Let $|\lambda| > \|b\|$. Then - in the same way as for the geometric series - it can be shown that the series

$$\sum_{n=0}^{\infty} b^n/\lambda^{n+1}\cdot 1$$

converges in B and that

$$\sum_{n=0}^{\infty} b^n/\lambda^{n+1}\cdot 1 = (\lambda\cdot 1-b)^{-1}$$

hence $\lambda \in \mathbb{C}\backslash\sigma(b)$.

iii) $\sigma(b)$ is nonempty.
If $\|b\| = 0$, then b is not invertible, hence $0 \in \sigma(b)$. Now let $\|b\| \neq 0$.
It follows easily from (**) in the proof of proposition (1.1.5) that

$$\mathbb{C}\backslash\sigma(b) \longrightarrow B,\ \lambda \longrightarrow (\lambda\cdot 1-b)^{-1},$$

is an analytic function. For $|\lambda| > k\|b\|$, $k \in \mathbb{N}$, we have

$$\|(\lambda\cdot 1-b)^{-1}\| = \|\sum_{n=0}^{\infty} b^n/\lambda^{n+1}\cdot 1\| \leq (1/k\|b\|)\sum_{n=0}^{\infty}(1/k)^n,$$

hence

(*) $\qquad \|(\lambda\cdot 1-b)^{-1}\| \longrightarrow 0$ for $|\lambda| \to \infty$.

Suppose that $\sigma(b) = \emptyset$. Then by Liouville's theorem and (*) $(\lambda\cdot 1-b)^{-1} = 0$, for all $\lambda \in \mathbb{C}$, and we get a contradiction. □

Remark. It follows from ii) of the proof above that $\sigma(b)$ is contained in the disc $\{\lambda : |\lambda| \leq \|b\|\}$.

(1.1.7) Corollary (Gelfand-Mazur). A Banach algebra B which is a field must be isometrically isomorphic to $\mathbb{C}$.

Proof. Let $b \in B$, $b \neq 0$ be an arbitrary element. By theorem (1.1.6) there exists $\lambda \in \sigma(b)$, this means $\lambda\cdot 1-b$ is not invertible. So $b = \lambda\cdot 1$, since B is a field, and we have proved that each element of B is a scalar multiple of 1. □

(For another proof cf. [REM].)

(1.1.8) Let B be a B-algebra, $b \in B$, then we call

$$r(b) = \sup\{\,|x| : x \in \sigma(b)\,\}$$

the spectral radius of b.

Note that $r(b) \leq \|b\|$ by remark (1.1.6).

Theorem. $r(b) = \lim_{n\to\infty} \|b^n\|^{1/n}$.

Proof. We have

$$(z\cdot 1-b)^{-1} = \sum_{n=0}^{\infty} b^n/z^{n+1}\cdot 1$$

for all z with $|z| > \|b\|$, cf. ii) of the proof of theorem (1.1.6). Moreover

$$\{|z|>r(b)\} \longrightarrow B,\ z \longrightarrow (z\cdot 1-b)^{-1},$$

is an analytic function, cf. (**) of the proof of (1.1.5).
Hence - by the uniqueness of the Laurent series -

$$(z\cdot 1-b)^{-1} = \sum_{n=0}^{\infty} b^n/z^{n+1}\cdot 1$$

on this domain. So for any particular z of this domain there is a constant M_z such that

$$\|b^n/z^{n+1}\cdot 1\| \le M_z \text{ for all } n \in \mathbb{N},$$

resp.

$$\limsup_{n\to\infty} \|b^n\|^{1/n} \le |z|.$$

Since the last inequality holds for all z with $|z| > r(b)$ we have

$$\limsup_{n\to\infty} \|b^n\|^{1/n} \le r(b).$$

Let $z \in \sigma(b)$ be an arbitrary element, then $z^n \in \sigma(b^n)$, since

$$b^n - z^n\cdot 1 = (b-z\cdot 1)(b^{n-1}+z\cdot b^{n-2}+\ldots+z^{n-1}\cdot 1).$$

So $|z^n| \le \|b^n\|$ by remark (1.1.6), hence

$$|z| \le \liminf_{n\to\infty} \|b^n\|^{1/n},$$

and

$$r(b) \le \liminf_{n\to\infty} \|b^n\|^{1/n},$$

since z was chosen arbitrarily. □

Remark. If B is a uB-algebra, then

$$r(b) = \lim_{n \to \infty} \|b^n\|^{1/n} = \lim_{n \to \infty} \|b^{2^n}\|^{1/2^n} = \|b\|, \text{ since } \|b^2\| = \|b\|^2.$$

(1.2) The spectrum of a B-algebra

(1.2.1) Let $\varphi : B \longrightarrow \mathbb{C}$ be a nontrivial algebra homomorphism, i.e. φ is a linear mapping and $\varphi(fg) = \varphi(f)\varphi(g)$ for all $f,g \in B$. Then $\varphi(1) = 1$, hence $1 = \varphi(b)\varphi(b^{-1})$ for all $b \in B^{-1}$, i.e. $\varphi(b) \neq 0$ for all invertible elements $b \in B$. Since $\varphi(\varphi(b)\cdot 1 - b) = 0$, we see that $\varphi(b) \in \sigma(b)$ for all $b \in B$, hence $|\varphi(b)| \le \|b\|$ for all $b \in B$ by remark (1.1.6) and we have proved that φ is continuous. We state this result explicitely.

Theorem. Let B be a B-algebra. Then every algebra homomorphism $\varphi : B \longrightarrow \mathbb{C}$ is automatically continuous.

(1.2.2) Definition. The set $M(B) = \{\varphi : B \longrightarrow \mathbb{C},\ \varphi$ is a nontrivial algebra homomorphism$\}$ is called the spectrum of B.

By the consideration above M(B) is contained in the closed unit disc of the dual space of B. Hence it makes sense to endow M(B) with the (relative) weak* topology. This topology is also referred to as the Gelfand topology. We shall show that M(B) is a compact space and that each maximal ideal in B is of the form

$$\ker \varphi = \{ b \in B : \varphi(b) = 0 \},\ \varphi \in M(B).$$

For each $b \in B$ define the mapping

$$\hat{b} : M(B) \longrightarrow \mathbb{C},\ \varphi \longrightarrow \varphi(b),$$

$\hat{b}$ is called the Gelfand transform of b and we set $\hat{B} = \{\hat{b} : b \in B\}$. Recall that the Gelfand topology is the coarsest topology on M(B) such that all elements of $\hat{B}$ are continuous, i.e. sets of the form

(*) $\quad \{\psi \in M(B) : |\hat{b}_i(\psi) - \hat{b}_i(\varphi)| < 1,\ i = 1,\dots,n\},\ n \in \mathbb{N},\ b_1,\dots,b_n \in B,$

give a basis of open neighbourhoods of an element $\varphi \in M(B)$. Note that $M(B)$ is a Hausdorff space.

(1.2.3) The next proposition follows from the consideration in (1.2.1).

Proposition. The map

$$\Gamma : B \longrightarrow C(M(B)),\ b \longrightarrow \hat{b},$$

is continuous, where $C(M(B))$ is endowed with the sup-norm $\|\cdot\|_{M(B)}$. In particular we have

$$\|\hat{b}\|_{M(B)} \leq \|b\| \text{ for all } b \in B.$$

(1.2.4) Examples. We determine the spectrum of the algebras $C(K)$ and $P(K)$ (cf. (1.1.3)).
First let K be a compact Hausdorff space, and let B be a closed subalgebra of $C(K)$ which contains the constants and which separates the points of K, i.e. for distinct points $x,y \in K$ there exists $f \in B$ so that $f(x) \neq f(y)$. Then B is a uniform Banach algebra. For $x \in K$ denote by φ_x the point evaluation homomorphism at x

$$\varphi_x : B \longrightarrow \mathbb{C},\ f \longrightarrow f(x).$$

Consider the map

$$j : K \longrightarrow M(B),\ x \longrightarrow \varphi_x.$$

Then j is injective since B separates the points of K and it follows from (1.2.2) (*) that j is continuous, hence $j: K \longrightarrow j(K)$ is a homeomorphism, since K is compact ($j(K)$ with the relative Gelfand topology). Note that in particular $M(B) \neq \emptyset$ for these algebras. (We shall show below that $M(B) \neq \emptyset$ for every B-algebra B.)

i) Claim: $M(C(K)) = K$,
more precisely $M(C(K)) = j(K)$, i.e. each $\varphi \in M(C(K))$ is a point evaluation at some point $x \in K$. Hence we can identify $M(C(K))$ and K as a

set and the Gelfand topology and the original topology on K coincide.

Suppose there exists $\varphi \in M(C(K))\setminus j(K)$. Then we find for each $x \in K$ an element $\tilde{f}_x \in C(K)$ such that $\varphi(\tilde{f}_x) \neq \tilde{f}_x(x)$. Set $f_x = \tilde{f}_x - \varphi(\tilde{f}_x)$, then

$$\varphi(f_x) = 0 \text{ and } f_x(x) \neq 0.$$

By continuity there exists an open neighbourhood $U(x)$ of x in K so that $f_x(y) \neq 0$ for all $y \in U(x)$. Since K is compact there are $x_1,\ldots,x_r$ such that

$$K \subset U(x_1) \cup \ldots \cup U(x_r).$$

Then

$$f = \sum_{i=1}^{r}(f_{x_i} \cdot \overline{f_{x_i}}) \in C(K),$$

where $\overline{f_{x_i}}$ denotes the complex conjugate of f_{x_i}, and $f(y) > 0$ for all $y \in K$. Hence $1/f \in C(K)$, but

$$1 = \varphi(1) = \varphi(f/f) = \varphi(1/f)\varphi\Big(\sum_{i=1}^{r} f_{x_i} \cdot \overline{f_{x_i}}\Big) =$$

$$= \varphi(1/f)\Big(\sum_{i=1}^{r} \varphi(f_{x_i})\varphi(\overline{f_{x_i}})\Big) = 0,$$

and we get a contradiction.

ii) If $K \subset \mathbb{C}^n$ is a compact set, then

$$\hat{K} = \{z \in \mathbb{C}^n : |p(z)| \le \|p\|_K \text{ for all polynomials } p\}$$

is called the polynomially convex hull of K. K is called polynomially convex if $K = \hat{K}$. Using Runge's theorem one can show that a compact set $K \subset \mathbb{C}$ is polynomially convex iff $\mathbb{C}\setminus K$ is connected ([STO], p.24). For $n>1$ no such purely topological description of polynomially compact sets holds.

We claim that $M(P(K)) = \hat{K}$ (as a topological space). Let $f \in P(K)$ and let $(p_n)_n$ be a sequence of polynomials such that $\|f-p_n\|_K \longrightarrow 0$ for $n \longrightarrow \infty$. Since

$$\|p_n - p_m\|_K = \|p_n - p_m\|_{\hat{K}} \quad \text{for all } n,m \in \mathbb{N},$$

we see that $(p_n|_{\hat{K}})_n$ is a Cauchy sequence in $C(\hat{K})$, hence $(p_n)_n$ converges uniformly on $\hat{K}$ to an extension $\tilde{f}$ of f. It follows that the restriction map $P(\hat{K}) \longrightarrow P(K)$, $f \longrightarrow f|_K$ is an isometrical isomorphism and it suffices to determine $M(P(\hat{K}))$. Let $\varphi \in M(P(\hat{K}))$ be an arbitrary element. Denote by $z_1,\dots,z_n$ the coordinate functions and set $x = (\varphi(z_1),\dots,\varphi(z_n))$. Then $\varphi(p) = p(x)$ for all polynomials p. Suppose that $x \in \mathbb{C}\backslash K$. Then we find a polynomial q such that

$$|\varphi(q)| = |q(x)| > \|q\|_{\hat{K}},$$

a contradiction to (1.2.3). So $x \in K$ and $\varphi(f) = f(x)$ for all $f \in P(\hat{K})$, since the polynomials lie dense in $P(\hat{K})$, and we have proved that $j(\hat{K}) = M(P(\hat{K}))$.

iii) The spectrum $M(C^k([0,1]))$ can be identified as a topological space with the unit interval $[0,1]$, cf. [GAM] p. 6, and

$$M(R(K)) = \{z \in \mathbb{C}^n : p(z) \in p(K) \text{ for all polynomials } p\},$$

in particular $M(R(K)) = K$ if K is a compact subset of the plane, cf. [GAM] p.69.

<u>(1.2.5)</u> Next we consider the maximal ideals of a B-algebra B. Recall that an ideal m in a commutative ring R is called maximal iff $m \neq R$ and m is not contained in another proper ideal of R. As it is well known $m \subset R$ is a maximal ideal iff R/m is a field. Moreover every ideal is contained in a maximal ideal.

<u>Lemma.</u> Let $m \subset B$ be a proper ideal, B a B-algebra.Then $\overline{m}$, the closure of m is again a proper ideal in B. In particular every maximal ideal $m \subset B$ is closed.

<u>Proof.</u> It is easy to show that $\overline{m}$ is an ideal.
To prove that $\overline{m}$ is a proper ideal we have to show that $1 \in B\backslash\overline{m}$. But

this is an immediate consequence of (1.1.5), since $1 \in B^{-1}$.
If m is a maximal ideal then $m \subset \overline{m} \subset B$ and $\overline{m} \neq B$, hence $m = \overline{m}$. □

(1.2.6) Let m be a maximal ideal. By lemma (1.2.5) m is closed, so B/m is a Banach space with respect to the quotient norm

$$\|f+m\|' = \inf \{ \|f+g\| : g \in m \}.$$

We have for arbitrary $h_1, h_2 \in m$, $f, g \in B$,

$$\|f+h_1\| \, \|g+h_2\| \geq \|(f+h_1)(g+h_2)\| = \|fg+h_1g+h_2f+h_1h_2\|.$$

It follows - since $h_1g+h_2f+h_1h_2 \in m$ - that

$$\|(f+m)(g+m)\|' \leq \|f+m\|'\|g+m\|'.$$

Let $g \in B$ be an arbitrary element such that $\|1-g\| < 1$. Then g is invertible by (*) of the proof of (1.1.5). Hence $\|1+h\| \geq 1$ for all $h \in m$, and

$$1 \leq \|1+m\|' = \inf\{\|1+h\| : h \in m\} \leq \|1+0\| = 1.$$

So 1+m is an identity for B/m such that $\|1+m\|' = 1$. It follows that B/m is a B-algebra. Since moreover B/m is a field, we have $B/m = \mathbb{C}$ by the theorem of Gelfand-Mazur.
Denote by $\Omega(B)$ the set of all maximal ideals of B. We can establish a one-to-one correspondence between the elements of M(B) and $\Omega(B)$.

Theorem. Let B be a B-algebra. The map

$$T: M(B) \longrightarrow \Omega(B), \; \varphi \longrightarrow \ker \varphi,$$

is bijective.

Proof. Let $\varphi \in M(B)$, then

$$B/\ker \varphi \longrightarrow \mathbb{C}, \; b + \ker \varphi \longrightarrow \varphi(b),$$

is a ring isomorphism, hence ker φ is a maximal ideal, i.e. T is well-defined. Clearly T is injective.
Now let $m \in \Omega(B)$ be an arbitrary element. The canonical map

$B \longrightarrow B/m$ induces a nontrivial homomorphism $\varphi_m: B \longrightarrow \mathbb{C}$ by the considerations above. We have $\ker \varphi_m = m$. So T is surjective, too. □

(1.2.7) In the next theorem we consider the connection between M(B) and the spectrum $\sigma(b)$ of an element $b \in B$.

Theorem. Let B be a B-algebra, and let $b \in B$ be an element. Then

$$\hat{b}(M(B)) = \sigma(b).$$

Proof. We have $\hat{b}(M(B)) \subset \sigma(b)$ by the considerations in (1.2.1). If $z \in \sigma(b)$, then $z \cdot 1 - b$ is not invertible, i.e. $z \cdot 1 - b$ is contained in a maximal ideal m. By theorem (1.2.6) there is $\varphi \in M(B)$ such that $m = \ker \varphi$, hence $\varphi(z \cdot 1 - b) = 0$, resp. $\hat{b}(\varphi) = z$. □

(1.2.8)Theorem. Let B be a B-algebra. Then M(B) is a nonempty compact space.

Proof. $M(B) \neq \emptyset$ by theorem (1.2.7) and theorem (1.1.6). By (1.2.1) M(B) is contained in E, the closed unit disc of the dual space of B. Since E is weak* compact it suffices to show that M(B) is a weak* closed subset of E.

Let $\rho \in E \setminus M(B)$. Hence there are $f, g \in B$ such that

$$\rho(fg) \neq \rho(f)\rho(g).$$

Then - for sufficiently small $\varepsilon > 0$ - we have $\delta(fg) \neq \delta(f)\delta(g)$ for all δ of the weak* open set

$$U = \{ \delta : |\delta(g) - \rho(g)| < \varepsilon, |\delta(f) - \rho(f)| < \varepsilon, |\delta(fg) - \rho(fg)| < \varepsilon \}.$$

Hence $U \cap M(B) = \emptyset$. □

(1.2.9) Theorem. Let B be a B-algebra, $f_1, \ldots, f_n \in B$. Then $(f_1, \ldots, f_n)$- the ideal which is generated by $f_1, \ldots, f_n$ - is a proper ideal of B iff

$$\{ \varphi \in M(B) : \varphi(f_1) = \ldots = \varphi(f_n) = 0 \} \neq \emptyset.$$

Proof. $(f_1,\dots,f_n)$ is a proper ideal iff there exists a maximal ideal m such that $(f_1,\dots,f_n) \subset m$. By theorem (1.2.6) there exists $\varphi \in M(B)$ such that $m = \ker \varphi$. □

(1.2.10) Notation. Let K be a compact Hausdorff space. By a uB-algebra (B,K) we mean a subalgebra of C(K), which separates the points of K and contains the constants.

We show that each uB-algebra is isomorphic to an algebra of this type.

Theorem. Let B be a uB-algebra. Then

$$\Gamma: B \longrightarrow C(M(B)),\ b \longrightarrow \hat{b},$$

is an isomorphism onto a closed pointseparating subalgebra of C(M(B)), and we have $\|b\| = \|\hat{b}\|_{M(B)}$ for each $b \in B$.

Proof. Clearly $\hat{B}$ separates the points of M(B) and contains the constants. We have

$$\hat{b}(M(B)) = \sigma(b)$$

for each $b \in B$ by theorem (1.2.7), so

$$\|\hat{b}\|_{M(B)} = r(b) = \|b\|$$

for each $b \in B$ by remark (1.1.8). □

(1.2.11) We shall show next that for certain B-algebras any two norms under which the algebra is a B-algebra are equivalent. It is clear that one cannot expect that such a result holds for all B-algebras, since one can make any Banach space into a B-algebra by defining all products to be zero and adjoining an identity.

Definition. Let A be a B-algebra. The intersection of all maximal ideals of A is called the radical of A and is denoted by rad A. A is called semisimple if rad A = {0}.

By theorem (1.2.6)

$$\text{rad } A = \bigcap_{\varphi \in \Omega(A)} \ker \varphi .$$

It follows that A is semisimple iff the Gelfand map

$$\Gamma : A \longrightarrow C(M(A)),\ f \longrightarrow \hat{f},$$

is injective. Note that in particular each uB-algebra is semisimple.

Theorem. Let A and B be B-algebras. If B is semisimple then each homomorphism $\varphi : A \longrightarrow B$ is automatically continuous.

Proof. We show that the graph of φ is closed and obtain our result from the closed graph theorem (B.2).
Let $(a_n, \varphi(a_n))_n$ be a sequence with $a_n \longrightarrow a$ and $\varphi(a_n) \longrightarrow b$ as $n \to \infty$. If $\psi \in M(B)$ then $\psi \circ \varphi \in M(A)$, hence

$$\psi \circ \varphi(a_n) \longrightarrow \psi \circ \varphi(a) \text{ as } n \to \infty$$

by theorem (1.2.1). Since $\psi \circ \varphi(a_n) \longrightarrow \psi(b)$ we have

$$\psi(b) = \psi(\varphi(a)) \text{ for all } \psi \in M(B).$$

Since B is semisimple we get $b = \varphi(a)$, as desired. □

As a corollary we get the uniqueness of the norm topology for semisimple B-algebras. This result is due to Gelfand.

(1.2.12) Corollary. Let A be a semisimple B-algebra with respect to a norm $\|\cdot\|$. If A is a B-algebra with respect to a second norm $\|\cdot\|_1$, then $\|\cdot\|$ and $\|\cdot\|_1$ are equivalent.

Proof. The map id : $(A, \|\cdot\|_1) \longrightarrow (A, \|\cdot\|)$, $a \longrightarrow a$ is bijective. It is continuous by the theorem (1.2.11) and hence a homeomorphism by the open mapping theorem (B.1). □

(1.2.13) We return to polynomial convexity and consider finitely generated B-algebras.

Definition. i) Let B be a B-algebra, and let $f_1,\ldots,f_n \in B$. The set

$$\sigma(f_1,\ldots,f_n) = \{(\varphi(f_1),\ldots,\varphi(f_n)) \in \mathbb{C}^n : \varphi \in M(A)\}$$

is called the joint spectrum of $f_1,\ldots,f_n$.
ii) B is called n-generated by $f_1,\ldots,f_n$ if the polynomials in $f_1,\ldots,f_n$ lie dense in B.

Theorem. Let B be an n-generated B-algebra with generating elements $f_1,\ldots,f_n$. Then $\sigma(f_1,\ldots,f_n) \subset \mathbb{C}^n$ is a polynomially convex compact set and the map

$$\hat{F} : M(A) \longrightarrow \sigma(f_1,\ldots,f_n),\ \varphi \longrightarrow (\hat{f}_1(\varphi),\ldots,\hat{f}_n(\varphi)),$$

is a homeomorphism.
B is topologically and algebraically isomorphic to $P(\sigma(f_1,\ldots,f_n))$, if B is moreover a uB-algebra.

Proof. Let $\hat{F}(\varphi) = \hat{F}(\psi)$, then $\varphi(p(f_1,\ldots,f_n)) = \psi(p(f_1,\ldots,f_n))$ for all polynomials p, hence $\varphi=\psi$ since B is n-generated by $f_1,\ldots,f_n$, i.e. $\hat{F}$ is injective.
$\hat{F}$ is clearly continuous. Hence $\sigma(f_1,\ldots,f_n) = \hat{F}(M(B))$ is compact by theorem (1.2.8) and $\hat{F}$ is a homeomorphism.
Let $z \in \hat{\sigma}(f_1,\ldots,f_n)$ (the polynomially convex hull of $\sigma(f_1,\ldots,f_n)$) be an arbitrary point, then

$$|p(z)| \le \|p\|_{\sigma(f_1,\ldots,f_n)} = \|p(f_1,\ldots,f_n)^\wedge\|_{M(B)} \le \|p(f_1,\ldots,f_n)\|$$

for all polynomials p, where $p(f_1,\ldots,f_n)^\wedge$ denotes the Gelfand transform of $p(f_1,\ldots,f_n)$. For the last inequality cf. (1.2.3). Hence

$$\varphi_z(p(f_1,\ldots,f_n)) = p(z)$$

defines a continuous homomorphism on the algebra of all polynomials in $f_1,\dots,f_n$. Since B is n-generated by $f_1,\dots,f_n$, φ_z extends to an element $\tilde{\varphi}_z$ of M(B) and

$$\sigma(f_1,\dots,f_n) = \hat{\sigma}(f_1,\dots,f_n)$$

since $\hat{F}(\tilde{\varphi}_z) = z$.
If B is a uB-algebra, then B is isomorphic to $\hat{B}$ by theorem (1.2.10) and it is easy to check that

$$P(\sigma(f_1,\dots,f_n)) \longrightarrow \hat{B},\ f \longrightarrow f\circ\hat{F}$$

is in fact an isometry.

□

Remark. If $f_1,\dots,f_n$ are arbitrary elements of a B-algebra B, then in general $\sigma(f_1,\dots,f_n)$ is not polynomially convex.
For example set $\partial D = \{x \in \mathbb{C} : |x| = 1\}$, $B = C(\partial D)$ and $z : \partial D \longrightarrow \mathbb{C}$, $x \longrightarrow x$, then $\sigma(z) = \partial D$ is not polynomially convex.

(1.3) The Shilov boundary

(1.3.1) Let K be a compact Hausdorff space, and let B be a pointseparating closed subalgebra of C(K), which contains the constants. By (1.2.4) K can be identified with $j(K) \subset M(B)$ and each $f \in B$ can be extended to an element $\hat{f} \in \hat{B} \subset C(M(B))$ by theorem (1.2.10). So M(B) can be interpreted as the "largest space" to which all elements of B can be extended in such a way that B is isomorphic to the algebra of the extended functions. (This will become clearer in (4.2)).
It is natural to ask whether there exists a "minimal" compact set $L \subset K$ such that B is isomorphic to $B|_L = \{f|_L : f \in B\}$.
This question can be answered by introducing the Shilov boundary.

Definition. Let B be a B-algebra. A closed subset $E \subset M(B)$ is a boundary for B if each $\hat{f} \in \hat{B}$ assumes its maximum modulus on E.

(1.3.2) Remark. i) In particular M(B) is a boundary for B.
ii) Let (B,K) be a uB-algebra. Then

$$\|f\|_K = \sup\{|f(x)| : x \in K\} = \|\hat{f}\|_{M(B)}$$

for all $f \in B$ by (1.2.10). Hence j(K) is a boundary for B, where j denotes the map introduced in (1.2.4).

Example. Denote by $\overline{D}$ the closed unit disc. Then each closed set $E \subset \overline{D}$ which contains the set $\{|x| = 1\}$ is a boundary for $P(\overline{D})$ by the maximum modulus principle.

(1.3.3) Theorem. The intersection of all (closed) boundaries for B is a boundary for B.

(1.3.4) Lemma. Let $f_1,\dots,f_n \in B$ and set

$$V = \{\varphi \in M(B) : |\hat{f}_i(\varphi)| < 1,\ i=1,\dots,n\}.$$

Then either V meets every boundary for B or $E\setminus V$ is a boundary for B for every boundary E for B.

Proof. Suppose there exists a boundary E for B such that $E\setminus V$ is not a boundary for B.
Hence there is $f \in B$ such that $\|\hat{f}\|_{M(B)} = 1 > \|\hat{f}\|_{E\setminus V}$. Replacing f by a suitable power f^m if necessary, we can assume that $\|\hat{f}\cdot\hat{f}_j\|_{E\setminus V} < 1$, $j = 1,\dots,n$.
If $\varphi \in E\cap V$, then

$$|\hat{f}(\varphi)\hat{f}_j(\varphi)| < 1,\ j=1,\dots,n,$$

by the definition of V. Since E is a boundary for B we get

$$(*) \qquad \|\hat{f}\cdot\hat{f}_j\|_{M(B)} = \|\hat{f}\cdot\hat{f}_j\|_E < 1,\ j=1,\dots,n.$$

Let $\varphi \in M(B)$ such that $|\hat{f}(\varphi)| = 1$. Suppose that $\varphi \in M(B)\setminus V$. Then there is $j \in \{1,\dots,n\}$ so that $|\hat{f}_j(\varphi)| \geq 1$, hence $|\hat{f}\cdot\hat{f}_j(\varphi)| \geq 1$, and we

get a contradiction to $\|\hat{f}\cdot\hat{f}_j\|_{M(B)} < 1$.
It follows that $\{\varphi \in M(B) : |\hat{f}(\varphi)| = 1\} \subset V$, so V meets every boundary for B.

□

Proof of theorem (1.3.3). Let $f \in B$ be an arbitrary element and set

$$S = \{\varphi \in M(B) : |\hat{f}(\varphi)| = \|\hat{f}\|_{M(B)}\}.$$

Denote by γB the intersection of all boundaries for B.
Suppose $S \cap \gamma B = \emptyset$. Then for each $\varphi \in S$, there exists a boundary E_φ for B such that $\varphi \in M(B)\setminus E_\varphi$. Since E_φ is closed we find by (1.2.2) (*) an open neighbourhood V_φ of φ of the form described in the lemma such that $V_\varphi \cap E_\varphi = \emptyset$. Since S is compact there are $\varphi_1,\ldots,\varphi_r$ so that

$$S \subset V_{\varphi_1} \cup \ldots \cup V_{\varphi_r}$$

Using the lemma r-times we see that $M(B)\setminus(V_{\varphi_1} \cup \ldots \cup V_{\varphi_r})$ is a boundary for B, a contradiction. Hence $S \cap \gamma B \neq \emptyset$, i.e. $\gamma B \neq \emptyset$ and γB is a boundary for B.

□

(1.3.5) Definition. The set γB is called the Shilov boundary of B.

Examples. i) Denote by $\overline{D}$ the closed unit disc, then $\{ |x| = 1 \} = \gamma P(\overline{D})$. More generally let $K \subset \mathbb{C}$ be a polynomially convex compact set. Then $\gamma P(K)$ equals the topological boundary of K, which we denote by ∂K ([GAM] p.27, [STO] p.277).
ii) If $K \subset \mathbb{C}$ is a compact set, then $\gamma R(K) = \partial K$ ([GAM] p.27, [STO] p.277).

(1.3.6) Proposition. Let A be a B-algebra. Then φ is contained in γA iff for each neighbourhood U of φ in M(A) there exists $f_U \in A$ so that

$$\|\hat{f}_U\|_U = \sup\{|\hat{f}_U(\psi)| : \psi \in U\} > \|\hat{f}_U\|_{M(A)\setminus U}.$$

Proof.i) Let $\varphi \in \gamma A$. Suppose there exists a neighbourhood U of φ such that

$$\|\hat{f}\|_U \leq \|\hat{f}\|_{M(A)\setminus U} \text{ for all } f \in A.$$

Then $M(A)\setminus U$ is a boundary for A, hence $U \cap \gamma A = \emptyset$, a contradiction.
ii) Suppose that for each neighbourhood U of φ there is $f_U \in A$ so that

$$\|\hat{f}_U\|_U > \|\hat{f}_U\|_{M(A)\setminus U}.$$

Then $\gamma A \cap U \neq \emptyset$ for all neighbourhoods U of φ. It follows that $\varphi \in \gamma A$, since γA is closed. □

(1.3.7) Lemma. Let A be a B-algebra, $f \in A$, then $\hat{f}(\gamma A)$ contains the topological boundary of $\hat{f}(M(A))$.

Proof. We denote by $\partial\hat{f}(M(A))$ the topological boundary of $\hat{f}(M(A))$. Suppose there exists $\lambda \in \partial\hat{f}(M(A))\setminus\hat{f}(\gamma A)$. Since $\hat{f}(\gamma A)$ is a compact subset, there is $\varepsilon > 0$ such that

$$\{z: |z-\lambda| < \varepsilon\} \cap \hat{f}(\gamma A) = \emptyset.$$

Choose $\lambda_o \in \mathbb{C}\setminus\hat{f}(M(A))$, so that $|\lambda-\lambda_o| < \varepsilon/4$. Then $f-\lambda_o\cdot 1$ is invertible in A by (1.2.7) and

$$((f-\lambda_o\cdot 1)^{-1})\hat{} = 1/(\hat{f}-\lambda_o),$$

where $((f-\lambda_o\cdot 1)^{-1})\hat{}$ denotes the Gelfand transform of $(f-\lambda_o\cdot 1)^{-1}$. Set $S = \hat{f}^{-1}(\lambda)$. By our assumption $S \cap \gamma A = \emptyset$. Let $\varphi \in \gamma A$ be an arbitrary element, then

$$|1/(\hat{f}(\varphi)-\lambda_o)| \leq 4/\varepsilon$$

but

$$|1/(\hat{f}(\psi)-\lambda_o)| > 4/\varepsilon \text{ for all } \psi \in S,$$

a contradiction to $S \cap \gamma A = \emptyset$. □

(1.3.8) Defintion. Let B be a B-algebra.
i) A point $\varphi \in M(B)$ is called strong boundary point (or p-point) for B if for every neighbourhood V of φ there exists an element $f \in B$ such that $\|\hat{f}\|_{M(B)} = \hat{f}(\varphi) = 1$ and $\|\hat{f}\|_{M(B)\setminus V} < 1$.
ii) φ is called a peak point for B iff there is $f \in B$ so that $\hat{f}(\varphi) = 1$ and $|\hat{f}(\psi)| < 1$ for $\psi \in M(B)\setminus\{\varphi\}$.

Example. Set $\partial D = \{x \in \mathbb{C}: |x| = 1\}$. Each point e^{it}, $0 \leq t < 2\pi$, is a peak point for $P(\overline{D})$ (consider the functions $(1 + z\cdot e^{-it})/2$).

(1.3.9) We first show that in general not every point of the Shilov boundary is a peak point.
Example. Set $K = \overline{D} \times \{0\} \cup \{0\} \times [0,1] \subset \mathbb{C}^2$.
Denote by z_1, z_2 the coordinate functions. We show that K is polynomially convex.
We prove more generally that if $K_i \subset \mathbb{C}^{n_i}$, $i=1,\dots,r$, are polynomially convex compact sets then

$$K = K_1 \times \dots \times K_r \subset \mathbb{C}^{n_1+\dots+n_r}$$

is polynomially convex, too.
Let $x = (x_1,\dots,x_r)$, $x_i \in \mathbb{C}^{n_i}$, be a point in $\mathbb{C}^{n_1+\dots+n_r} \setminus K$.

W.l.o.g. $x_1 \in \mathbb{C}^{n_1} \setminus K_1$. Hence there exists a polynomial p so that $|p(x_1)| > \|p\|_{K_1}$. Set

$$q(y_1,\dots,y_{n_1+\dots+n_r}) = p(y_1,\dots,y_{n_1}),$$

then $|q(x)| > \|q\|_K$.

Consider the algebra $P(K)$. We have $M(P(K)) = K$ by (1.2.4) ii). Each point of $\partial\overline{D} \times \{0\}$ is a peak point for $P(K)$ (cf. example (1.3.8) and replace z by z_1). The same is true for all points $(0,y)$, $0<y\leq 1$ (consider for example $1-(z_2-y)^2$). Hence

$$\partial D \times \{0\} \cup \{0\} \times [0,1] \subset \gamma P(K)$$

since $\gamma P(K)$ is a closed subset of K.

Now let $f \in P(K)$ be an arbitrary element. Then $f_{|\overline{D}\times\{0\}}$ can be interpreted as an element of $P(\overline{D})$ and it follows from the maximum modulus principle that in fact

$$\partial D \times \{0\} \cup \{0\} \times [0,1] = \gamma P(K)$$

and that (0,0) is not a peak point for P(K).

(1.3.10) Before we establish the connection between peak points, strong boundary points and the Shilov boundary, we introduce the notion of representing measures.

Defintion. Let (B,X) be a uB-algebra. A positive regular Borel measure μ on X is said to be a representing measure for $x \in X$ provided

$$f(x) = \int_X f d\mu \text{ for all } f \in B.$$

Note that the point mass at x is a representing measure for x. The next result shows that there exists always a representing measure, which is concentrated on the Shilov boundary.

Theorem. Let B be a B-algebra, and let $x \in M(B)$. Then there exists a positive measure μ on γB such that

$$\hat{f}(x) = \int_{\gamma B} \hat{f}\, d\mu \text{ for all } f \in B.$$

(cf. [STO], p. 46.)

(1.3.11) We now return to peak points and strong boundary points.

Theorem. Let (B,X) be a uB-algebra, and let $y \in X$. Then the following conditions are equivalent:

i) For each neighbourhood V of y there exists $f \in B$ such that

$\|f\|_X = 1$, $|f(y)|>3/4$ and $\|f\|_{X\setminus V} < 1/4$.

ii) The point y is a strong boundary point for B.

iii) The only representing measure for y is the point mass at y.

(In i) and ii) we have identified K and $j(K) \subset M(B)$, cf. (1.2.4).)
([STO] p. 47, [GAM] p. 54, p. 59).

<u>(1.3.12) Theorem.</u> Let B be a uB-algebra.
i) The Shilov boundary for B is the closure of the set of all strong boundary points.
ii) A point $x \in M(B)$ is a peak point iff it is a strong boundary point and a G_δ (i.e. x is the intersection of countable many open subsets of M(B)). In particular if M(B) is a metric space, then x is a peak point iff it is a strong boundary point.
([STO] p. 53, p. 54).

<u>Example.</u> Let $K \subset \mathbb{C}$ be a polynomially convex compact set. Then $M(P(K)) = K$ by (1.2.4)ii). Each point of ∂K, the topological boundary of K, admits a unique representing measure, since P(K) is a Dirichlet algebra. Hence each point of ∂K is a peak point for P(K).
([GAM] p. 37, [STO] p. 285).

<u>(1.3.13)</u>The fact that P(K) is a Dirichlet algebra is also used for the proof of Mergelyan's theorem:

<u>Theorem.</u> Let $K \subset \mathbb{C}$ be a polynomially convex compact set. Let f be a continuous function on K which is holomorphic on the interior of K. Then $f \in P(K)$.
([GAM] p.48, [STO] p.287).

<u>(1.3.14)Definition.</u> Let A be a B-algebra, and let $\varphi \in M(A)$. We say that D is a point derivation on A at φ if D is a linear functional on

A, and if

$$D(fg) = D(f)\varphi(g)+D(g)\varphi(f)$$

whenever $f,g \in A$.
For example if $\overline{D}$ denotes the closed unit disc, then

$$P(\overline{D}) \longrightarrow \mathbb{C},\ f \longrightarrow f'(0),$$

where $f'(0)$ denotes the derivative of f at 0, defines a point derivation on $P(\overline{D})$. Nontrivial point derivations may occur even in the absence of analytic structure. We shall need the following result on the nonexistence of point derivations.

Theorem. Let A be a uB-algebra and let $\varphi \in M(A)$ be a peak point, then there exists no nonzero point derivation on A at φ.
([BRW], p. 78).

Remark. Let $K \subset \mathbb{C}$ be a polynomially convex compact set, and let x be a boundary point of K. It follows by example (1.3.12) and the theorem that there exists no nonzero point derivation for P(K) at x.

(1.4) The holomorphic functional calculus

(1.4.1) Let f be a holomorphic function on the plane

$$f(z) = \sum_{n=0}^{\infty} a_n z^n.$$

Let b be an element of a B-algebra B. Then it is easy to see that the series

$$\sum_{n=0}^{\infty} a_n b^n$$

converges in B to an element a and that

$$\hat{a}(\varphi) = f(\hat{b}(\varphi)) \text{ for all } \varphi \in M(B).$$

This result is a special case (cf. remark (1.4.4) i) below) of a theorem which is called the holomorphic functional calculus.

Before we state the theorem we recall the definition of holomorphic functions in several variables. Let $G \subset \mathbb{C}^n$ be an open set. A function $f: G \longrightarrow \mathbb{C}$ is called holomorphic if for each fixed $(a_1,\ldots,a_n) \in G$ and each $j \in \{1,\ldots,n\}$ the function of one variable determined by the assignment

$$z \longrightarrow f(a_1,\ldots,a_{j-1},z,a_{j+1},\ldots,a_n)$$

is holomorphic in the one variable sense. We denote by Hol(G) the algebra of all holomorphic functions on G. Hol(G) endowed with the compact open topology becomes a Fréchet algebra (in particular a locally convex space).

Next we introduce the algebra H(S), S an arbitrary subset of $\mathbb{C}^n$. Denote by K(S) the collection of all open sets in $\mathbb{C}^n$ which contain S. We define an equivalence relation "$\sim$" on the disjoint union of the algebras Hol(U), $U \in K(S)$, in the following way:
Let $U_1, U_2 \in K(S)$, $F \in \mathrm{Hol}(U_1)$, $G \in \mathrm{Hol}(U_2)$, then $F \sim G$ provided there is $U_3 \in K(S)$ such that $F|_{U_3} = G|_{U_3}$. If $F \in \mathrm{Hol}(U)$, $U \in K(S)$, then we denote by $\tilde{F}$ the equivalence class of F ($\tilde{F}$ is called the germ of F). The collection of all equivalence classes is denoted by H(S). For $U \in K(S)$ set

$$i_U : \mathrm{Hol}\,(U) \longrightarrow H(S),\ F \longrightarrow \tilde{F},$$

Note that $\tilde{F}(z)$ is well-defined for $z \in S$.
H(S) can be naturally endowed with an algebra structure:
Let $\tilde{F},\tilde{G} \in H(S)$, $\lambda,\mu \in \mathbb{C}$ be arbitrary elements. Choose representatives $F,G \in \mathrm{Hol}(U)$ and define $\tilde{F}\cdot\tilde{G} = i_U(F\cdot G)$ resp. $\lambda\tilde{F}+\mu\tilde{G} = i_U(\lambda F+\mu G)$. It can be easily seen that the definition of the algebraic operations is independent of the chosen representatives. We endow H(S) with the strongest locally convex topology so that the maps i_U are continuous for each $U \in K(S)$. (H(S) is the locally convex inductive limit of the system $(\mathrm{Hol}(U),r_V^U)_{U,V \in K(S)}$, where K(S) is directed as fol-

lows: $U \le V$ if $U \subset V$ and r_U^V is just the restriction mapping

$$\mathrm{Hol}(V) \longrightarrow \mathrm{Hol}(U),\ f \longrightarrow f|_U,\ \text{for } U \le V\ .)$$

Let F be another locally convex space. A linear map $\tau : H(S) \longrightarrow F$ is continuous iff $\tau \circ i_U : \mathrm{Hol}(U) \longrightarrow F$ is continuous for each $U \in K(S)$. ([JAR], p. 110).

(1.4.2) We now state the holomorphic functional calculus in a version due to Zame [ZAM].

Theorem. Let B be a B-algebra, and let $b = \{b_1,\dots,b_n\} \subset B$. Let $U \subset \mathbb{C}^n$ be an open neighbourhood of $\sigma(b_1,\dots,b_n)$. Then there is a unique continuous homomorphism $\vartheta_b : \mathrm{Hol}(U) \longrightarrow B$ such that

i) $\vartheta_b(z_i) = b_i$, $1 \le i \le n$, (z_i denote the coordinate functions)

and

ii) $\varphi(\vartheta_b(g)) = g(\hat{b}_1(\varphi),\dots,\hat{b}_n(\varphi))$ for all $g \in \mathrm{Hol}(U)$, $\varphi \in M(B)$.

(1.4.3) Corollary. There exists a unique continuous homomorphism $\vartheta_b : H(\sigma(b_1,\dots,b_n)) \longrightarrow B$ such that

i) $\vartheta_b(\tilde{z}_i) = b_i$, $1 \le i \le n$,

and

ii) $\varphi(\vartheta_b(\tilde{g})) = \tilde{g}(\hat{b}_1(\varphi),\dots,\hat{b}_n(\varphi))$ for all $\tilde{g} \in H(\sigma(b_1,\dots,b_n))$, $\varphi \in M(B)$.

For a weaker version of the holomorphic functional calculus, cf. [GAM] p. 77, [STO] p. 63.

(1.4.4) Remarks. i) Let $b \in B$, $f \in \mathrm{Hol}(\mathbb{C})$

$$f(z) = \sum_{n=0}^{\infty} a_n z^n\ .$$

Then clearly

$$\vartheta_b\Big(\sum_{n=0}^{k} a_n z^n\Big) = \sum_{n=0}^{k} a_n b^n.$$

Hence by continuity

$$\vartheta_b(f) = \sum_{n=0}^{\infty} a_n b^n.$$

ii) We often write f(b) instead of $\vartheta_b(f)$. It follows from i) that

$$\exp(b) = \sum_{n=0}^{\infty} b^n/n!.$$

It is not hard to show that

$$\exp(a+b) = \exp(a)\exp(b)$$

for a,b $\in$ B. We note for later use that the mapping

$$B \longrightarrow B,\ b \longrightarrow \exp(b),$$

is continuous, since

$$\|\exp(a+b) - \exp(a)\| \le \|\exp(a)\| \Big(\sum_{n=1}^{\infty} \|b\|^n/n!\Big)$$

and the last expression tends to zero as $\|b\|$ tends to zero.

(1.4.5) We give some applications of the holomorphic calculus.

Theorem (Oka-Weil). Let $K \subset \mathbb{C}^n$ be a polynomially convex compact set. Then every function holomorphic in a neighbourhood of K can be approximated uniformly on K by polynomials.
If $K \subset \mathbb{C}^n$ is a rationally convex compact set, i.e. M(R(K)) = K, then $F|_K \in R(K)$ for each function F which is holomorphic in an open neighbourhood of K.
([GAM] p. 84, [STO] p. 368).

(1.4.6) Shilov idempotent theorem . Let B be a B-algebra. Suppose that $E \subset M(B)$ is an open and closed set. Then there is a unique $f \in B$ such that $f^2 = f$ and $\hat{f}$ is the characteristic function of E.

([GAM] p. 88, [STO] p. 75).

Remark. Let B be a uB-algebra such that $M(B)$ is a totally disconnected space. Let x, y, $x \neq y$, be two points of $M(B)$. Choose an open and closed neighbourhood U of x such that $y \in M(B)\setminus U$. Then there exists $\hat{f} \in \hat{B}$ such that $\hat{f}|_U = 1$ and $\hat{f}|_{M(B)\setminus U} = 0$, i.e. the real-valued functions in $\hat{B}$ separate the points of $M(B)$ and we get $C(M(B)) = \hat{B}$ by the theorem of Stone-Weierstrass.

(1.4.7) Theorem. Let B be a B-algebra, $f \in B$ and $h \in C((M(B))$ so that $\hat{f} = \exp(h)$. Then there exists $g \in B$ such that $f = \exp(g)$ and $h = \hat{g}$. (Cf. the proof of corollary 8.22, [STO] p. 80).

(1.4.8) Theorem (Arens-Royden). Let B be a B-algebra, and let $h \in C(M(B))^{-1}$. Then there exist $f \in B^{-1}$ and $g \in C(M(B))$ such that $h = \hat{f}\exp(g)$.
([GAM] p. 89).

(1.4.9) The next theorem, due to Rossi, is also based on deep methods of the theory of several complex variables.

Definition. Let B be a B-algebra.
i) A closed subset $E \subset M(B)$ is called local peak set provided there is $\hat{f} \in \hat{B}$ and an open neighbourhood U of E such that $\hat{f}|_E = 1$, while $|\hat{f}| < 1$ on $U \setminus E$.
ii) E is called peak set if U can be taken to be $M(B)$.

Theorem (Local maximum modulus theorem). Local peak sets are peak sets.
([GAM] p. 91, [STO] p. 90).

Corollary. If $U \subset M(B)$ is an open set, then

$$\|\hat{f}\|_{\overline{U}} = \|\hat{f}\|_{(\gamma B \cap U) \cup \partial U} \quad \text{for all } \hat{f} \in \hat{B},$$

where $\overline{U}$ (resp. ∂U) denotes the closure (resp. boundary) of U. In particular

$$\|\hat{f}\|_{\overline{U}} = \|\hat{f}\|_{\partial U}$$

if $\gamma B \cap U = \emptyset$.
([GAM] p. 92, [STO] p. 96).

(1.5) Analytic structure in spectra.

The last section of the first chapter is devoted to some results which deal with the problem of finding analytic structure in the spectrum of a uB-algebra. In particular we introduce the notion of Gleason parts.

(1.5.1) Definition. Let B be a uB-algebra, and let $\varphi, \psi \in M(B)$. We say that φ and ψ lie in the same Gleason part for B ($\varphi \sim \psi$) if

$$\sup\{|\varphi(f)| : f \in B, \|f\| = 1, \psi(f) = 0\} < 1.$$

Remark. The relation "$\sim$" is an equivalence relation.

Theorem. i) Let B be a uB-algebra, let E be a Gleason part for B, and let $f \in B$. If $|\varphi(f)| = \|f\|$ for some $\varphi \in E$, then $\hat{f}|_E$ is constant.
ii) If $K \subset E$ is compact then

$$\hat{K} = \{\varphi \in M(A) : |\varphi(f)| \leq \|\hat{f}\|_K \text{ for all } f \in B\}$$

is contained in E.
([STO] p. 167, p. 168).

(1.5.2) Theorem. Let B be a uB-algebra, and let $\varphi \in M(B)$. If E, the Gleason part for B which contains φ, consists of more than one point and if there is a unique representing measure for φ supported on γB, then there is a one-to-one continuous map h from D, the open unit disc in $\mathbb{C}$, onto E such that $\hat{f} \circ h$ is holomorphic on D for all $\hat{f} \in \hat{B}$. ([GAM] p. 158, [STO] p. 170).

<u>(1.5.3)</u> It was hoped that Gleason parts would provide analytic structure in the spectrum but it turned out that they can be quite pathological.

<u>Theorem.</u> Let B be a uB-algebra, then there exists a uB-algebra A with a Gleason part E homeomorphic to M(B) such that $A|_E$ is isomorphic to B.
([STO] p. 192).

<u>(1.5.4)</u> Next we consider nontrivial Gleason parts of the algebra R(K), K $\subset \mathbb{C}$ a compact set.
Denote by λ the Lebesgue measure on the plane and set

$$D_{x,\varepsilon} = \{ z \in \mathbb{C} : |z-x| < \varepsilon \} \text{ for } x \in \mathbb{C},\ \varepsilon > 0.$$

<u>Theorem.</u> Let $x \in K$ such that the Gleason part π for R(K) which contains x is nontrivial, then

$$\lim_{\varepsilon \to 0} \frac{\lambda(D_{x,\varepsilon} \cap \pi)}{\lambda(D_{x,\varepsilon})} = 1. \text{ ([STO], p.312).}$$

<u>(1.5.5)</u> Finally we state a result, due to Gleason, which provides analytic structure in the spectrum.

<u>Definition.</u> i) Let $G \subset \mathbb{C}^n$ be a domain. A closed subset $X \subset G$ is called analytic subset of G if for each $z \in X$ there is a neighbourhood W of z in G and a family Φ of holomorphic functions on W such that

$$X \cap W = \{y \in W : f(y) = 0 \text{ for all } f \in \Phi\}.$$

ii) A continuous function f on X is said to be holomorphic provided that for all $y \in X$ there is neighbourhood U of z and a holomorphic function F on U such that $F|_{U\cap X} = f|_{U\cap X}$.

We denote the set of all holomorphic functions on X by Hol(X).

<u>Theorem.</u> Let B be a B-algebra, and let $\varphi \in M(B)$. Suppose there are $g_1,\ldots,g_n \in B$ such that every $f \in \ker \varphi$ is of the form $\sum_{i=1}^{n} f_i g_i$, $f_1,\ldots,f_n \in B$. Then there is a neighbourhood W of φ and an analytic subset X of a neighbourhood U of 0 in $\mathbb{C}^n$ such that

i) the map $\hat{G}: W \longrightarrow X$, $\psi \longrightarrow (\hat{g}_1(\psi),\ldots,\hat{g}_n(\psi))$, is a homeomorphism, and

ii) $\hat{f}\circ\hat{G}^{-1} \in \mathrm{Hol}(X)$ for all $\hat{f} \in \hat{B}$.

Moreover $\ker \psi$ is algebraically generated by $g_1-\psi(g_1),\ldots,g_n-\psi(g_n)$ for all $\psi \in W$.

([GAM] p. 154, [STO] p. 144).

CHAPTER 2

THE ALGEBRA OF HOLOMORPHIC FUNCTIONS

Before we start with the general theory of uniform Fréchet algebras we consider one of the most important examples of this class, the algebra of all holomorphic functions on a reduced complex space X, which we denote by Hol(X). It can be easily seen that Hol(X) is not a Banach algebra but a Fréchet algebra. Several new aspects appear, which are not known from the theory of Banach algebras. For example Hol(X), when endowed with the compact open topology, is a Montel space. For a Banach algebra this can only happen if it is a finite dimensional vector space. Another point is the observation that the spectrum of Hol(X) is "never" a compact space.
For Fréchet algebras we shall not only ask for the existence of analytic structure in the spectrum but shall also seek for properties which characterize Hol(X) (for certain complex spaces X) within the class of Fréchet algebras. So it is natural to collect first properties of Hol(X). We shall discuss them later in the general setting of Fréchet algebras. In particular we shall keep busy with the question to what extent they are sufficient for the desired characterizations.

As in the first section on Banach algebras we omit most of the proofs of the stated results, since they can be found in every textbook on several complex variables. We mention - beside others - the books of Hörmander [HÖR], Gunning/Rossi [G/R], Krantz [KRN], Grauert/Remmert [GR/R 2] and Kaup/Kaup [K/K]. A good source for results on the continuation problem for holomorphic functions is the survey article of S. Hayes [HAY 1].

(2.1) General theory

(2.1.1) Definition. Let $G \subset \mathbb{C}^n$ be an open set. A function $f: G \longrightarrow \mathbb{C}$ is called holomorphic if for each fixed $(a_1,\ldots,a_n) \in G$ and each $j \in \{1,\ldots,n\}$ the function of one variable determined by the assignment

$$z \longrightarrow f(a_1,\ldots,a_{j-1}, z, a_{j+1},\ldots,a_n)$$

is holomorphic in the one variable sense. The set of all holomorphic functions on G will be denoted by Hol(G).

Remark. Let $p(z_1,\ldots,z_n)$ be a polynomial then $p|_G \in \mathrm{Hol}(G)$.

(2.1.2) Due to a nontrivial theorem of Hartogs a holomorphic function is necessarily continuous.

We state some results which can be carried over relatively easily from the one variable theory.

(2.1.3) We first introduce some notations.

We set

$$\mathbb{N} = \{1,2,\ldots\},\ \mathbb{N}_o = \{0\} \cup \mathbb{N},$$

$$\mathbb{R}_+^n = \{(r_1,\ldots,r_n) \in \mathbb{R}^n: r_i > 0,\ i=1,\ldots,n\}.$$

Denote by x_i (resp. y_i) the real (resp. imaginary) part of z_i $(i=1,\ldots,n)$. Define partial differential operators on $\mathbb{C}^n$ by

$$\frac{\partial}{\partial z_j} = \frac{1}{2}\left(\frac{\partial}{\partial x_j} - i\frac{\partial}{\partial y_j}\right),\ j = 1,\ldots,n,$$

$$\frac{\partial}{\partial \bar{z}_j} = \frac{1}{2}\left(\frac{\partial}{\partial x_j} + i\frac{\partial}{\partial y_j}\right),\ j = 1,\ldots,n.$$

Note that $f: G \longrightarrow \mathbb{C}$ is holomorphic iff

$$\frac{\partial f}{\partial \bar{z}_j} = 0 \text{ for } j = 1,\ldots,n.$$

For $z = (z_1,\dots,z_n) \in \mathbb{C}^n$ and $r = (r_1,\dots,r_n) \in \mathbb{R}^n_+$, define

$$D_{z,r} = \{(y_1,\dots,y_n) \in \mathbb{C}^n : |y_i - z_i| < r_i,\ i=1,\dots,n\}.$$

For $v = (v_1,\dots,v_n) \in \mathbb{N}^n_o$, $z = (z_1,\dots,z_n) \in \mathbb{C}^n$ and $f \in \mathrm{Hol}(G)$ set

$$v! = v_1!\cdot\ldots\cdot v_n!,\quad z^v = z_1^{v_1}\cdot\ldots\cdot z_n^{v_n},$$

$$D^v f = \frac{\partial^{v_1+\dots+v_n} f}{\partial_{z_1}^{v_1}\dots\partial_{z_n}^{v_n}} .$$

We can now state Cauchy's integral formula:

Let $w = (w_1,\dots,w_n) \in G$, $r = (r_1,\dots,r_n) \in \mathbb{R}^n_+$ such that the closure of $D_{w,r}$ is contained in G, then

$$D^v f(z) = \frac{v!}{(2\pi i)^n} \int_{|\xi_n - w_n| = r_n} \dots \int_{|\xi_1 - w_1| = r_1} \frac{f(\xi_1,\dots,\xi_n)}{(\xi_1 - z_1)^{v_1+1}\dots(\xi_n - z_n)^{v_n+1}}\, d\xi_1\dots d\xi_n$$

for all $v = (v_1,\dots,v_n) \in \mathbb{N}^n_o$, $z = (z_1,\dots,z_n) \in D_{w,r}$ and $f \in \mathrm{Hol}(G)$.

<u>(2.1.4)</u> As a consequence of Cauchy's integral formula we get the following estimate: Let $K \subset G$ be a compact subset, and let $r = (r_1,\dots,r_n) \in \mathbb{R}^n_+$ such that the compact set

$$K_r = \{(z_1,\dots,z_n) \in \mathbb{C}^n : \text{There exists } (y_1,\dots,y_n) \in K \text{ such that } |y_i - z_i| \le r_i \text{ for } i=1,\dots,n\}$$

is contained in G, then

$$\|D^v f\|_K \le v!\|f\|_{K_r} / r^v$$

for all $v \in \mathbb{N}_o^n$, $f \in Hol(G)$, where $\|f\|_{K_r} = \sup\{|f(x)|: x \in K_r\}$.

<u>(2.1.5)</u> As in the case n=1 we have a power series representation of holomorphic functions.

For $x = (x_1,\dots,x_n) \in \mathbb{C}^n$ set $|x| = (|x_1|,\dots,|x_n|)$.

Let $\sum_{v\in\mathbb{N}_o^n} a_v z^v$ be a power series ($a_v \in \mathbb{C}$ for all $v \in \mathbb{N}_o^n$)

which converges at x with respect to a fixed order of summation. Then $\sum a_v z^v$ converges absolutely and uniformly on each compact subset of $D_{o,|x|}$ to a limit that is independent of the order of summation.

If $|x| \in \mathbb{R}_+^n$, then $\sum a_v z^v$ defines an element of $Hol(D_{o,|x|})$. Vice versa if f is holomorphic on a polydisc $D_{x,r}$ ($x \in \mathbb{C}^n$, $r \in \mathbb{R}_+^n$), then the Taylor series of f

$$\sum_{v\in\mathbb{N}_o^n} \frac{1}{v!} D^v f(x)(z-x)^v$$

converges uniformly on each compact subset K of $D_{x,r}$ to f.

<u>(2.1.6) Identity Theorem.</u> Let $G \subset \mathbb{C}^n$ be a domain. If $f \in Hol(G)$ vanishes on a nonempty open subset of G then f vanishes identically.

<u>Remark</u> Note that in the case n>1 f must not vanish on G if f vanishes on a sequence $(a_n)_n$ in G, which converges to a point $a \in G$. Take for example $G = \mathbb{C}^2$, $a_n = (0,1/n)$ and $f = z_1$, the first coordinate function.

<u>(2.1.7)</u> We denote by C(G) the algebra of all complex-valued continuous functions on G. We have $Hol(G) \subset C(G)$ by (2.1.2).

<u>Theorem.</u> Let $f \in C(G)$ and let $(f_m)_m \subset Hol(G)$ be a sequence which converges compactly to f, i.e. $(f_m)_m$ converges uniformly on each

compact set $K \subset G$ to f, then $f \in \mathrm{Hol}(G)$.

(2.1.8) Open-mapping Theorem. Let $G \subset \mathbb{C}^n$ be a domain, and let $f \in \mathrm{Hol}(G)$ be a nonconstant function. Then f(G) is an open subset of $\mathbb{C}$.

(2.1.9) Maximum principle. Let $G \subset \mathbb{C}^n$ be a domain, $f \in \mathrm{Hol}(G)$ and let $K \subset G$ be a compact subset such that there exists a point $a \in \mathrm{int}(K)$ with $|f(a)| = \|f\|_K = \sup\{|f(z)| : z \in K\}$. Then f is constant.

Remark. We conclude from the maximum principle that

$$\|f\|_K = \|f\|_{\partial K}$$

for each compact set $K \subset G$ and each $f \in \mathrm{Hol}(G)$, where ∂K denotes the boundary of K.

(2.1.10) As in the one dimensional case sums, products and complex multiples of holomorphic functions are holomorphic. Hol(G) is thus a commutative algebra whose unit element is the function with constant value 1.

We consider the algebra Hol(G) in greater detail. First we endow Hol(G) with the compact open topology. If $K \subset G$ is a compact set, then

$$\|f+g\|_K \le \|f\|_K + \|g\|_K,$$

$$\|\lambda f\|_K = |\lambda| \, \|f\|_K,$$

$$\|fg\|_K \le \|f\|_K \, \|g\|_K \text{ for all } f,g \in \mathrm{Hol}(G), \quad \lambda \in \mathbb{C},$$

i.e the mapping

$$\|\cdot\|_K : \mathrm{Hol}(G) \longrightarrow \mathbb{R}, \; f \longrightarrow \|f\|_K,$$

defines a submultiplicative seminorm (cf. (3.1.4)).

An open neighbourhoodbasis of $f \in Hol(G)$ (with respect to the compact open topology) is given by all sets of the form

$$\{g \in Hol(G) : \|f-g\|_K < \varepsilon\},$$

K a compact subset of G, $\varepsilon > 0$.

<u>(2.1.11)</u> We can describe the compact open topology in another way. Let $...K_m \subset K_{m+1} \subset ...$ be a countable compact exhaustion of G such that $K_m \subset \operatorname{int} K_{m+1}$ for all $m \in \mathbb{N}$ (where $\operatorname{int} K_{m+1}$ denotes the interior of K_{m+1}). Then the sequence of seminorms $(\|\cdot\|_{K_m})_m$ generates a topology on Hol(G). A neighbourhoodbasis of $f \in Hol(G)$ is given by the sets

$$\{g \in Hol(G) : \|f-g\|_{K_m} < 1/m\}, \ m \in \mathbb{N}.$$

It is easily seen that $(\|\cdot\|_{K_m})_m$ generates just the compact open topology . Moreover

$$d(f,g) = \sum_{m=1}^{\infty} 2^{-m}\|f-g\|_{K_m}/(1+\|f-g\|_{K_m})$$

defines a metric on Hol(G) which generates the compact open topology (cf. [KÖT], p. 205).

The operations

$$\mathbb{C} \times Hol(G) \longrightarrow Hol(G), \ (\lambda,f) \longrightarrow \lambda f,$$
$$Hol(G) \times Hol(G) \longrightarrow Hol(G), \ (f,g) \longrightarrow f+g,$$
$$Hol(G) \times Hol(G) \longrightarrow Hol(G), \ (f,g) \longrightarrow fg,$$

are continuous, so Hol(G) is a metrizable topological algebra (cf. (3.1.5) for the definition).

Let $(f_m)_m$ be a Cauchy sequence in Hol(G), that means that for each $k \in \mathbb{N}$ $(f_m|_{K_k})_m$ is a Cauchy sequence in $C(K_k)$.

Since $(C(K_k),\|\cdot\|_{K_k})$ is a complete vector space and since

$K_k \subset \text{int } K_{k+1}$ for all $k \in \mathbb{N}$ it follows that $(f_m)_m$ converges compactly to a continuous function f, hence $f \in \text{Hol}(G)$ by (2.1.7). So Hol(G) is a complete metrizable topological algebra. Such algebras are called Fréchet algebras (cf. (3.1.5)).

(2.1.12) Since the coordinate functions are holomorphic, we find for each pair of points $a,b \in G$, $a \neq b$, an element $f \in \text{Hol}(G)$ with $f(a) \neq f(b)$. We say that Hol(G) separates the points of G.
Pointseparating subalgebras of C(X), X a σ-compact and locally compact space, which are complete with respect to the compact open topology will be called uniform Fréchet algebras, cf. (4.1.2). Hence Hol(G) is a uniform Fréchet algebra.

(2.1.13) Using the maximum principle we show that Hol(G) is not a Banach algebra. For simplicity assume that G is a domain. Recall that a subset B of a topological linear space is called bounded if for each neighbourhood U of zero there exists $\rho > 0$ such that

$$B \subset \rho U = \{\rho f : f \in U\}.$$

Suppose that Hol(G) is a Banach algebra, i.e. there is a norm $\|\cdot\|$ which generates the compact open topology. Set

$$B = \{f \in \text{Hol}(G) : \|f\| < 1\}.$$

Then B is a bounded subset of Hol(G).
Since B is a neighbourhood of zero, there exists a compact set $K \subset G$ and $\varepsilon > 0$ so that

$$(*) \qquad U = \{f : \|f\|_K < \varepsilon\} \subset B$$

according to the definition of the compact open topology.
Choose a compact set $L \subset G$ with $K \subset \text{int } L$ and let $f \in \text{Hol}(G)$ be a nonconstant function, w.l.o.g. $\|f\|_K = 1$. Then $\|f\|_L > 1$ by the maximum principle. For $n \in \mathbb{N}$ set

$$g_n = f^n \cdot \varepsilon/2.$$

Then $g_n \in B$ for all $n \in \mathbb{N}$ by (*). Now consider

$$W = \{f : \|f\|_L < 1\}$$

W is a neighbourhood of zero but there exists no $\rho>0$ such that $B \subset \rho W$, since $\|g_n\|_L$ tends to ∞ for $n \longrightarrow \infty$. Hence B cannot be bounded.

(2.1.14) Montel's Theorem. Let $F \subset \mathrm{Hol}(G)$ be a bounded set, i.e. for each compact set $K \subset G$ there exists $M_K > 0$ such that $\sup \{\|f\|_K : f \in F\} \le M_K$. Then F is a relatively compact subset of Hol(G).

Remark. A Fréchet space is called Montel space if each bounded subset is relatively compact.
So Hol(G) is a Montel space. Since the unit ball in a normed space E is relatively compact iff E is finite dimensional, we conclude that there are no infinite dimensional normed Montel spaces. We could have used this observation to show that Hol(G) is not a Banach algebra.

(2.1.15) We use Montel's Theorem to show that Hol(G) is a Schwartz algebra. We start with the definition of a Schwartz space.

Let E be a Fréchet space and let $(p_n)_n$ be a sequence of seminorms which defines the topology of E. W.l.o.g. we can assume that $p_n(a) \le p_{n+1}(a)$ for all $a \in E$ and all $n \in \mathbb{N}$, cf. [KÖT] p. 205. Set

$$N_k = \{a \in E : p_k(a) = 0\}$$

and let E_k be the completion of E/N_k with respect to the norm

$$p_k'(a + N_k) = p_k(a).$$

Each E_k is a Banach space. Since $p_k \le p_{k+1}$ we get for $m > k$ natural linear maps

$$\Phi_{m,k}: E/N_m \longrightarrow E_k,\ a + N_m \longrightarrow a + N_k.$$

Since $\Phi_{m,k}$ is continuous we can extend it to E_m.
We denote the extended map again by $\Phi_{m,k}$. E will be called a Fréchet Schwartz space iff for each $k \in \mathbb{N}$ there is $m(k) > k$ such that $\Phi_{m(k),k}$ is a compact operator, i.e. the image of every bounded subset of $E_{m(k)}$ is relatively compact in E_k.

Now let $G \subset \mathbb{C}^n$ be an open set. Choose a compact exhaustion $\dots \subset K_m \subset K_{m+1} \subset \dots$ of G so that $K_m \subset \operatorname{int} K_{m+1}$ for all $m \in \mathbb{N}$. Then $(\|\cdot\|_{K_m})_m$ defines the topology of Hol(G) by (2.1.11).
Claim: $\Phi_{k+1,k}: \mathrm{Hol}(G)_{k+1} \longrightarrow \mathrm{Hol}(G)_k$ is a compact operator, hence Hol(G) is a Fréchet Schwartz algebra.
We first note that $f|_{\operatorname{int} K_{k+1}} \in \mathrm{Hol}(\operatorname{int} K_{k+1})$ for all $f \in \mathrm{Hol}(G)_{k+1}$ by (2.1.7). Let $(f_m)_m$ be an arbitrary bounded sequence in $\mathrm{Hol}(G)_{k+1}$.
W.l.o.g. we can assume that each f_m is the restriction of an element of Hol(G).
Then a subsequence $(f_{m_l}|_{\operatorname{int} K_{k+1}})_l$ converges compactly to $f \in \mathrm{Hol}(\operatorname{int} K_{k+1})$ by Montel's theorem. Hence $\Phi_{k+1,k}(f_{m_l})_l$ converges in $\mathrm{Hol}(G)_k$ and our claim is proved.

(2.2) Analytic continuation

(2.2.1) We now turn to a phenomenon in the theory of several complex variables, which is unknown in the one dimensional theory. As was first recognized by Hartogs there are domains G in $\mathbb{C}^n$ such that each holomorphic function on G can be extended to a holomorphic function on a larger domain. As for B-algebras we shall introduce the notion of the spectrum of the algebra Hol(G) and discuss the connection with the extension problem.

Consider the domain $X = \mathbb{C}^2\backslash\{0\}$. Let $f \in \mathrm{Hol}(X)$ be an arbitrary function. For $(z_1,z_2) \in D_{0,1} = \{(z,w) : |z|<1, |w|<1\}$ set

$$F(z_1,z_2) = \frac{1}{2\pi i} \int_{|w|=1} \frac{f(z_1,w)}{w-z_2}\, dw .$$

Then $F \in \mathrm{Hol}(D_{0,1})$ since the integration and differentiation with respect to $\bar{z}_j$ (j=1,2) may be carried out in any order. By the Cauchy integral formula for holomorphic functions of one variable $F = f$ on the set $D_{0,1} \backslash\{(z,w) : z=0\}$.
Hence $F = f$ on $D_{0,1}\backslash\{0\}$ by (2.1.6) and we see that

$$\tilde{f}(z) = \begin{cases} F(z) & \text{if} \quad z \in D_{0,1} \\ f(z) & \text{if} \quad z \in \mathbb{C}^2\backslash D_{0,1} \end{cases}$$

defines an element of $\mathrm{Hol}(\mathbb{C}^2)$ such that $\tilde{f}|_{\mathbb{C}^2\backslash\{0\}} = f$.

Denote by r the restriction mapping

$$r : \mathrm{Hol}(\mathbb{C}^2) \longrightarrow \mathrm{Hol}(\mathbb{C}^2\backslash\{0\}),\ f \longrightarrow f|_{\mathbb{C}^2\backslash\{0\}}.$$

Then r is a continuous algebra homomorphism. It is surjective, as noticed above, and it is injective by (2.1.6). Hence r is a topological and algebraical isomorphism by the open mapping theorem for Fréchet spaces (cf. B 1).
We say that a domain $\tilde{G}$ containing G is a holomorphic extension of G if the restriction mapping r is a (topological) isomorphism.
It is natural to ask whether to a given domain G there exists a "largest" holomorphic extension $\tilde{G}$ (the holomorphic envelope of G), resp. to characterize those domains $G \subset \mathbb{C}^n$ which admit no proper

holomorphic extension. Such domains are called domains of holomorphy.

Remark. Each domain $G \subset \mathbb{C}$ is a domain of holomorphy. (Consider the set $S = \{1/(z-x) : x \in \partial G\} \subset \mathrm{Hol}(G)$. Let $\tilde{G} \neq G$ be an arbitrary domain containing G. Then clearly S is not contained in $r(\mathrm{Hol}(\tilde{G}))$.)

(2.2.2) Before we answer the questions posed above we recall the definition of a complex manifold.
Let $U \subset \mathbb{C}^n$ be an open set. We say that a mapping $f = (f_1,\dots,f_m): U \longrightarrow \mathbb{C}^m$ is holomorphic if $f_i \in \mathrm{Hol}(U)$ for $i = 1,\dots,m$.

Definition. a) Let X be a Hausdorff space. Then X is called a complex manifold of dimension n if there is a family $(U_i,\varphi_i,V_i)_{i\in I}$ such that

i) $(U_i)_{i\in I}$ is an open covering of X and each V_i is an open subset of $\mathbb{C}^n$,

ii) $\varphi_i : U_i \longrightarrow V_i$ is a homeomorphism for all $i \in I$,

iii) $\varphi_i \circ \varphi_j^{-1}: \varphi_j(U_{ij}) \longrightarrow \varphi_i(U_{ij})$ is a holomorphic map, whenever $U_{ij} = U_i \cap U_j \neq \emptyset$ for $i,j \in I$. The system $(U_i,\varphi_i,V_i)_{i\in I}$ is called an atlas for X.

b) A complex structure on X is a maximal atlas $A = (U_i,\varphi_i,V_i)_{i\in I}$. Elements of A are called coordinate systems.
c) A function $f: X \longrightarrow \mathbb{C}$ is called holomorphic provided that for each $x \in X$ there exists a coordinate system (U_i,φ_i,V_i) such that $x \in U_i$ and $f \circ \varphi_i^{-1} \in \mathrm{Hol}(V_i)$. The set of all holomorphic functions on X will be denoted by $\mathrm{Hol}(X)$.
d) Let X resp. Y be an n-dimensional resp. m-dimensional complex manifold with atlas $(U_i,\varphi_i,V_i)_{i\in I}$ resp $(\tilde{U}_j,\tilde{\varphi}_j,\tilde{V}_j)_{j\in J}$.
A continuous function $f: X \longrightarrow Y$ is said to be holomorphic if

$$\tilde{\varphi}_j \circ f \circ \varphi_i^{-1} : \varphi_i(U_i \cap f^{-1}(\tilde{U}_j)) \longrightarrow \mathbb{C}^m$$

is holomorphic for all $i \in I$, $j \in J$ whenever $f^{-1}(\tilde{U}_j) \cap U_i \neq \emptyset$.
e) A connected complex manifold X is called Riemann domain over $\mathbb{C}^n$ if there exist $p_1,\dots,p_n \in \mathrm{Hol}(X)$ so that the map

$$p : X \longrightarrow \mathbb{C}^n,\ x \longrightarrow (p_1(x),\dots, p_n(x))$$

is a local homeomorphism. In this case p is called realization map.

Remarks. i) Let $U \subset \mathbb{C}^n$, $W \subset \mathbb{C}^m$ be open subsets. Suppose there is a biholomorphic map $f: U \longrightarrow W$, then $n = m$. It follows that the dimension of a complex manifold is well-defined.
ii) If (X,p) is a Riemann domain over $\mathbb{C}^n$ then the set of all triples $(U, p|_U, p(U))$ such that $p|_U$ is a homeomorphism is an atlas for X. Note that in particular each domain of $\mathbb{C}^n$ is a Riemann domain over $\mathbb{C}^n$.
iii) We shall only consider complex manifolds X which have a countable topology and which are holomorphically separable, i.e. for each pair of distinct points $x,y \in X$ there is $f \in \mathrm{Hol}(X)$ so that $f(x) \neq f(y)$. In this case Hol(X) is again a uniform Fréchet algebra, when endowed with the compact open topology and (2.1.6) - (2.1.15) remain valid.

If X is connected and holomorphically separable then X has a countable topology, cf. [K/K] p. 228. In particular each holomorphically separable Riemann domain has a countable topology. Every domain $G \subset \mathbb{C}^n$ is holomorphically separable since distinct points of G can be separated by the coordinate functions. We remark that not every Riemann domain over $\mathbb{C}^n$ is holomorphically separable.
iv) By (2.1.8) all holomorphic functions on a compact complex manifold are necessarily constant, hence Riemann domains over $\mathbb{C}^n$ cannot be compact spaces.

(2.2.3) We return to the questions posed in (2.2.1). As known from the one variable theory there are holomorphic functions on domains $G \subset \mathbb{C}$ which have no largest domain of definition in $\mathbb{C}$. Take for example $\sqrt{z}$ on $\mathbb{C}\setminus\{Re(z)\geq 0,\ Im(z)=0\}$. This leads to the definition of Riemann surfaces. For $n>1$ it may even happen that every holomorphic function on a domain $G \subset \mathbb{C}^n$ can be extended to a holomorphic function on a Riemann domain over $\mathbb{C}^n$, which cannot be realized as domain in $\mathbb{C}^n$. For an example cf. for instance [G/R], p. 43. Hence it makes sense to consider the extension problem in the class of Riemann domains over $\mathbb{C}^n$.

<u>Definition</u> a) A Riemann domain (Y,p) over $\mathbb{C}^n$ is a holomorphic extension of a Riemann domain (X,q) over $\mathbb{C}^n$ if there is a holomorphic map $\alpha: X \longrightarrow Y$ so that

$$\alpha^* : Hol(Y) \longrightarrow Hol(X),\ f \longrightarrow f\circ\alpha,$$

is bijective.

b) A Riemann domain $(E(X),\tilde{q})$ over $\mathbb{C}^n$ is called an envelope of holomorphy of (X,p)

i) if there exists $\alpha: X \longrightarrow E(X)$, so that $(E(X),\tilde{q})$ is a holomorphic extension of X with respect to α;

ii) if $\beta: X \longrightarrow Y$ defines another holomorphic extension of X, there is a uniquely determined holomorphic map $\gamma: Y \longrightarrow E(X)$ so that the diagram

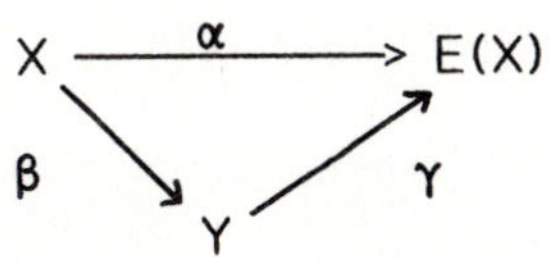

is commutative.

iii) (X,π) is called a domain of holomorphy if $X = E(X)$.

<u>Remarks.</u> i) The envelope of holomorphy $E(X)$ is a holomorphic extension of every other holomorphic extension of X. The uniqueness of γ guarantees that $E(X)$ is uniquely determined up to biholomorphism.

ii) If Y is a holomorphic extension of X via the map α, then α^* is a topological algebra homomorphism by the open mapping theorem for Fréchet spaces.

iii) $\mathbb{C}^2$ is a holomorphic extension of $\mathbb{C}^2\backslash\{0\}$ via the map

$$\alpha : \mathbb{C}^2\backslash\{0\} \longrightarrow \mathbb{C}^2,\ x \longrightarrow x,$$

as was proved in (2.2.1).

iv) We shall consider in (4.2) the extension problem in the general setting of uF-algebras.

(2.2.4) It was shown by Thullen that every Riemann domain (X,π) over $\mathbb{C}^n$ has an envelope of holomorphy E(X).

One is inclined to ask how to construct E(X). This is the point where the spectrum of the algebra Hol(X) comes into account. Let Y be a holomorphic extension of the Riemann domain (X,p) over $\mathbb{C}^n$ via the map α. Then for each $y \in Y$

$$\alpha_y : \mathrm{Hol}(X) \longrightarrow \mathbb{C},\ f \longrightarrow (\alpha*)^{-1}(f)(y),$$

defines a continuous nonzero algebra homomorphism on Hol(X). Hence we get a map from Y into M(Hol(X)), the space of all nonzero continuous complex-valued algebra homomorphisms on Hol(X), by

$$Y \longrightarrow M(\mathrm{Hol}(X)),\ y \longrightarrow \alpha_y,$$

and it is reasonable to conjecture that E(X) = M(Hol(X)) if we can give M(Hol(X)) the structure of a Riemann domain over $\mathbb{C}^n$.

Definition. i) Let X be a holomorphically separable complex manifold with a countable topology. Then we denote by M(Hol(X)) the set of all nonzero complex-valued continuous algebra homomorphisms on Hol(X). We endow M(Hol(X)) with the relative weak* topology.Note that M(Hol(X)) is a subset of the topological dual space of Hol(X).

ii) Let $f \in$ Hol(X), then

$$\hat{f} : M(\mathrm{Hol}(X)) \longrightarrow \mathbb{C},\ \varphi \longrightarrow \varphi(f),$$

is called the Gelfand transform of f. The algebra of all Gelfand transforms of Hol(X) will be denoted by $\text{Hol}(X)^{\wedge}$.

Remarks. i) The weak* topology on M(Hol(X)) is the weakest topology such all elements of $\text{Hol}(X)^{\wedge}$ become continuous functions.
ii) The map

$$j : X \longrightarrow M(\text{Hol}(X)),\ x \longrightarrow \varphi_x,$$

with $\varphi_x(f) = f(x)$ for all $f \in \text{Hol}(X)$ is injective and continuous and

$$j* : \text{Hol}(X)^{\wedge} \longrightarrow \text{Hol}(X),\ \hat{f} \longrightarrow \hat{f} \circ j,$$

is a topological algebra homomorphism, where $\text{Hol}(X)^{\wedge}$ is endowed with the compact open topology. The inverse mapping is given by

$$\text{Hol}(X) \longrightarrow \text{Hol}(X)^{\wedge},\ f \longrightarrow \hat{f},\ \text{cf. (4.1.3).}$$

(2.2.5) Let (X,p) be a Riemann domain over $\mathbb{C}^n$ (with countable topology). Recall that if $p = (p_1,\dots,p_n)$, then $p_i \in \text{Hol}(X)$ for $i=1,\dots,n$. It was shown by Igusa/Remmert/Iwahashi/Forster that E(X) is topologically just the spectrum M(Hol(X)). Rossi introduced an analytic structure into M(Hol(X)) making it to a Riemann domain over $\mathbb{C}^n$. The realization map is given by

$$\hat{p} : M(\text{Hol}(X)) \longrightarrow \mathbb{C}^n,\ \varphi \longrightarrow (\hat{p}_1(\varphi),\dots,\hat{p}_n(\varphi)),$$

where $\hat{p}_i$ denotes the Gelfand transform of p_i $(i=1,\dots,n)$.
With this structure we have M(Hol(X)) = E(X). Moreover the holomorphic maps on E(X) are just the Gelfand transforms of Hol(X).
So we have answered the first question of (2.2.1).

(2.2.6) We turn to the second question: Which properties characterize the Riemann domains (X,π) which agree with their envelope of holomorphy?

Defintion. Let X be a complex manifold.

i) X is called holomorphically convex provided that for each compact subset K the holomorphically convex hull

$$\hat{K}_{Hol(X)} = \{x \in X : |f(x)| \leq \|f\|_K \text{ for all } f \in Hol(X)\}$$

is a compact subset of X.

ii) X is called Stein (manifold) iff X is holomorphically separable and holomorphically convex.

Remarks. i)Let $K \subset \mathbb{C}^n$ be an arbitrary compact subset. Then $\hat{K}_{Hol(\mathbb{C}^n)}$ is a closed subset of $\mathbb{C}^n$ by the definition. Choose r>0 so that

$$K \subset D_{o,r} = \{(x_1,\dots,x_n) \in \mathbb{C} : |x_i|<r,\ i=1,\dots,n\}$$

and let $y = (y_1,\dots,y_n)$ be an arbitrary point of $\mathbb{C} \setminus D_{o,r}$, w.l.o.g. $|y_1| \geq r$. Then $|z_1(y)| > \|z_1\|_K$ where z_1 denotes the first coordinate function, hence $\hat{K}_{Hol(X)} \subset D_{o,r}$, i.e. $\hat{K}_{Hol(X)}$ is a compact subset of $\mathbb{C}^n$ and $\mathbb{C}^n$ is holomorphically convex.

It was proved by Oka that the envelope of holomorphy of a Riemann domain over $\mathbb{C}^n$ is Stein. Vice versa, if (X,π) is Stein then E(X) = X. Even stronger: If (X,π) is holomorphically convex, then (X,π) is holomorphically separable, hence Stein, and E(X) = X.

Remark. It follows from this theorem and our earlier observations that $\mathbb{C}^2$ is the envelope of holomorphy of $\mathbb{C}^2\setminus\{0\}$.

(2.2.7) Theorem. Let (X,π) be a Riemann domain over $\mathbb{C}^n$. Then the following statements are equivalent:

i) X is a domain of holomorphy;

ii) X ist Stein;

iii) X is holomorphically convex;

iv) the natural map $j: X \longrightarrow M(Hol(X))$, cf. remark (2.2.4)ii) is a homeomorphism;

v) for every sequence $(x_m)_m$ without accumulation point there exists $f \in \mathrm{Hol}(X)$ such that $\sup\{|f(x_m)|: m \in \mathbb{N}\} = \infty$;

vi) X is holomorphically separable and for every finite set of functions $f_1,\dots,f_r \in \mathrm{Hol}(X)$ without common zero on X there exist $g_1,\dots,g_r \in \mathrm{Hol}(X)$ such that

$\sum_{i=1}^{r} g_i f_i = 1.$

In this case we say that the weak Nullstellensatz is valid for Hol(X).

Remark. We shall consider later the properties mentioned in iii) - vi) for uniform Fréchet algebras, cf. chapter 4 and (7.3).

Examples. We give two important examples of domains of holomorphy.

i) We say that a domain $G \subset \mathbb{C}^n$ is polynomially convex provided that the polynomially convex hull $\hat{K}$ (cf. (1.2.4)ii)) of every compact subset $K \subset G$ is contained in G. Since each polynomial (restricted to G) is a holomorphic function on G, we see that every polynomially convex domain is in particular holomorphically convex.

ii) A domain $G \subset \mathbb{C}^n$ is called a complete Reinhardt domain if for each $(x_1,\dots,x_n) \in G$ the set

$$\{(y_1,\dots,y_n) \in \mathbb{C}^n: |y_i| \le |x_i|,\ i = 1,\dots,n\}$$

is contained in G. A complete Reinhardt domain is called logarithmically convex if the set

$$\{(\log|x_1|,\dots,\log|x_n|): (x_1,\dots,x_n) \in G,\ x_i \neq 0 \text{ for } i = 1,\dots,n\}$$

is a convex subset of $\mathbb{R}^n$.

Let G be a complete Reinhardt domain. Denote by $\tilde{G}$ the intersection of all logarithmically convex complete Reinhardt domains which contain it. Then $\tilde{G}$ is a logarithmically convex complete Reinhardt domain and the Taylor series

$$\sum_{\nu\in\mathbb{N}_0^n} \frac{1}{\nu!} D^\nu f(0) z^\nu,$$

of each $f \in Hol(G)$ converges uniformly on compact subsets of $\tilde{G}$ to an element $\tilde{f} \in Hol(\tilde{G})$. Since $\tilde{f}|_G = f$ by (2.1.5) and (2.1.6) we see that $\tilde{G}$ is a holomorphic extension of G. In fact a complete Reinhardt domain is a domain of holomorphy iff it is logarithmically convex resp. iff it is polynomially convex.
Finally we state that the domain of convergence of a power series $\sum a_\nu z^\nu$ is a logarithmically convex complete Reinhardt domain.

(2.3) Stein spaces

(2.3.1) Next we consider complex spaces, which in general have no longer the structure of a complex manifold.
(To keep the text elementary we avoid sheaf theory throughout this book.)

Definition. a) Let $G \subset \mathbb{C}^n$ be a domain. A closed subset X of G is called an analytic subset of G if for each $x \in X$ there exists a neighbourhood U of x in G and a family F of holomorphic functions on U so that

$$X\cap U = V(F) = \{ x \in U: f(x) = 0 \text{ for all } f \in F\}.$$

b) A continuous function $f: X \longrightarrow \mathbb{C}$ is called holomorphic if for each $x \in X$ there exists a neighbourhood U of x in G and $\tilde{f} \in Hol(U)$ such that $\tilde{f}_{|X\cap U} = f_{|X\cap U}$.
We denote by Hol(X) the algebra of all holomorphic functions on X.
Let $X \subset G \subset \mathbb{C}^n$ and $X' \subset G' \subset \mathbb{C}^m$ be two analytic subsets and let $f = (f_1,\dots f_m): X \longrightarrow X'$ be a continuous function.
Then f is called holomorphic if $f_i \in Hol(X)$, $i = 1,\dots,m$.
c) Let X be a Hausdorff space. Then X is called a reduced complex space if there exists a family $(U_i,\varphi_i,X_i)_{i\in I}$ so that

i) $(U_i)_{i\in I}$ is an open covering of X and each X_i is an analytic subset of a domain $G_i \subset \mathbb{C}^{n_i}$.

ii) $\varphi_i: U_i \longrightarrow X_i$ is a homeomorphism for all $i \in I$,

iii) $\varphi_i \circ \varphi_j^{-1}: \varphi_j(U_{ij}) \longrightarrow \varphi_i(U_{ij})$

is a holomorphic function whenever $U_{ij} = U_i \cap U_j \neq \emptyset$ for $i,j \in I$.

The algebra of all holomorphic functions on X is denoted by Hol(X).

Remarks. i) Note that if $G \subset \mathbb{C}^n$ is a domain then G itself is an analytic subset of G, hence a complex manifold is in particular a reduced complex space.

ii) Consider the analytic subset $X = \{(z,w) \in \mathbb{C}^2: z\cdot w = 0\}$. Then X looks in most points - precisely in all points of $X\backslash\{(0,0)\}$ - like a complex manifold. This statement remains true for arbitrary reduced complex spaces X.

A point $x \in X$ is called regular if there exists a neighbourhood U of x in X such that U is biholomorphically equivalent to a domain in some $\mathbb{C}^n$, otherwise x is called singular.

The set of all singular points of X is a nowhere dense (analytic) subset of X.

iii)As in the case of manifolds we only consider holomorphically separable (analogous definition as in remark (2.2.2)iii)) reduced complex spaces which have a countable topology. If X is holomorphically separable and connected then X has automatically a countable topology, cf. [K/K] p. 228. Hol(X), when endowed with the compact open topology, is again a uniform Fréchet algebra.

We note that the identity theorem (2.1.6) does not hold for arbitrary analytic subsets. For example let X be as in ii) and consider the holomorphic function $z: X \longrightarrow \mathbb{C}$, $(x,y) \longrightarrow x$. Then z vanishes on an open subset of X but does not vanish identically.

However the identity theorem holds for irreducible analytic subsets. (An analytic subset $X \subset G \subset \mathbb{C}^n$ is called irreducible if whenever X is the union of two analytic subsets X_1, X_2 then either $X = X_1$ or $X = X_2$.)

The analogous statements to (2.1.7), (2.1.8), remark (2.1.9) and (2.1.13) - (2.1.15) remain true. (A proof of the analogous statement of (2.1.15) will be given in (2.3.8) below.)

(2.3.2) We shortly consider the extension problem for reduced complex spaces. If we replace in (2.2.6) the term " complex manifold" by "reduced complex space", then we get the definitions of a holomorphically convex (resp. Stein) reduced complex space.
The results differ essentially from that for Riemann domains over $\mathbb{C}^n$. We first mention that for an arbitrary reduced complex space an envelope of holomorphy does not exist in general (for an example cf. [KRA 5]). Even if the envelope of holomorphy exists it must not be a Stein space. The connection between the envelope of holomorphy E(X) and the spectrum M(Hol(X)) has not been established in the general case.
The theorem of Igusa/Remmert/Iwahashi/Forster however remains true for arbitrary reduced complex spaces X, i.e. X is Stein iff the map

$$j: X \longrightarrow M(\mathrm{Hol}(X)),\ x \longrightarrow \varphi_x ,$$

with $\varphi_x(f) = f(x)$ for all $f \in \mathrm{Hol}(X)$, is a homeomorphism. If X is Stein then E(X) exists and coincides with X.

We cannot discuss the importance of the property Stein in detail but state some results, cf. for example [K/K] § 57.

(2.3.3) Let A be an analytic subset (analogous definition as in (2.3.1)) of a Stein space X, and let $f \in \mathrm{Hol}(A)$ be an arbitrary element. Then there exists $\tilde{f} \in \mathrm{Hol}(X)$ such that $\tilde{f}|_A = f$.

(2.3.4) The following statements are equivalent for a maximal ideal $m \subset Hol(X)$, X a Stein space:

i) m is closed in Hol(X);

ii) m is finitely generated;

iii) there exists $x \in X$ so that $m = \{f \in Hol(X): f(x) = 0\}$.

It was shown by Cartan that each finitely generated ideal $I \subset Hol(X)$ is closed, while the converse statement is false ([FOR 3], p. 314).

Opposite to the situation for B-algebras (cf. (1.2.5)) there are "always" maximal ideals $m \subset Hol(X)$, which are not closed. In connection with this observation there appears the following question: Is every nontrivial algebra homomorphism $\varphi: Hol(X) \longrightarrow \mathbb{C}$ automatically continuous? This question can be answered in the positive if X is a finite dimensional Stein space (for a generalization cf. [EPH]).

We shall return to this question for Fréchet algebras - known as Michael's problem - in chapter 10.

(2.3.5) Let Y be a reduced complex space, X a Stein space, and let $\varphi: Hol(X) \longrightarrow Hol(Y)$ be a continuous algebra homomorphism. Then there exists a unique holomorphic map $f: Y \longrightarrow X$ so that $\varphi(g) = g \circ f$ for all $g \in Hol(X)$.

(2.3.6) Let X be a Stein space, and let $Y \subset X$ be an open subset which is Hol(X)-convex, i.e. $\hat{K}_{Hol(X)}$ is contained in Y for each compact subset $K \subset Y$, then the restriction mapping

$$Hol(X) \longrightarrow Hol(Y),\ f \longrightarrow f|_Y.$$

has dense range. ((X,Y) is called a Runge pair.)

(2.3.7) We state an approximation result, cf. [FOR 2]. Let X be a Stein space and let $A \subset Hol(X)$ be a subalgebra, such that X is

- A-separable, i.e. for each pair of distinct points x, y in X there exists $f \in A$ so that $f(x) \neq f(y)$,

- A-convex, i.e.

$$\hat{K}_A = \{y \in X: |f(y)| \leq \sup\{|f(x)|: x \in K\} \text{ for all } f \in A\}$$

is compact for each compact subset $K \subset X$,

- A-regular, i.e. for each $x \in X$ there are $f_1^x,\dots,f_r^x \in A$, an open neighbourhood U of x, and an analytic subset Y of an open set in $\mathbb{C}^r$ such that

$U \longrightarrow Y$, $t \longrightarrow (f_1^x(t),\dots,f_r^x(t))$,

is a biholomorphic mapping.

Then A is dense in Hol(X).

(2.3.8) Finally - using (2.3.3) and permanence properties of Schwartz spaces - we show that Hol(X) (X a reduced complex space) is a Schwartz space.

We mention that subspaces, quotients and products of Schwartz spaces are again Schwartz spaces (cf. [JAR]).

First let X be an analytic subset of a polycylinder $P \subset \mathbb{C}^n$. One easily proves that X is holomorphically convex and hence Stein. Set

$$I = \{f \in \mathrm{Hol}(P): f_{|X} = 0\}$$

Then I is a closed subspace of Hol(P) and Hol(P)/I becomes a Fréchet space when endowed with the quotient topology. Hence Hol(P)/I is a Fréchet Schwartz space. The map

$$\mathrm{Hol}(P)/I \longrightarrow \mathrm{Hol}(X),\ f+I \longrightarrow f|_X$$

is a continuous and injective linear mapping. It is also surjective by (2.3.3) and hence a topological isomorphism by the open mapping theorem for Fréchet spaces. We conclude that Hol(X) is a Fréchet Schwartz space.

Now let X be an arbitrary analytic subset of a domain $G \subset \mathbb{C}^n$. For each $x \in X$ choose a polycylinder $P_x \subset G$ such that $x \in P_x$. Since $\mathrm{Hol}(X \cap P_x)$ is a Fréchet Schwartz space for every $x \in X$ the same is true for the product $\prod_{x\in X} \mathrm{Hol}(X\cap P_x)$. Since we can represent Hol(X) as a closed subspace of $\prod_{x\in X} \mathrm{Hol}(X\cap P_x)$, via the map

$$\mathrm{Hol}(X) \longrightarrow \prod_{x\in X} \mathrm{Hol}(X\cap P_x),\ f \longrightarrow (f_{|P_x\cap X})_{x\in X},$$

Hol(X) is a Fréchet Schwartz space, too.

Let X be a reduced complex space, and let $(U_i,\varphi_i,X_i)_{i\in I}$ be as in definition (2.3.1)c). Then Hol(X) can be represented as a closed subspace of the Schwartz space $\prod_{i\in I} \mathrm{Hol}(X_i)$ via the map

$$\mathrm{Hol}(X) \longrightarrow \prod_{i\in I} \mathrm{Hol}(X_i),\ f \longrightarrow (f\circ \varphi_i^{-1})_{i\in I}\ .$$

Hence Hol(X) is a Fréchet Schwartz space.

Remark. In a similar way one can prove the stronger statement that Hol(X) is a nuclear space (for the definition cf. [JAR]).

PART 2

GENERAL THEORY OF FRECHET ALGEBRAS

This part is devoted to the introduction of Fréchet algebras and uniform Fréchet algebras. Important examples of (uniform) Fréchet algebras are

- Hol(X), the algebra of all holomorphic functions on a hemicompact reduced complex space;
- C(X), the algebra of all continuous functions on a hemicompact k-space.

It is an important tool in our theory to represent each Fréchet algebra as the projective limit of a sequence of Banach algebras. This representation enables us to prove many theorems for Fréchet algebras which are known for Banach algebras, e.g. the theorem of Gelfand-Mazur, the holomorphic functional calculus, Shilov's idempotent theorem etc..

However the question as whether each multiplicative linear functional on a Fréchet algebra is necessarily continuous is still unanswered. Hence in some cases we have to develope new proofs for theorems, which in the Banach algebra case are a consequence of the automatic continuity of characters. For example we mention the proof of the uniqueness of topology for semisimple Fréchet algebras.

One significant difference appears in connection with the fact that the spectrum of a Fréchet algebra in general is no longer a compact but a hemicompact space. The topology of the spectrum is not quite well understood. For example there are Fréchet algebras whose spectrum is not a k-space. We also mention that Rossi's maximum modulus principle fails to be true for Fréchet algebras.

CHAPTER 3

THEORY OF FRECHET ALGEBRAS, BASIC RESULTS

We always consider associative and commutative algebras over the field of complex numbers. Moreover - with the exception of (3.1) and (3.2) - we shall assume that all considered algebras have an identity 1.

(3.1) Fréchet algebras

In this section we introduce Fréchet algebras. By a Fréchet algebra A we mean an algebra which - considered as a topological vector space - is complete metrizable and which has a neighbourhoodbasis $(U_i)_{i\in\mathbb{N}}$ of zero consisting of convex sets U_i such that $U_i \cdot U_i \subset U_i$ for all $i \in \mathbb{N}$. It turns out that the topology of A can always be defined by a sequence of multiplicative seminorms.
At the end of this section we shall discuss several examples. In particular we shall deal with the question when C(X) - the algebra of all continuous functions on a space X, endowed with the compact open topology - is a Fréchet algebra.

(3.1.1) Recall that by a Banach algebra A we mean a Banach space A which is also an algebra, in which multiplication and the norm are linked by the inequality

$$\|fg\| \le \|f\|\|g\|, \; f,g \in A.$$

Before we generalize this notion in a suitable way for Fréchet algebras, we recall some definitions and results which can be found in any textbook on topological linear spaces, for instance in the books of Köthe [KÖT] or Jarchow [JAR].

Definition. Let E be a complex vector space. A subset $M \subset E$ is said to be convex (resp. absolutely convex) if whenever it contains x and y it contains all $\tau x + \mu y$ where $\tau,\mu \geq 0$ and $\tau+\mu=1$ (resp. $\tau,\mu \in \mathbb{C}$ and $|\tau|+|\mu|\leq 1$). The absolutely convex hull $\Gamma(M)$ of a set $M \subset E$ is the intersection of all the absolutely convex sets which contain M.

Remark $\Gamma(M)$ consists of all the terms of the form $\sum_{i=1}^{n} a_i x_i$, $a_i \in \mathbb{C}$, $\sum_{i=1}^{n} |a_i| \leq 1$, $x_i \in M$ ([JAR] p. 101, p. 102).

(3.1.2) Definition. A complex vector space is called topological vector space if a Hausdorff topology τ is defined on E such that the mappings

$$E \times E \longrightarrow E, \ (x,y) \longrightarrow x+y,$$

and

$$\mathbb{C} \times E \longrightarrow E, \ (\lambda,x) \longrightarrow \lambda x,$$

are continuous.

Remarks. i) If E is a topological vector space, $x \in E$ and $\Phi = \{U_\alpha\}$ is a basis of neighbourhoods of zero, then the sets $x+U_\alpha$, $U_\alpha \in \Phi$, form a basis for the neighbourhood filter of x.

ii) A topological vector space is said to be locally convex if it has a basis of neighbourhoods $\Phi = \{U_\alpha\}$ of zero consisting of convex sets U_α.

iii) A subset M of a vector space E is said to be absorbent, if for each element $x \in E$ a suitable multiple ρx, $\rho>0$, lies in M. Note that every neighbourhood of zero in a topological vector space is absorbent.

(3.1.3) Definition. Let E be a vector space. A map $p: E \longrightarrow [0,\infty)$ is called seminorm if

i) $p(x+y) \le p(x)+p(y)$ for all $x,y \in E$;

ii) $p(\lambda x) = |\lambda| p(x)$ for all $x \in E$, $\lambda \in \mathbb{C}$.

If p is a seminorm, then the set

$$V_p = \{x \in E: p(x) \le 1\}$$

is an absolutely convex and absorbent set. Vice versa if $U \subset E$ is an absolutely convex and absorbent set, then the gauge functional

$$p_U(x) = \inf\{\rho>0: x \in \rho U\}$$

defines a seminorm on E ([KÖT], p. 182).

(3.1.4) Definition. i) Let A be an algebra. A seminorm $p: A \longrightarrow [0,\infty)$ is said to be (sub-)multiplicative if

$$p(xy) \le p(x)p(y) \text{ for all } x,y \in A.$$

ii) A subset $U \subset A$ is called multiplicative if $U^2 = UU \subset U$.

Remark. Let $p_1,\dots,p_n$ be multiplicative seminorms on A. Then

$$A \longrightarrow [0,\infty),\ x \longrightarrow \sup\{p_1(x),\dots,p_n(x)\}$$

defines a seminorm on A which is easily seen to be multiplicative.

The next proposition gives the connection between multiplicative sets and multiplicative seminorms.

Proposition. i) If p is a multiplicative seminorm on A, then

$$V_p = \{x \in A: p(x) \le 1\}$$

is an absorbent, absolutely convex and multiplicative set.

ii) If U is an absolutely convex, absorbent and multiplicative set then

$$p_U(x) = \inf\{\ \rho>0: x \in \rho U\}$$

defines a multiplicative seminorm.

Proof. i) By (3.1.3) we only have to show that V_p is multiplicative. This is the case, since if $x,y \in V_p$, then

$$p(xy) \le p(x)p(y) \le 1.$$

ii) By (3.1.3) it remains to show that p_U is multiplicative. Let $x,y \in A$ and $a,b>0$ such that $x \in aU$ and $y \in bU$. Then

$$xy \in abU^2 \subset abU,$$

hence

$$p_U(xy) \le ab,$$

and the result follows.

□

(3.1.5) Definition. i) A topological algebra is an algebra, which is a topological vector space, such that the multiplication

$$A \times A \longrightarrow A, \ (a,b) \longrightarrow ab$$

is a continuous mapping.

ii) A topological algebra is called a locally multiplicatively convex algebra (LMC-algebra) if there is a basis of neighbourhoods of zero consisting of multiplicative and convex sets.

iii) A Fréchet algebra (F-algebra) is an LMC-algebra which is moreover a complete metrizable topological vector space.

Remark. Since we want to concentrate on F-algebras in this book we do not deal with the more general concept of LMC-algebras. We refer the reader, who is interested in the theory of LMC-algebras, to the books of Beckenstein et al. [BNS], Mallios [MAL] and Zelazko [ZEL 1].

(3.1.6) We now want to show that the topology of an F-algebra can be generated by a sequence of multiplicative seminorms.

i) Let A be an algebra, and let $(p_n)_n$ be a sequence of multiplicative seminorms on A such that

a) $p_n(x) \le p_{n+1}(x)$ for all $n \in \mathbb{N}$, $x \in A$;

b) for each $x \in A$, $x \neq 0$, there is $n \in \mathbb{N}$ such that $p_n(x) \neq 0$.

Then $(p_n)_n$ generates a topology on A in the following way. For $x \in A$, $n \in \mathbb{N}$ set

$$U_n(x) = \{y \in A: p_n(x-y) < 1/n\}.$$

Then $\Phi=(U_n(x))_n$ is a basis of neighbourhoods of x for a (locally convex) topology on A and A is metrizable ([JAR], p. 40). By proposition (3.1.4) each $U_n(0)$ is an absolutely convex and multiplicative set. It is easy to show that A is an LMC-algebra, i.e. the multiplication is continuous, so A becomes an F-algebra if A is complete.

ii) On the other hand every F-algebra arises in this way.

Let A be an F-algebra. Since A is metrizable we can choose a basis of neighbourhoods $(U_n)_n$ of zero consisting of multiplicative and convex sets (cf. [JAR], p. 40). Fix $n \in \mathbb{N}$. Since A is in particular a locally convex space we can choose an absolutely convex neighbourhood of zero such that $W \subset U_n$ ([JAR], p.108). Choose $k \in \mathbb{N}$ so that $U_k \subset W$. Let $x,y \in \Gamma(U_k)$ be two arbitrary elements of the absolutely convex hull of U_k, i.e.

$$x= \sum_{i=1}^{m} \rho_i x_i, \quad \sum_{i=1}^{m} |\rho_i| \le 1, \; x_i \in U_k$$

and

$$y = \sum_{i=1}^{l} \mu_i y_i, \quad \sum_{i=1}^{l} |\mu_i| \le 1, \; y_i \in U_k \text{ (cf. (3.1.1))}.$$

Then

$$xy = \sum_{i=1}^{m} \sum_{j=1}^{l} \rho_i \mu_j x_i y_j .$$

It follows that $\Gamma(U_k)$ is multiplicative, since

$$\sum_{i=1}^{m} \sum_{j=1}^{l} |\rho_i \mu_j| \le 1 \text{ and } x_i y_j \in U_k^2 \subset U_k .$$

Since $\Gamma(U_k) \subset W$, A has a countable basis of neighbourhoods of zero $\ldots \supset W_k \supset W_{k+1} \supset \ldots$, consisting of absolutely convex and multiplicative sets. By proposition (3.1.4) and (3.1.2)iii) every W_k corresponds with a multiplicative seminorm p_{W_k}. We have

$$p_{W_k} \le p_{W_{k+1}} \text{ for all } k \in \mathbb{N}$$

because $W_k \supset W_{k+1}$. If $x \in A$, $x \neq 0$, there is k such that $x \in A \backslash W_k$, hence $p_{W_k}(x) > 0$. Finally one sees that the seminorms $(p_{W_k})_k$ generate the original topology of A.

<u>(3.1.7)</u> Let A be an algebra with identity 1, and let $(p_n)_n$ be a sequence of multiplicative seminorms on A so that $p_n(1) \neq 0$ for all $n \in \mathbb{N}$. Set

$$q_n(f) = \sup\{p_n(fg) : p_n(g) = 1\},$$

then each q_n defines a seminorm on A and $q_n(1) = 1$. Moreover

$$q_n(f) \le p_n(f)$$

and

$$p_n(f) \le p_n(1)q_n(f) \text{ for all } n \in \mathbb{N},\ f \in A,$$

since

$$p_n(f \cdot 1/p_n(1)) \le q_n(f).$$

It follows that $(q_n)_n$ generates the same topology on A as $(p_n)_n$. Each q_n is multiplicative, since for $f,g,h \in A$

$$p_n(fgh) \le p_n(gh)q_n(f),$$

hence

$$q_n(fg) = \sup\{p_n(fgh) : p_n(h) = 1\}$$
$$\le \sup\{q_n(f)p_n(gh) : p_n(h) = 1\} \le q_n(f)q_n(g).$$

By remark (3.1.4)

$$q'_n(f) = \sup\{q_1(f),\dots,q_n(f)\}$$

defines a multiplicative seminorm on A. Clearly

$$q'_n \le q'_{n+1}, \text{ for } n \in \mathbb{N}.$$

It follows from [KÖT], p. 203 that $(q'_n)_n$ defines the same topology on A as $(q_n)_n$. Hence the following definition makes sense.

Definition. Let A be an F-algebra. By a generating (or defining) sequence of seminorms for A we mean a sequence of multiplicative seminorms $(p_n)_n$ on A which generates the topology of A (in the above described way) such that

$$p_n \le p_{n+1}, \text{ for all } n \in \mathbb{N},$$

and

$$p_n(1) = 1, \text{ for all } n \in \mathbb{N}, \text{ if A has an identity.}$$

(3.1.8) There is a canonical way of adjoining an identity to an F-algebra A. We denote by A* the direct sum

$$A^* = \mathbb{C} \oplus A = \{(\lambda,a) : \lambda \in \mathbb{C},\ a \in A\}$$

and define addition and multiplication in A* by

$$(\lambda,a) + (\mu,b) = (\lambda+\mu,a+b)$$

and

$$(\lambda,a)(\mu,b) = (\lambda\mu,\mu a+\lambda b+ab).$$

If $(p_n)_n$ is a generating sequence of multiplicative seminorms for A then we endow A* with the topology which is generated by the multiplicative seminorms

$$q_n(\lambda,a) = p_n(a)+|\lambda|,\ n \in \mathbb{N}.$$

It is easily shown that A* becomes an F-algebra and that (1,0) is an identity. We shall refer to A* as the algebra obtained by adjoining an identity to A.

(3.1.9) Examples. i) Of course each B-algebra is an F-algebra.
ii) Denote by $C^\infty([0,1])$ the algebra of all infinitely differentiable functions on the unit interval with pointwise operations. For $n \in \mathbb{N}$ define

$$p_n(f) = 2^{n-1}\sup\{|f^{(k)}(x)|: x \in [0,1], k = 0,1,\dots,n-1\},$$

where $f^{(k)}$ denotes the k-th derivative of f. It can be verified that each p_n is a seminorm. By using Leibniz's rule for computing the k-th derivative it can be shown that p_n is moreover multiplicative. Clearly $p_1(f) = 0$ iff $f \equiv 0$. We endow $C^\infty([0,1])$ with the topology which is generated by the sequence $(p_n)_n$.

Let $(f_m)_m$ be a Cauchy sequence in $C^\infty([0,1])$, i.e. $(f_m)_m$ is a Cauchy sequence with respect to each seminorm p_n. It follows for n=1 that $(f_m)_m$ converges uniformly on [0,1] to a continuous function g. Since $(f_m)_m$ is a Cauchy sequence with respect to p_2, the sequence of the derivatives $(f_m^{(1)})_m$ converges uniformly on [0,1] to a continuous function g_1 and - as is well known - g is differentiable and $g^{(1)}=g_1$. Continuing in this manner we see that $g \in C^\infty([0,1])$ and $(f_m^{(k)})_m$ converges uniformly on [0,1] to $g^{(k)}$, i.e. the topology generated by the sequence $(p_n)_n$ is complete, i.e. $C^\infty([0,1])$ becomes an F-algebra.

We want to show that $C^\infty([0,1])$ endowed with the above introduced topology is not a B-algebra.

Recall first that a set U in a topological vector space E is called bounded if for each neighbourhood V of zero there is $\rho>0$ so that $U \subset \rho V$. If $(E,\|\cdot\|)$ is a normed space - e.g. a B-algebra - then $\{f \in E: \|f\| < 1\}$ is clearly a bounded neighbourhood of zero. Suppose there exists a bounded neighbourhood U of zero in $C^\infty([0,1])$. Then there is $k \in \mathbb{N}$ such that

$$\{f: p_k(f) < 1/k\} \subset U.$$

For $n \in \mathbb{N}$ set

$$f_n: [0,1] \longrightarrow \mathbb{R},\ x \longrightarrow 1/k \cdot 2^{-k} \cdot 10^{-n(k-1)} \sin(10^n x).$$

Then $f_n \in C^\infty([0,1])$ and

$$p_l(f_n) = 1/k \cdot 2^{l-k-1} \cdot 10^{n(l-k)},\ l \in \mathbb{N},$$

i.e. $f_n \in U$ for all $n \in \mathbb{N}$, but $(p_l(f_n))_{n\in\mathbb{N}}$ is unbounded for every $l>k$. Hence U cannot be a bounded neighbourhood of zero, since for example there exists no $\rho>0$ such that

$$(f_n)_n \subset \rho\cdot\{f: p_{k+1}(f) < 1\}.$$

iii) The algebra C(X).

Let X be a Hausdorff space. We denote by C(X) the algebra of all continuous complex-valued functions on X (with pointwise operations).

If $K \subset X$ is a compact subset we define

$$\|f\|_K = \sup\{|f(x)|: x \in K\}.$$

Obviously $\|\cdot\|_K$ is a multiplicative seminorm on C(X). We always endow C(X) with the topology which is generated by the seminorms $(\|\cdot\|_K)_{K\subset X \text{ compact}}$.

It is called the compact open topology or topology of uniform convergence on compact sets. A neighbourhoodbasis of zero is given by all sets of the form

$$\{f \in C(X): \|f\|_K < \varepsilon\},\ \varepsilon>0,\ K \subset X \text{ compact}.$$

We want to determine a class of spaces such that C(X) becomes an F-algebra. We first recall some definitions and results.

A topological space X is called regular if for each point x and each neighbourhood U of x there is a closed neighbourhood V of x such that $V \subset U$. X is called completely regular if for each $x \in X$ and each neighbourhood U of x there exists a continuous function f on X to the closed unit interval such that f(x) = 0 and f is identically one on X\U. A completely regular space is regular.

By a normal space X we mean a space such that for each disjoint pair of closed sets, A and B, there are disjoint open sets U and V so that $A \subset V$ and $B \subset U$. By Urysohn's lemma we find for each disjoint pair of closed subsets A,B of a normal space X a continuous

function f on X to the unit interval such that f is zero on A and one on B. Hence normal spaces are completely regular.
By a Lindelöf space we mean a topological space so that each open cover has a countable subcover. Each regular Lindelöf space is normal, in particular compact spaces or topological spaces satisfying the second axiom of countability are normal ([KEL], p. 113).

Tietze's Extension Theorem. Let X be a normal space, E be a closed subset, and let f be a continuous function on E to the closed interval $[a,b] \subset \mathbb{R}$. Then f has a continuous extension g to X such that $g(X) \subset [a,b]$.
([KEL] p. 242).

As a consequence of Tietze's theorem we get another extension theorem for completely regular spaces.
Theorem . Suppose K is a compact subset of the completely regular space X. If $f \in C(K)$, then there is $g \in C(X)$ extending f such that

$$\sup\{|g(x)| : x \in X\} = \|f\|_K.$$

(Cf. [JAR], p.29).

A Hausdorff space is called a k-space if every subset intersecting each compact subset in a closed set is itself closed. Note that a complex-valued function f on a k-space X is continuous iff it is continuous on each compact subset of X. Examples of k-spaces are locally compact and first countable spaces ([KEL], p. 231).

Definition. A Hausdorff space X is called hemicompact if there is a countable compact exhaustion $\ldots K_n \subset K_{n+1} \subset \ldots$ of X such that for each compact subset $K \subset X$ there is $n \in \mathbb{N}$ so that $K \subset K_n$. We call such an exhaustion $(K_n)_n$ admissible.

Remark. Each hemicompact space is a Lindelöf space.

Theorem. Let X be a completely regular space. Then $C(X)$ is an F-algebra iff X is a hemicompact k-space.

Proof. i) Suppose X is a hemicompact k-space. Let $(K_n)_n$ be an admissible exhaustion of X. The compact open topology is generated by the sequence of seminorms $(\|\cdot\|_{K_n})_n$, i.e. $C(X)$ is metrizable ([KÖT], p. 205). Let $(f_n)_n$ be a Cauchy sequence in $C(X)$, i.e. $(f_n)_n$ is a Cauchy sequence with respect to each seminorm $\|\cdot\|_{K_m}$. Since $C(K_m)$ is complete with respect to the norm $\|\cdot\|_{K_m}$, $(f_n|_{K_m})_n$ converges uniformly on K_m to a continuous function $\tilde{f}_m$. Set

$$f: X \longrightarrow \mathbb{C},\ x \longrightarrow \tilde{f}_m(x), \text{ if } x \in K_m.$$

Then f is well-defined and $f|_{K_m} = \tilde{f}_m$, i.e. $f|_{K_m}$ is continuous for each $m \in \mathbb{N}$. Since X is a k-space this implies that $f \in C(X)$.

ii) Now let $C(X)$ be an F-algebra.

Since $C(X)$ is metrizable, there is a countable neighbourhoodbasis of zero, hence there are a sequence of positive numbers $(c_n)_n$ and a sequence of compact subsets $(K_n)_n$ such that

$$(*) \qquad \{f \in C(X): \|f\|_{K_n} < c_n\}$$

is a neighbourhoodbasis of zero. We claim that $(K_n)_n$ is an admissible exhaustion of X, i.e. X is a hemicompact space.

To see this choose an arbitrary compact subset $K \subset X$. By $(*)$ there is $n \in \mathbb{N}$ such that

$$V = \{f \in C(X): \|f\|_{K_n} < c_n\} \subset W = \{f \in C(X): \|f\|_K < 1\}.$$

Suppose there is a point $x \in K \setminus K_n$. Since X is completely regular there is $g \in C(X)$ such that $g(x) = 1$ and $g|_{K_n} \equiv 0$. So $g \in V$ but $g \in C(X)\setminus W$, a contradiction and hence $K \subset K_n$.

It follows that the seminormsystem $(\|\cdot\|_{K_n})_n$ determines the topology of C(X). W.l.o.g. we can assume that $K_n \subset K_{n+1}$ for all $n \in \mathbb{N}$.

(**) Next we claim that a complex-valued function f on X is continuous iff $f|_{K_m} \in C(K_m)$ for each $m \in \mathbb{N}$.

Let f be a function such that $f|_{K_m}$ is continuous for each $m \in \mathbb{N}$. By the extension theorem we find for each $m \in \mathbb{N}$ an extension $f_m \in C(X)$ of $f|_{K_m}$. Since

$$f_m|_{K_n} = f|_{K_n} \text{ for } m \geq n$$

$(f_m)_m$ is a Cauchy sequence with respect to every seminorm $\|\cdot\|_{K_n}$, hence $(f_m)_m$ converges to $g \in C(X)$. Clearly $g = f$ and our claim is proved.

Now let $E \subset X$ be a subset such that $E \cap K_l$ is closed for each $l \in \mathbb{N}$. It remains to show that E is closed.

Let $x \in X \setminus E$ be an arbitrary point, w.l.o.g. $x \in K_1$. Choose a decreasing sequence of positive numbers $(a_n)_n$ so that $1 > a_i$, $a_i > a_{i+1}$ and $a_i > 1/2$ for all $i \in \mathbb{N}$.

Since K_1 is a normal space we find $f_1 \in C(K_1)$ such that

$$1 \geq f_1 \geq 0,\ f_1(x) = 0 \text{ and } f_{1|E \cap K_1} \equiv 1.$$

Suppose that we have already constructed $f_i \in C(K_i)$ $(i=1,\ldots,n-1)$ such that

α) $f_i \geq 0$;

β) $f_i|_{E \cap K_i} \geq a_i$;

γ) $f_i|_{K_{i-1}} = f_{i-1}$ for $i=2,\ldots,n-1$.

According to Tietze's theorem we can choose an extension $\tilde{f}_n$ of f_{n-1} to K_n such that $\tilde{f}_n \geq 0$. If $\tilde{f}_n(x) \geq a_n$ for all $x \in E \cap K_n$ then set $f_n = \tilde{f}_n$. Otherwise K_{n-1} and $E \cap K_n \cap \{x \in K_n : \tilde{f}_n(x) \leq a_n\}$ are disjoint closed subsets of K_n (recall that $a_{n-1} > a_n$). Hence we find $g \in C(K_n)$, $1 \geq g \geq 0$, such that

$$g|_{K_{n-1}} \equiv 0 \text{ and } g|_{E \cap K_n \cap \{\tilde{f}_n \leq a_n\}} \equiv 1.$$

Set $f_n = \tilde{f}_n + g$. Then

$$f_n \geq 0,\ f_n|_{K_{n-1}} = f_{n-1} \text{ and } f_n|_{E \cap K_n} \geq a_n.$$

Set

$$f: X \longrightarrow \mathbb{C},\ x \longrightarrow f_n(x) \text{ if } x \in K_n.$$

Then f is well-defined and $f \in C(X)$ by (**), since $f|_{K_n} = f_n$ for each $n \in \mathbb{N}$.

Moreover $f(x) = 0$ and $f(y) \geq 1/2$ for $y \in E$. Hence $\{y \in X: f(y) < 1/4\}$ is an open neighbourhood of x which does not intersect E, i.e. E is closed since x was chosen arbitrarily.

□

<u>(3.1.10) Remarks .</u> i) Note that the hypothesis "completely regular space" was only needed to prove that X is hemicompact if C(X) is an F-algebra. In particular we have shown that C(X) is an F-algebra if X is a hemicompact k-space.

ii) Let X be a hemicompact k-space, let $x \in X$ be an arbitrary point, and let U be an open neighbourhood of x. Repeating the last part of the preceeding proof we find $f \in C(X)$ so that $f \geq 0$, $f(x)=0$ and $f(y) \geq 1/2$ for all $y \in X \setminus U$. Hence $\{y \in X: f(y) \leq 1/4\}$ is a closed neighbourhood of x which is contained in U, i.e. X is regular, hence X is a normal space since it is also a Lindelöf space ([KEL] p. 113).

iii) For a detailed study of the algebra C(X) - X a regular space - we refer the reader to Beckenstein et al. [BNS].

(3.2) The spectrum of a Fréchet algebra

As for B-algebras we define the spectrum M(A) of an F-algebra A to be the set of all nonzero complex-valued and continuous algebra homomorphisms endowed with the Gelfand topology. Recall that the spectrum of a B-algebra (with identity) is a nonempty compact space.

The main result of this section shows that in general the spectrum of an F-algebra is not a compact but a nonempty hemicompact space. As a consequence we get Gelfand-Mazur's theorem, i.e. we prove that an F-algebra which is a field is isomorphic to $\mathbb{C}$. This enables us to exhibit a one-to-one correspondence between closed maximal ideals and the elements of M(A). Finally we determine the spectrum of A*, the algebra obtained by adjoining an identity.

(3.2.1) Definition. Let A be an F-algebra.

i) We denote by S(A) the set of all nonzero complex-valued algebra homomorphisms and by M(A) the set of all continuous members of S(A). We call M(A) the spectrum of A.

ii) For $f \in A$, we define the function

$$\hat{f}: S(A) \longrightarrow \mathbb{C},\ \varphi \longrightarrow \varphi(f),$$

$\hat{f}$ is called the Gelfand transform of f. We set

$$\hat{A} = \{\hat{f}|_{M(A)} : f \in A\}.$$

iii) We always endow S(A) resp. M(A) with the coarsest topology such that all Gelfand transforms are continuous functions on S(A) resp. M(A). This topology is called the Gelfand topology.

(3.2.2) Remarks. i) If $\varphi \in S(A)$ (resp. $\varphi \in M(A)$), then sets of the form

$$\{\psi \in S(A): |\psi(f_i)-\varphi(f_i)|<1,\ i=1,\dots,r\},\ f_1,\dots,f_r \in A,$$

resp.

$$\{\psi \in M(A): |\psi(f_i)-\varphi(f_i)|<1,\ i=1,\dots r\},\ f_1,\dots,f_r \in A$$

are a basis of open neighbourhoods of φ for the Gelfand topology on S(A) (resp. M(A)).

ii) Let A,B be F-algebras with generating sequences of seminorms

$(p_n)_n$ resp. $(q_n)_n$, and let $\varphi: A \longrightarrow B$ be a linear mapping. Then φ is continuous iff for each $k \in \mathbb{N}$ there exists $n(k) \in \mathbb{N}$ and a constant $c_k>0$ such that

$$q_k(\varphi(f)) \le c_k p_{n(k)}(f) \text{ for all } f \in A.$$

Let φ be a continuous algebra homomorphism, and suppose that

$$q_n(g^2) = q_n(g)^2 \text{ for all } g \in B,\ n \in \mathbb{N}.$$

Then

$$q_k(\varphi(f))^{2^m} = q_k(\varphi(f^{2^m})) \le c_k p_{n(k)}(f^{2^m}) \le c_k(p_{n(k)}(f))^{2^m}$$

for all $m \in \mathbb{N}$, $f \in A$. Thus we may choose each $c_k=1$. In particular $\varphi \in S(A)$ is an element of $M(A)$ iff there is $n \in \mathbb{N}$ such that

$$|\varphi(f)| \le p_n(f) \text{ for all } f \in A.$$

iii) We remark that $\varphi \in S(A)$ implies that $\varphi(1)=1$ if A has an identity.
iv) It is an unsolved problem whether $S(A)=M(A)$ for every F-algebra A, i.e. whether each homomorphism $\varphi: A \longrightarrow \mathbb{C}$ is automatically continuous. We shall return to this question later in chapter 10.

Example . We determine the spectrum of $C(\mathbb{R})$ by elementary calculations. We show that $S(C(\mathbb{R})) = M(C(\mathbb{R})) = \mathbb{R}$, more precisely: if $\varphi: C(\mathbb{R}) \longrightarrow \mathbb{C}$ is a nontrivial algebra homomorphism, then there exists $x \in \mathbb{R}$ such that $\varphi(f)=f(x)$ for all $f \in C(\mathbb{R})$.
(The following proof is due to Aron and Fricke [A/F]. We shall prove in chapter 10 that $M(C(X)) = S(C(X)) = X$ for each hemicompact k-space X.)

Proof. Set $z: \mathbb{R} \longrightarrow \mathbb{C}$, $t \longrightarrow t$.
Let $\varphi \in S(C(\mathbb{R}))$ be an arbitrary element. Set $c=\varphi(z)$. We claim that $\varphi(f)=f(c)$ for all $f \in C(\mathbb{R})$.
If $f \in C(\mathbb{R})$, then $g=f-f(c) \in C(\mathbb{R})$, $g(c)=0$ and $\varphi(g)=\varphi(f)-f(c)$, since $\varphi(1)=1$, i.e. our assertion is proved if $\varphi(g)=0$ for all $g \in C(\mathbb{R})$ such

that $g(c)=0$.

Case 1: Assume there exists $\varepsilon>0$ such that $f(t)=0$ for $c-\varepsilon\le t\le c+\varepsilon$. Define

$$g(t) = \begin{cases} f(t)/(t-c), & \text{if } t \ne c \\ 0 & \text{, if } t = c. \end{cases}$$

Then $g \in C(\mathbb{R})$, $g(t)(z(t)-c)=f(t)$ for all $t \in \mathbb{R}$, hence $\varphi(f) = 0$.

Case 2: Suppose there is $f \in C(\mathbb{R})$, $f(c)=0$ so that $\varphi(f) \ne f(c)$, w.l.o.g. $\varphi(f) = 1$.

Choose $\varepsilon>0$ so that $|f(t)| < 1/2$ for $c-\varepsilon \le t \le c+\varepsilon$ and set

$$h(t) = \begin{cases} f(c+\varepsilon), & \text{if } t \ge c+\varepsilon \\ f(t), & \text{if } c-\varepsilon < t < c+\varepsilon \\ f(c-\varepsilon), & \text{if } t \le c-\varepsilon. \end{cases}$$

Then $h \in C(\mathbb{R})$ and $\varphi(h)-\varphi(f) = \varphi(h-f) = 0$ by case 1, hence $\varphi(h) = 1$. The function $1-h$ is invertible in $C(\mathbb{R})$ by construction, so

$$1 = \varphi(1) = \varphi(1-h)\varphi((1-h)^{-1}) = 0,$$

and we get a contradiction.

(3.2.3) We shall frequently use the following simple observation.

Proposition. Let A be an F-algebra with identity 1. Let $\varphi_1,\dots,\varphi_k$ be distinct points of $M(A)$, and let $c_1,\dots,c_k \in \mathbb{C}$. Then there is $f \in A$ so that $\hat{f}(\varphi_i) = c_i$, $i=1,\dots,k$.

Proof. For $i,j \in \{1,\dots,k\}$, $i \ne j$, choose $f_{ij} \in \ker \varphi_j = \{f\in A: \varphi_j(f)=0\}$ such that $\hat{f}_{ij}(\varphi_i) \ne 0$. Then $\hat{h}_{ij}(\varphi_i) = 1$ and $\hat{h}_{ij}(\varphi_j) = 0$ for

$$h_{ij} = f_{ij}/\varphi_i(f_{ij})\cdot 1 \in A.$$

Set

$$f = \sum_{i=1}^{k} c_i\Big(\prod_{j=1,j\ne i}^{k} h_{ij}\Big).$$

□

<u>(3.2.4)</u> Let A be an F-algebra without unit, $(p_n)_n$ be a defining sequence of seminorms for A, and let A* be the algebra obtained by adjoining an identity (with the sequence of seminorms $(q_n)_n$ defined as in (3.1.8)). Set

$$\varphi_o : A^* \longrightarrow \mathbb{C},\ (\lambda,f) \longrightarrow \lambda.$$

Then $\varphi_o \in M(A^*)$.

For $\psi \in S(A)$ define

$$\tilde{\psi} : A^* \longrightarrow \mathbb{C},\ (\lambda,f) \longrightarrow \lambda+\psi(f),$$

and for $\varphi \in S(A^*)\backslash\{\varphi_o\}$ set

$$\varphi' : A \longrightarrow \mathbb{C},\ f \longrightarrow \varphi(0,f).$$

It can be easily checked that $\tilde{\psi} \in S(A^*)\backslash\{\varphi_o\}$ (resp. $\varphi' \in S(A)$) for all $\psi \in S(A)$ (resp. $\varphi \in S(A^*)\backslash\{\varphi_o\}$) and that $(\tilde{\psi})' = \psi$ (resp. $(\tilde{\varphi'}) = \varphi$), i.e. the map

$$i: S(A) \longrightarrow S(A^*)\backslash\{\varphi_o\},\ \psi \longrightarrow \tilde{\psi},$$

is bijective.

Let $\varphi \in S(A)$, and let $n \in \mathbb{N}$ so that

$$|\varphi(f)| \leq p_n(f) \text{ for all } f \in A,$$

(i.e. $\varphi \in M(A)$ by (3.2.2)ii)) then

$$|\tilde{\varphi}(\lambda,f)| \leq |\varphi(f)| + |\lambda| \leq p_n(f) + |\lambda| = q_n(\lambda,f)$$

for all $(\lambda,f) \in A^*$, i.e. $\tilde{\varphi} \in M(A^*)\backslash\{\varphi_o\}$.

Vice versa if $\varphi \in S(A^*)\backslash\{\varphi_o\}$ and

$$|\varphi(\lambda,f)| \leq q_n(\lambda,f) \text{ for all } (\lambda,f) \in A^*,$$

then

$$|\varphi'(f)| = |\varphi(0,f)| \leq q_n(0,f) = p_n(f)$$

for all $f \in A$, hence

$$i(M(A)) = M(A^*)\backslash\{\varphi_o\}.$$

Finally choose arbitrary $f_1,\dots,f_r \in A$, $\lambda_1,\dots,\lambda_r \in \mathbb{C}$ and $\mu \in S(A)$, then

$$i(\{\psi \in S(A)\colon |\mu(f_i)-\psi(f_i)| < 1,\ i=1,\dots,r\}) =$$
$$= \{\varphi \in S(A*)\setminus\{\varphi_o\}\colon |\tilde{\mu}(0,f_i)-\varphi(0,f_i)| < 1,\ i=1,\dots,r\} =$$
$$= \{\varphi \in S(A*)\setminus\{\varphi_o\}\colon |\tilde{\mu}(\lambda_i,f_i)-\varphi(\lambda_i,f_i)| < 1,\ i=1,\dots,r\}.$$

It follows from this and (3.2.2)i) that

$$i\colon S(A) \longrightarrow S(A*)\setminus\{\varphi_o\}$$

resp.

$$i\colon M(A) \longrightarrow M(A*)\setminus\{\varphi_o\}$$

are homeomorphisms.

<u>IMPORTANT REMARK.</u> In the sequel we only consider F-algebras with identity 1.

<u>(3.2.5)</u> Let A,B be F-algebras, and let $T\colon A \longrightarrow B$ be a continuous nontrivial algebra homomorphism, i.e. $T(1) = 1$. The adjoint spectral map T^* is defined by

$$T^*\colon M(B) \longrightarrow M(A),\quad \varphi \longrightarrow \varphi\circ T.$$

<u>Lemma.</u> Let A,B and T be as above. Then

i) T^* is continuous;

ii) T^* is injective if T has dense range;

iii) $(S\circ T)^* = T^*\circ S^*$ if C is another F-algebra and $S\colon B \longrightarrow C$ is a continuous nontrivial algebra homomorphism;

iv) T^* is a homeomorphism if T is bijective.

<u>Proof.</u> Let $\varphi \in M(B)$ and let $f_1,\dots,f_r \in A$. Then

$$(T^*)^{-1}(\{\psi \in M(A)\colon |T^*(\varphi)(f_i)-\psi(f_i)| < 1,\ i=1,\dots,r\}) =$$
$$= \{\omega \in M(B)\colon |\varphi(T(f_i))-\omega(T(f_i))| < 1,\ i=1,\dots,r\}.$$

The last set is open in M(B), hence $(T^*)^{-1}(U)$ is open for every open set $U \subset M(A)$ (cf. remark (3.2.2)i)).

ii) Let $T^*(\varphi) = T^*(\psi)$ for $\varphi,\psi \in M(B)$,

then

$$\varphi \circ T(f) = \psi \circ T(f) \text{ for all } f \in A,$$

hence

$$\varphi(g) = \psi(g) \text{ for all } g \in B,$$

since T(A) is dense in B, thus $\varphi = \psi$.

iii) This is obvious.

iv) By the open mapping theorem for Fréchet spaces T is a homeomorphism. We have

$$(T^{-1})^* \circ T^* = id_{M(B)} \text{ and } T^* \circ (T^{-1})^* = id_{M(A)},$$

so T^* is bijective and $(T^*)^{-1} = (T^{-1})^*$ is continuous by i). □

(3.2.6) We shall need lemma (3.2.5) to determine the spectrum of an F-algebra.

Let A be an F-algebra with defining sequence of seminorms $(p_n)_n$ (recall that $p_n \leq p_{n+1}$ for all $n \in \mathbb{N}$). For each $n \in \mathbb{N}$ denote by I_n the ideal

$$I_n = \ker p_n = \{a \in A : p_n(a) = 0\},$$

and denote by A_n the completion of the algebra A/I_n with respect to the norm

$$p_n'(a+I_n) = p_n(a),$$

so A_n is a B-algebra by definition. Note that A_n has an identity since A has an identity by our general assumption on F-algebras. We set

$$\pi_n : A \longrightarrow A_n, \; a \longrightarrow a+I_n.$$

Then π_n is a continuous homomorphism which has dense range, hence

π_n^* is continuous and injective by lemma (3.2.5). This implies - since $M(A_n)$, as the spectrum of a B-algebra with identity, is compact - that

$$\pi_n^*\colon M(A_n) \longrightarrow \pi_n^*(M(A_n)) \subset M(A)$$

is a homeomorphism.

Let φ be an arbitrary element of $M(A)$, then there exists $n \in \mathbb{N}$ such that

$$|\varphi(f)| \le p_n(f) \text{ for all } f \in A \text{ (cf. (3.2.2.)ii)).}$$

If we define

$$\tilde{\varphi}\colon A/I_n \longrightarrow \mathbb{C},\ f+I_n \longrightarrow \varphi(f),$$

then $\tilde{\varphi}$ is a well-defined continuous homomorphism on A/I_n and if we denote the unique extension of $\tilde{\varphi}$ to A_n again by $\tilde{\varphi}$ then

$$\pi_n^*(\tilde{\varphi}) = \varphi.$$

Conversely, if $\varphi \in M(A_n)$, then

$$|\pi_n^*(\varphi)(f)| = |\varphi\circ\pi_n(f)| \le p'_n(f+I_n) = p_n(f),$$

(for the inequality cf. (1.2.1)) hence

$$(*) \qquad \pi_n^*(M(A_n)) = \{\varphi\in M(A)\colon |\varphi(f)| \le p_n(f) \text{ for all } f \in A\}$$

So

$$\pi_n^*(M(A_n)) \subset \pi_m^*(M(A_m)) \text{ for } m\ge n$$

and

$$M(A) = \bigcup_{n\in\mathbb{N}} \pi_n^*(M(A_n)).$$

(3.2.7) In the sequel we always identify $M(A_n)$ with $\pi_n^*(M(A_n))$. By (3.2.6) we already know that $\ldots M(A_n) \subset M(A_{n+1}) \subset \ldots$ is a countable compact exhaustion of $M(A)$, i.e. $M(A)$ is σ-compact. In fact we can improve this result.

Recall that a countable compact exhaustion $\ldots K_n \subset K_{n+1} \subset \ldots$ of a

Hausdorff space X is called admissible provided that for each compact set $K \subset X$ there is $n \in \mathbb{N}$ such that $K \subset K_n$. A space with an admissible exhaustion is called hemicompact.

Remark. There are σ-compact spaces which are not hemicompact. Take for instance $\mathbb{Q}$, the set of all rational numbers, endowed with the induced euclidean topology. $\mathbb{Q}$ is clearly σ-compact and satisfies the first axiom of countability. Since hemicompact spaces which satisfy the first axiom of countability are locally compact, cf. (4.4.1), $\mathbb{Q}$ cannot be hemicompact.

We want to prove that $(M(A_n))_n$ is an admissible exhaustion of M(A).

To show this let $K \subset M(A)$ be an arbitrary compact set. For each $\varphi \in K$ the set $\{f \in A: |\varphi(f)| \le 1\}$ is closed, hence

$$K^o = \bigcap_{\varphi \in K} \{f \in A: |\varphi(f)| \le 1\} = \{f \in A: \|\hat{f}\|_K \le 1\}$$

is closed, too.

Since - by the definition of the Gelfand topology - each Gelfand transform $\hat{f}$ is continuous on M(A), $\hat{f}$ is bounded on the compact set K, hence

$$A = \bigcup_{n=1}^{\infty} nK^o.$$

By Baire's theorem (cf. B.3) there is $n \in \mathbb{N}$ such that nK^o - and thus K^o - has nonempty interior. So we can choose $k \in \mathbb{N}$, $\varepsilon > 0$ and $g \in K^o$ such that

$$\{f \in A: p_k(f-g) < \varepsilon\} \subset K^o.$$

Let $h \in A$ so that $p_k(h) \ne 0$, and let $\varphi \in K$, then

$$g+(\varepsilon h/2p_k(h)) \in K^o$$

hence

$$|\varphi(g+(\varepsilon h/2p_k(h)))| \le 1.$$

From this follows

(**) $|\varphi(h)| \le 4p_k(h)/\varepsilon$, since $|\varphi(g)| \le 1$.

If $p_k(h) = 0$, then $g+c\cdot h \in K^o$ for all $c>0$, hence $|\varphi(h)| \le 2/c$ for all $c>0$, i.e.

(***) $0 = |\varphi(h)| = p_k(h)$.

Thus $\varphi \in M(A_k)$ by (*) of (3.2.6), (**), (***) and (3.2.2)ii).

We can now summarize our considerations above. (Recall once more that we consider only algebras with identity.)

(3.2.8) Theorem ([MIC]). Let A be an F-algebra, and let A_n be defined as above. Then (in a natural) sense

$$M(A) = \bigcup_{n\in\mathbb{N}} M(A_n)$$

and $(M(A_n))_n$ is an admissible exhaustion of M(A), i.e. M(A) is a hemicompact space.

Corollary. The spectrum of an F-algebra is never empty.

Proof. $M(A_1)$ is not empty by (1.2.8). □

(3.2.9) As an immediate consequence we get the theorem of Gelfand-Mazur for F-algebras.

Theorem. If A is an F-algebra which is a field, then $A = \mathbb{C}$.

Proof. The complex multiples of the identity form a subalgebra of A which is isomorphic to $\mathbb{C}$. It suffices to show that every $f \in A$ is a complex multiple of the identity.
Let $f \in A$. By the corollary M(A) is not empty. Choose an arbitrary $\varphi \in M(A)$, then $f-\varphi(f)\cdot 1$ is not invertible. (Otherwise there exists $g \in A$ such that

$$(f-\varphi(f)\cdot 1)g = 1,$$

hence

$$1 = \varphi(1) = \varphi(f-\varphi(f)\cdot 1)\varphi(g) = 0\cdot\varphi(g) = 0.$$

This is absurd.) Since A is a field we have $f-\varphi(f)\cdot 1 = 0$. □

<u>(3.2.10)</u> Let A be an F-algebra, whose topology is defined by the sequence of seminorms $(p_n)_n$. Let I be a proper closed ideal of A. Then A/I endowed with the quotient topology is a Fréchet space and the topology is defined by the seminorms

$$q_n(f+I) = \inf\{p_n(f+g)\colon g \in I\} \text{ (cf. [JAR], p. 76).}$$

It is easy to show that multiplication is continuous, i.e. A/I is a topological algebra. Let $f,g \in A$, $h_1,h_2 \in I$ be arbitrary elements. Then

$$p_n(f+h_1)p_n(g+h_2) \geq p_n(fg+fh_2+gh_1+h_1h_2).$$

It follows that

$$q_n(f+I)q_n(g+I) \geq q_n(fg+I), \text{ since } fh_2+gh_1+h_1h_2 \in I,$$

i.e. each q_n is multiplicative and A/I becomes an F-algebra. Next we want to determine the spectrum of A/I. Set

$$V(I) = \{\varphi \in M(A)\colon \varphi(f) = 0 \text{ for all } f \in I\}.$$

Since the natural map

$$j\colon A \longrightarrow A/I,\ f \longrightarrow f+I,$$

is a surjective and continuous homomorphism, the map

$$j^*\colon M(A/I) \longrightarrow M(A),\ \varphi \longrightarrow \varphi\circ j$$

is continuous and injective (lemma (3.2.5)). We have

$$j^*(y)(f) = 0 \text{ for each } f \in I \text{ and each } y \in M(A/I)$$

hence

$$j^*(M(A/I)) \subset V(I).$$

For $\varphi \in V(I)$ define

$$\tilde{\varphi}: A/I \longrightarrow \mathbb{C},\ f+I \longrightarrow \varphi(f).$$

Then $\tilde{\varphi}$ is a well-defined homomorphism. By (3.2.2)ii) there is $n \in \mathbb{N}$ so that

$$|\varphi(f)| \leq p_n(f) \text{ for all } f \in A.$$

Let $h \in I$ be an arbitrary element. Then

$$|\tilde{\varphi}(f+I)| = |\varphi(f)| = |\varphi(f+h)| \leq p_n(f+h),$$

for each $f \in A$, hence

$$|\tilde{\varphi}(f+I)| \leq q_n(f+I) \text{ for all } f \in I,$$

thus $\tilde{\varphi} \in M(A/I)$ by (3.2.2)ii). Clearly $j^*(\tilde{\varphi}) = \varphi$, hence

$$j^*(M(A/I)) = V(I).$$

Using (3.2.2)i) it is easy to show that

$$j^*: M(A/I) \longrightarrow V(I)$$

is in fact a homeomorphism, $V(I)$ endowed with the relative Gelfand topology.

<u>(3.2.11)</u> An ideal I of A is called maximal if $I \neq A$ and I is contained in no other proper ideal of A. I is maximal iff A/I is a field. Hence - by the theorem of Gelfand-Mazur - A/I is isomorphic to $\mathbb{C}$, if A is an F-algebra and $I \subset A$ is a closed maximal ideal. So we can interpret the natural projection $j: A \longrightarrow A/I$ as an element of $M(A)$ with kernel I.
On the other hand $\ker \varphi$ is a maximal closed ideal of A if $\varphi \in M(A)$. This proves the following theorem.

<u>Theorem</u> . The correspondence $\varphi \longrightarrow \ker \varphi$, is a one-to-one correspondence between the elements of $M(A)$ and the maximal closed ideals in A.

<u>Remark</u>. Every maximal ideal in a B-algebra is closed (1.2.5), hence

there is a one-to-one correspondence between the maximal ideals of A and the elements of M(A) in this case (1.2.6). Opposite to this result maximal ideals of F-algebras are in general not closed, as we shall see next.

Example. We have $M(C(\mathbb{R})) = \mathbb{R}$ by example (3.2.2), i.e. each maximal closed ideal $m \subset C(\mathbb{R})$ is of the form

$$I_x = \{f \in C(\mathbb{R}) : f(x) = 0\}$$

for some $x \in \mathbb{R}$ by the theorem above. Consider the following ideal

$$I = \{f \in C(\mathbb{R}) : \text{There exists } k \in \mathbb{N} \text{ so that } f(x) = 0 \text{ for } x \geq k\}.$$

By Zorn's lemma I is contained in a maximal ideal m. Since there exists no $x \in \mathbb{R}$ so that $I \subset I_x$, the same is true for m, i.e. m is not closed.

(3.2.12) We close this section with a consideration of $\hat{A}$, the algebra of the Gelfand transforms.
Let A be an F-algebra with a defining sequence of seminorms $(p_n)_n$. By definition $\hat{A}$ is a subalgebra of C(M(A)) and

$$\Gamma : A \longrightarrow C(M(A)),\ f \longrightarrow \hat{f},$$

is an algebra homomorphism. We endow C(M(A)) with the compact open topology. Recall from (3.2.8) that $(M(A_n))_n$ is an admissible exhaustion of M(A). It follows that the compact open topology is generated by the seminorms

$$\|f\|_{M(A_n)} = \sup\{|f(x)| : x \in M(A_n)\}.$$

For $f \in A$ we have

$$\|\hat{f}\|_{M(A_n)} \leq p_n(f) \text{ by (3.2.6) } (*),$$

hence Γ is continuous.

(3.3) Projective limits

It is an important tool in our theory to represent each F-algebra as the projective limit of a sequence of B-algebras. This representation enables us to carry over many results which are known for B-algebras as we shall see later.

(3.3.1) Let $(E_n,d_n)_n$ be a sequence of metric spaces and assume that for each $n \in \mathbb{N}$ a continuous map

$$\varphi_n : E_{n+1} \longrightarrow E_n$$

is given. We say that this constitutes a projective system of metric spaces

$$\cdots \longrightarrow E_{n+1} \xrightarrow{\varphi_n} E_n \longrightarrow \cdots \quad .$$

If $\varphi_n(E_{n+1})$ is dense in E_n for all $n \in \mathbb{N}$ then we speak of a dense projective system.

Definition. The subset

$$\varprojlim E_n = \{(f_n)_n \in \prod_{n\in\mathbb{N}} E_n : \varphi_n(f_{n+1}) = f_n \quad \text{for all } n \in \mathbb{N}\}$$

endowed with the relative product topology is called the projective limit of the projective system.

(3.3.2) Lemma [ARE 1]. Let $\varprojlim E_n$ be the projective limit of a dense projective system

$$(*) \qquad \cdots \longrightarrow E_{n+1} \xrightarrow{\varphi_n} E_n \longrightarrow \cdots$$

of complete metric spaces. Then the map

$$\pi_k : \varprojlim E_n \longrightarrow E_k,\ (f_n)_n \longrightarrow f_k,$$

has dense range for all $k \in \mathbb{N}$.

Proof. It is sufficient to prove our assertion for k=1. We shall denote the composite of any m successive mappings in (*) by φ^m. Let $e_1 \in E_1$ be an arbitrary point, and let $\varepsilon>0$. We shall construct an element $(c_n)_n \in \varprojlim E_n$ such that

$$d_1(\pi_1((c_n)_n),e_1) < \varepsilon.$$

Choose a sequence $(\varepsilon_n)_n$ of positive numbers such that $\sum_{n\in\mathbb{N}} \varepsilon_n < \varepsilon/2$. Since φ has dense range we find $e_2 \in E_2$ so that

$$d_1(\varphi(e_2),e_1) < \varepsilon_1.$$

Next we choose $e_3 \in E_3$ such that

$$d_2(\varphi(e_3),e_2) < \varepsilon_2 \text{ and } d_1(\varphi^2(e_3),\varphi(e_2)) < \varepsilon_2.$$

This is possible because φ is continuous and has dense range. If we continue we get elements $e_i \in E_i$ such that

$$d_{i-k}(\varphi^k(e_i),\varphi^{k-1}(e_{i-1})) < \varepsilon_{i-1} \text{ for all } k<i.$$

For each $n \in \mathbb{N}$ the sequence

$$c_{n,p} = \varphi^{p-n}(e_p) \qquad (p = n+1,n+2,\ldots)$$

is a Cauchy sequence in E_n, since

$$d_n(c_{n,p+1},c_{n,p}) < \varepsilon_p.$$

We denote the limit of this sequence (in E_n) by c_n. We have

$$\varphi(c_{n+1}) = c_n,$$

since

$$\varphi(c_{n+1,p}) = \varphi(\varphi^{p-(n+1)}(e_p)) = \varphi^{p-n}(e_p) = c_{n,p},$$

hence

$$(c_n)_n \in \varprojlim E_n.$$

Moreover

$$d_1(c_1,e_1) \le d_1(c_1,\varphi^{p-1}(e_p)) + d_1(e_1,\varphi^{p-1}(e_p))$$

$$\leq d_1(c_1,c_{1,p}) + d_1(e_1,\varphi(e_2)) + \ldots + d_1(\varphi^{p-2}(e_{p-1}),\varphi^{p-1}(e_p))$$
$$< d_1(c_1,c_{1,p}) + \varepsilon/2$$

for all p. For $p \longrightarrow \infty$ we obtain the desired result. □

Remark. This lemma is usually referred to as the abstract version of the Mittag-Leffler theorem. Indeed the classical theorem of Mittag-Leffler for meromorphic functions as well as the Baire category theorem can be deduced from it (cf. [EST]).

(3.3.3) We now consider dense projective systems

$$\ldots \longrightarrow B_{n+1} \xrightarrow{\varphi_n} B_n \longrightarrow \ldots$$

where each B_n is a B-algebra with norm $\|\cdot\|_n$ and each φ_n is a continuous algebra homomorphism.

$\prod B_n$ is an algebra under coordinatewise defined algebraic operations.

Since the product of complete spaces is complete ([KEL],p. 194) and the topology on $\prod B_n$ is generated by the sequence of multiplicative seminorms

$$p_k^*((f_n)_n) = \max\{(\|f_j\|_j,\ j\leq k\},\ k \in \mathbb{N},$$

$\prod B_n$ is an F-algebra.

It is not hard to prove that $\varprojlim B_n$ is a closed subalgebra of $\prod B_n$, so $\varprojlim B_n$ is an F-algebra, too. Let $(f_n)_n \in \varprojlim B_n = A$.

For $k \in \mathbb{N}$ there is $c_k>0$ such that

$$\|f_k\|_k \leq p_k^*((f_n)_n) \leq c_k\|f_k\|_k$$

since

$$\|f_{k-i}\|_{k-i} = \|\varphi_{k-i}\circ\ldots\circ\varphi_{k-1}(f_k)\|_{k-i},\ i=1,\ldots,k-1.$$

Therefore the seminormsystem

(*) $$p_k((f_n)_n) = \|f_k\|_k$$

defines an equivalent topology on $A = \varprojlim B_n$.

As in (3.2.6) set $I_k = \{(f_n)_n \in A: \rho_k((f_n)_n)=0\}$ and denote by A_k the completion of A/I_k with respect to the norm

$$\rho_k'((f_n)_n + I_k) = \rho_k((f_n)_n) = \|f_k\|_k$$

The map

$$T: A/I_k \longrightarrow B_k,\ (f_n)_n + I_k \longrightarrow f_k$$

is an injective and continuous algebra homomorphism. By lemma (3.3.2) T has dense range, hence we can extend T to a topological algebra homomorphism $\tilde{T}: A_k \longrightarrow B_k$.

<u>(3.3.4) Theorem.</u> Let

$$\ldots \longrightarrow B_{n+1} \xrightarrow{\varphi_n} B_n \longrightarrow \ldots$$

be a dense projective system of B-algebras. Then $A = \varprojlim B_n$ is an F-algebra, M(A) can be identified in a natural way with $\cup M(B_n)$ and $(M(B_n))_n$ is an admissible exhaustion of M(A).

<u>Proof.</u> Use the proof of theorem (3.2.8) and the considerations above. □

<u>(3.3.5) Example.</u> For $k \in \mathbb{N}$ denote by $C^k([0,1])$ the algebra of all functions on [0,1] which are k-times continuously differentiable. $C^k([0,1])$ becomes a B-algebra with respect to the norm

$$p_{k+1}(f) = 2^k \sup\{|f^{(i)}(x)|: x \in [0,1],\ i=0,\ldots,k\},$$

where $f^{(i)}$ denotes the i-th derivative of f. We have

$$M(C^k([0,1])) = [0,1],$$

more precisely: Each element of $M(C^k([0,1]))$ is the point evaluation homomorphism at some point $x \in [0,1]$, cf. [GAM], p. 6.

We claim that Gelfand topology and induced euclidean topology coin-

cide on $[0,1]$. Set $z: [0,1] \longrightarrow \mathbb{C}$, $x \longrightarrow x$.
Let U be an open subset of $[0,1]$ with respect to the euclidean topology, and let $x \in U$. Then there is $\varepsilon>0$ such that

$$\{y \in [0,1]: |z(x)-z(y)| < \varepsilon\} \subset U,$$

i.e. U is open with respect to the Gelfand topology (cf. (3.2.2)i)).
Vice versa let $x \in [0,1]$, and let $f_1,\dots,f_r \in C^k([0,1])$. Then

$$\{y \in [0,1]: |f_i(x)-f_i(y)| < 1,\ i=1,\dots,r\}$$

is open with respect to the euclidean topology, since $f_1,\dots,f_r$ are elements of $C([0,1])$. Now it follows again from (3.2.2)i) that both topologies coincide on $[0,1]$.
Clearly

$$\dots \longrightarrow C^{k+1}([0,1]) \xrightarrow{r_k} C^k([0,1]) \longrightarrow \dots$$

constitutes a dense projective system of B-algebras with respect to the natural restriction mappings

$$r_k: C^{k+1}([0,1]) \longrightarrow C^k([0,1]),\ f \longrightarrow f.$$

Recall the definition of $C^\infty([0,1])$ from (3.1.9). We have

$$C^\infty([0,1]) = \varprojlim C^k([0,1])$$

and hence

$$M(C^\infty([0,1])) = [0,1] \text{ by (3.3.4).}$$

The Gelfand and the euclidean topology coincide on $[0,1]$ and we can identify $C^\infty([0,1])$ with $C^\infty([0,1])^\wedge$, the algebra of all Gelfand transforms of $C^\infty([0,1])$.

<u>(3.3.6)</u> We have proved above that $\varprojlim B_n$ is an F-algebra if it is the projective limit of a dense projective system. We now show that each F-algebra arises in this way, i.e. if A is an F-algebra then A is the projective limit of a dense projective system of B-algebras.
Let A be an F-algebra with defining sequence of seminorms $(p_n)_n$,

cf. (3.1.7).

As in (3.2.6) we denote by A_n the completion of the algebra $A/\ker p_n$ with respect to the norm

$$p'_n(f + \ker p_n) = p_n(f)$$

and by π_n the canonical mapping

$$A \longrightarrow A_n, \; f \longrightarrow f+\ker p_n.$$

For $m \geq n$ we define

$$\pi_{m,n}: A_m \longrightarrow A_n$$

to be the extension of the mapping

$$A/\ker p_m \longrightarrow A/\ker p_n, \; f + \ker p_m \longrightarrow f + \ker p_n.$$

π_n (resp. $\pi_{m,n}$) are (well-defined) continuous algebra homomorphisms which have dense range, and

$$(*) \qquad p'_n(\pi_{m,n}(f)) \leq p'_m(f) \text{ for all } f \in A_m,$$

since $p_n \leq p_m$.

Note that

$$\pi_{m,n} = \pi_{n+1,n} \circ \pi_{n+2,n+1} \circ \cdots \circ \pi_{m,m-1}.$$

We consider the dense projective system of B-algebras

$$\cdots \longrightarrow A_{n+1} \xrightarrow{\pi_{n+1,n}} A_n \longrightarrow \cdots$$

Let

$$T: A \longrightarrow \varprojlim A_n, \; f \longrightarrow (\pi_n(f))_n.$$

Then T is an algebra homomorphism. We have

$$\rho_k^*(T(f)) = \rho_k(T(f)) = p_k(f) \text{ for all } k \in \mathbb{N}, \; f \in A,$$

hence T is injective and continuous.

To show that T is surjective let $(f_n)_n \in \varprojlim A_n$ be an arbitrary element. For each $k \in \mathbb{N}$ we can choose $g_k \in A$ such that

$$p'_k(f_k - \pi_k(g_k)) < 1/k,$$

since π_k has dense range. So for $k \geq n$

$$(**) \qquad p'_n(f_n - \pi_n(g_k)) = p'_n(\pi_{k,n}(f_k) - \pi_{k,n}(\pi_k(g_k)))$$
$$\leq p'_k(f_k - \pi_k(g_k)) < 1/k.$$

For the first inequality use (*).

Fix $m \in \mathbb{N}$, and let $k \geq l \geq m$, then

$$p_m(g_k - g_l) \leq p_l(g_k - g_l) = p'_l(\pi_l(g_k - g_l))$$
$$\leq p'_l(\pi_l(g_k) - f_l) + p'_l(f_l - \pi_l(g_l)) \leq 1/k + 1/l \text{ by } (**).$$

It follows that $(g_n)_n$ is a Cauchy sequence in A. Set $g = \lim g_n$, then for $l \geq k$

$$\rho_k(T(g) - (f_n)_n) = p'_k(\pi_k(g) - f_k)$$
$$\leq p'_k(\pi_k(g) - \pi_k(g_l)) + p'_k(\pi_k(g_l) - f_k)$$
$$\leq p_k(g - g_l) + 1/l.$$

Hence we get

$$\rho_k(T(g) - (f_n)_n) = 0 \text{ for all } k \in \mathbb{N}, \text{ i.e.}$$

$T(g) = (f_n)_n$ and T is surjective, hence a topological homomorphism by the open mapping theorem for Fréchet spaces.

The next theorem follows from our considerations above.

(3.3.7) Theorem. The following statements are equivalent for an algebra A:

i) A is an F-algebra;

ii) A is the projective limit of a dense projective system of B-algebras.

CHAPTER 4

GENERAL THEORY OF UNIFORM FRECHET ALGEBRAS

(4.1) Uniform Fréchet algebras

In an analogous way to the theory of Banach algebras we introduce uniform Fréchet algebras and give equivalent characterizations. Important examples are the algebra of all holomorphic functions on a hemicompact reduced complex space and the algebra of all continuous functions on a hemicompact k-space.

(4.1.1) First recall the notion of a uniform Banach algebra (uB-algebra). A B-algebra A with norm $\|\cdot\|$ is called uB-algebra if

$$\|f\|^2 = \|f^2\| \text{ for all } f \in A.$$

In this case we have

$$\|f\| = \|\hat{f}\|_{M(A)} = \sup\{|\hat{f}(x)|: x \in M(A)\} \text{ for all } f \in A$$

and

$\Gamma: A \longrightarrow \hat{A}$, $f \longrightarrow \hat{f}$, is a topological homomorphism, i.e. $(\hat{A}, \|\cdot\|_{M(A)})$ is a uB-algebra (cf. (1.2.10)).

Since M(A) is a compact nonempty space (cf. (1.2.8)) and $\hat{A}$ separates the points of M(A) we see that each uB-algebra is topologically and algebraically isomorphic to a closed pointseparating subalgebra of C(K), K a compact nonempty Hausdorff space. Vice versa each closed pointseparating subalgebra A of C(K) is a uB-algebra, since clearly $\|f^2\|_K = \|f\|_K^2$ for each $f \in A$.

(4.1.2) Definition. Let A be an F-algebra. Then A is called uniform

Fréchet algebra (uF-algebra) if there is a defining sequence of seminorms $(p_n)_n$ such that

$$p_n(f^2) = p_n(f)^2 \text{ for all } f \in A,\ n \in \mathbb{N}.$$

<u>(4.1.3) Theorem</u>.a) The following statements are equivalent for an algebra A:

i) A is a uF-algebra;

ii) A is the projective limit of a dense projective system of uB-algebras;

iii) A contains the constants and is topologically and algebraically isomorphic to a pointseparating and complete subalgebra of C(X), where X is a hemicompact space and C(X) is endowed with the compact open topology.

b) If A is a uF-algebra, then

$$\Gamma: A \longrightarrow \hat{A},\ f \longrightarrow \hat{f},$$

defines a topological and algebraical isomorphism.

<u>Proof.</u> i) ⇒ iii) By theorem (3.2.8) M(A) is a hemicompact space. We shall show that

$$\Gamma: A \longrightarrow \hat{A},\ f \longrightarrow \hat{f}$$

is a topological and algebraical homomorphism.

Let $(p_n)_n$ be a defining sequence of seminorms for A so that

$$p_n(f^2) = p_n(f)^2 \text{ for all } f \in A,\ n \in \mathbb{N}.$$

Recall from (3.2.6) the definition of A_n, π_n and p'_n. Note that

$$p'_n(g^2) = p'_n(g)^2 \text{ for all } g \in A_n,$$

hence each A_n is a uB-algebra and

$$\|\hat{g}\|_{M(A_n)} = p'_n(g) \text{ for all } g \in A_n \text{ (cf. (1.2.10)).}$$

So we have for every $f \in A$, $n \in \mathbb{N}$

$$p_n(f) = p'_n(\pi_n(f)) = \|(\pi_n(f))^\wedge\|_{M(A_n)} = \|\hat{f}\|_{\pi_n^*(M(A_n))}$$
$$= \|\hat{f}\|_{M(A_n)}.$$

Here we have identified $M(A_n)$ and $\pi_n^*(M(A_n))$ (cf. (3.2.7)) and denote by $(\pi_n(f))^\wedge$ the Gelfand transform of $\pi_n(f)$.
Since the topology of $\hat{A}$ is generated by the seminorms $\|\cdot\|_{M(A_n)}$ (cf. (3.2.12)) Γ is a topological algebra homomorphism.
Clearly $\hat{A}$ separates the points of M(A) and contains the constants, since we consider only algebras with identity.
iii) ⇒ ii) Let $(K_n)_n$ be an admissible exhaustion of X. By our hypothesis A is an F-algebra with defining sequence of seminorms $(\|\cdot\|_{K_n})_n$. By (3.3.6)

$$T\colon A \longrightarrow \varprojlim A_n\,,\; f \longrightarrow (\pi_n(f))_n$$

defines a topological homomorphism. Since

$$\|f^2\|_{K_n} = \|f\|^2_{K_n} \quad \text{for all } f \in A,\ n \in \mathbb{N},$$

each A_n is a uB-algebra.
ii) ⇒ i) Let

$$\cdots \longrightarrow B_{n+1} \xrightarrow{\varphi_n} B_n \longrightarrow \cdots$$

be a dense projective sytem of uB-algebras. By (3.3.3) $\varprojlim B_n$ is an F-algebra with defining sequence of seminorms

$$\rho_k^*((f_n)_n) = \max\{\ \|f_j\|_j,\ j\le k\ \},\ k \in \mathbb{N}.$$

Since

$$\|f^2\|_j = \|f\|_j^2, \text{ for all } f \in B_j \text{ (cf. 4.1.1)}$$

we have

$$\rho_k^*(((f_n)_n)^2) = \rho_k^*((f_n)_n)^2 \text{ for all } k \in \mathbb{N},\ (f_n)_n \in \varprojlim B_n. \qquad \square$$

<u>(4.1.4) Examples.</u> i) Each uB-algebra is a uF-algebra.
ii) Recall that a Hausdorff space X is called a k-space if every subset intersecting each compact set in a closed set is itself closed.

If X is a hemicompact k-space then C(X) is complete in the compact open topology (cf. (3.1.9)iii)), hence C(X) is a uF-algebra by iii) of theorem (4.1.3).

iii) Let X be a hemicompact holomorphically separable reduced complex space, then Hol(X) - the algebra of all holomorphic functions on X - is complete in the compact open topology (cf. (2.3.1)iii)), i.e. Hol(X) is a uF-algebra by iii) of theorem (4.1.3).

iv) The algebra $A = C^{\infty}([0,1])$ is not a uF-algebra.

By (3.3.5) M(A) = [0,1], Gelfand and euclidean topology coincide on [0,1] and we can identify A with $\hat{A}$.

Suppose that A is a uF-algebra, then

$$\Gamma: A \longrightarrow \hat{A},\ f \longrightarrow \hat{f},$$

defines a topological algebra isomorphism, where $\hat{A}$ is endowed with the compact open topology, i.e. with the topology which is generated by the norm $\|\cdot\|_{[0,1]}$. Then, by the theorem of Stone-Weierstraß, $\hat{A} = C([0,1])$ and we get a contradiction, since for example $f: [0,1] \longrightarrow \mathbb{C},\ x \longrightarrow |x-(1/2)|$, is not contained in $\Gamma(A)$.

(4.1.5) Notation. By a uF-algebra (A,X) we mean a uF-algebra A, which is given as in iii) of theorem (4.1.3).

The reason that we introduce this notation is that we shall often want to study the carrier space X.

Let (A,X) be a uF-algebra, and let $K \subset X$ be a compact subset of X. We define A_K to be the completion of the algebra

$$A|_K = \{f|_K : f \in A\}$$

with respect to the norm $\|\cdot\|_K$. A_K is a uB-algebra. If $M \subset X$ (endowed with the relative topology) is a hemicompact subset of X and $\ldots \subset K_n \subset K_{n+1} \subset \ldots$ is an admissible exhaustion of M we consider the dense projective system of uB-algebras

$$\ldots A_{K_{n+1}} \xrightarrow{r_n} A_{K_n} \longrightarrow \ldots$$

with the natural restriction mappings

$$r_n : A_{K_{n+1}} \longrightarrow A_{K_n},\ f \longrightarrow f|_{K_n}.$$

We define

$$A_M = \varprojlim A_{K_n}.$$

By theorem (4.1.3) A_M is a uF-algebra.

We can interpret each element of A_M as a complex-valued function on M which can be approximated uniformly on each K_n by elements of A. If moreover M is a k-space, then $A_M \subset C(M)$, since in this case $f \in C(M)$ iff $f|_{K_n} \in C(K_n)$ for all $n \in \mathbb{N}$.

We note that $(A_{M_2})_{M_1} = A_{M_1}$ for $M_1 \subset M_2 \subset X$, M_1, M_2 hemicompact subsets of X.

<u>(4.1.6)</u> It is useful to reformulate theorem (3.3.4) and theorem (4.1.3) for uF-algebras (A,X).

<u>Theorem.</u> Let (A,X) be a uF-algebra and let $(K_n)_n$ be an admissible exhaustion of X. Then

i) A is topologically and algebraically isomorphic to $\varprojlim A_{K_n}$, the projective limit of the dense projective system

$$\ldots \longrightarrow A_{K_{n+1}} \xrightarrow{r_n} A_{K_n} \longrightarrow \ldots$$

with the restriction mappings

$$r_n : A_{K_{n+1}} \longrightarrow A_{K_n},\ f \longrightarrow f|_{K_n}.$$

ii) The spectrum M(A) can be identified as a set with $\cup M(A_{K_n})$ and $(M(A_{K_n}))_n$ is an admissible exhaustion of M(A).

(4.1.7) Examples. i) Let (X,μ) be a hemicompact space, and let $(K_n)_n$ be an admissible exhaustion of X. We define on X a new topology τ in the following way. A set $S \subset X$ is τ-closed if $S \cap K_n$ is μ-closed for each $n \in \mathbb{N}$. Then τ is finer as μ, both topologies coincide on each K_n, $(K_n)_n$ is an admissible exhaustion of X with respect to the τ-topology and (X,τ) is a k-space. So $C(X,\tau)$ becomes a uF-algebra by (4.1.4)ii). The map

$$\varprojlim C(K_n) \longrightarrow C(X,\tau), \ (f_n)_n \longrightarrow f,$$

where $f(x) = f_n(x)$ if $x \in K_n$, defines a topological and algebraical isomorphism, and

$$M(C(X,\tau)) = \bigcup K_n = X \text{ (as sets)}$$

by theorem (4.1.6) and the fact that $M(C(K_n)) = K_n$ (cf. (1.2.4)i)). Hence we can identify $C(X,\tau)$ and $C(X,\tau)^\wedge$, the algebra of the Gelfand transforms of $C(X,\tau)$. It follows from the definition that the Gelfand topology is coarser than the τ-topology. To show that both topologies coincide let S be a τ-closed subset, and let $p \in X\setminus S$. Since (X,τ) is completely regular by (3.1.10)ii), we find $f \in C(X,\tau)$, $1 \geq f \geq 0$, such that $f(p) = 0$ and $f|_S \equiv 1$, hence $\{y \in X: f(y) < 1/2\}$ is a Gelfand open neighbourhood of p, which does not intersect S. It follows that S is Gelfand closed.

If (X,μ) is a hemicompact k-space, then $(X,\mu) = (X,\tau)$ and we have proved the next theorem.

Theorem. Let X be a hemicompact k-space. Then M(C(X)) can be identified as a topological space with X.

ii) Let $G \subset \mathbb{C}$ be a domain. Choose an admissible exhaustion $(K'_n)_n$ of G. Denote by K_n the union of K'_n with all bounded components of $\mathbb{C}\setminus K'_n$ which are relatively compact in G.

Set

$$R(K_n) = \{f \in C(K_n): f \text{ can be approximated uniformly on } K_n \text{ by rational functions with poles off } K_n\}.$$

The restriction mapping

$$r_n: R(K_{n+1}) \longrightarrow R(K_n),\ f \longrightarrow f|_{K_n},$$

has dense range for each $n \in \mathbb{N}$ by Runge's theorem (cf. [NAR 1], p. 107) and (1.4.5). Hence

$$\ldots \longrightarrow R(K_{n+1}) \xrightarrow{r_n} R(K_n) \longrightarrow \ldots$$

constitutes a dense projective system of uB-algebras. Let $(f_n)_n$ be an arbitrary element of $\varprojlim R(K_n)_n$, then

$$f: G \longrightarrow \mathbb{C},\ x \longrightarrow f_n(x), \text{ if } x \in K_n,$$

defines an element of Hol(G). It follows from this and (1.4.5) that the map

$$\mathrm{Hol}(G) \longrightarrow \varprojlim R(K_n),\ f \longrightarrow (f|_{K_n})_n$$

is a topological homomorphism.
$M(R(K_n))$ can be identified with K_n (cf. (1.2.4)iii)) for all $n \in \mathbb{N}$, so by theorem (4.1.6)

$$M(\mathrm{Hol}(G)) = \cup\ M(R(K_n)) = \cup\ K_n = G$$

and it follows at once from (3.2.2)i) that the Gelfand topology coincides with the euclidean topology on G.
Since each domain $G \subset \mathbb{C}$ is a domain of holomorphy (remark (2.2.1)), this result is a special case of (2.2.7).

(4.2) Extension of a uF-algebra

We have seen in (4.1.3) that an F-algebra A is uniform iff A is isomorphic to a complete subalgebra of the continuous functions on a hemicompact space X. We shall prove in this section that for a given uF-algebra (A,X) the spectrum M(A) is the "largest" space to which all functions $f \in A$ can be extended.

(4.2.1) Let (A,X) be a uF-algebra. For each point $x \in X$ we get a continuous algebra homomorphism

$$\varphi_x: A \longrightarrow \mathbb{C},\ f \longrightarrow f(x).$$

Lemma. The structure map

$$j: X \longrightarrow M(A),\ x \longrightarrow \varphi_x,$$

is injective and continuous.

Proof. i) The map is injective since A separates the points of X.
ii) Let x be a point of X, and let $U \subset M(A)$ be an open neighbourhood of j(x). By (3.2.2)i) there are $f_1,\dots,f_r \in A$ such that

$$\{\varphi \in M(A): |\hat{f}_i(\varphi)-\hat{f}_i(\varphi_x)| < 1,\ i=1,\dots,r\} \subset U.$$

The continuity of $f_1,\dots,f_r$ implies that

$$S = \{y \in X: |f_i(x)-f_i(y)| < 1,\ i=1,\dots,r\}$$

is an open neighbourhood of x in X. Clearly $j(S) \subset U$. □

Remark. The map j has in general no other "good" properties. To explain this statement we give an example where j is bijective but not a homeomorphism.
Denote by $\overline{D}$ the closed unit disc in the plane. We define on $\overline{D}$ a new topology τ: A subset $U \subset \overline{D}$ is τ-open if $U \cap (\overline{D}\setminus\{0\})$ is open with respect to the euclidean topology. So the induced τ-topology on $\overline{D}\setminus\{0\}$ coincides with the euclidean topology, while 0 is an isolated point of $(\overline{D},\tau)$. The new topology is finer than the euclidean topology and $(\overline{D},\tau)$ is no longer a compact, but a locally compact and hemicompact space.
Recall that $P(\overline{D})$ denotes the algebra of all functions, which can be approximated uniformly on $\overline{D}$ by polynomials. We have $P(\overline{D}) \subset C(\overline{D},\tau)$

(complex-valued continuous functions on $(\overline{D},\tau)$) and it is easy to see that the topology which is induced on $P(\overline{D})$ as a subspace of $C(\overline{D},\tau)$ coincides with the usual topology on $P(\overline{D})$ - given by the sup-norm $\|\cdot\|_{\overline{D}}$. Hence $(P(\overline{D}),(\overline{D},\tau))$ is a uF-algebra - in fact a uB-algebra - and $M(P(\overline{D})) = \overline{D}$ by (1.2.4)ii). The map

$$j: (\overline{D},\tau) \longrightarrow \overline{D},\ x \longrightarrow x,$$

is bijective but not a homeomorphism.

<u>(4.2.2)</u> Recall from (2.2) that there are domains $G \subset \mathbb{C}^n$ such that all holomorphic functions on G can be extended to a larger domain. This leads to the definition of the envelope of holomorphy E(G), the "largest" domain to which all $f \in \mathrm{Hol}(G)$ can be analytically extended (cf. (2.2.3)). As mentioned in (2.2.5) E(G) is topologically just the spectrum M(Hol(G)).
Now we return to our general setting. Let (A,X) be a uF-algebra. Since the diagram

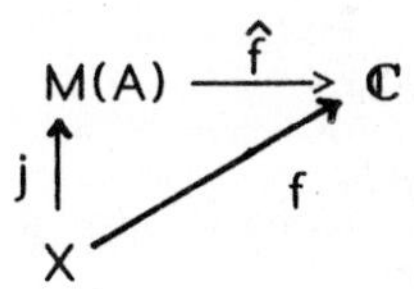

is commutative for each $f \in (A,X)$ we can interpret $\hat{f}$ as an extension of f on M(A) and in analogy to above it is natural to ask whether there exists a "largest" space Y such that
i) each f can be extended to a continuous function $\tilde{f}$ on Y;
ii) (A,X) is topologically and algebraically isomorphic to the algebra of all extended functions $\tilde{f}$.

<u>Definition.</u> i) A uF-algebra (B,Y) is called a morphic extension of (A,X) if there is a continuous function $f: X \longrightarrow Y$ such that the map

$$T_f: B \longrightarrow C(X),\ b \longrightarrow b \circ f,$$

induces a topological algebra isomorphism between B and A.

ii) A morphic extension (B,Y) of (A,X) is called morphic hull of (A,X) if for any other morphic extension (C,Z) of (A,X) there is a unique continuous function $h\colon Z \longrightarrow Y$ so that the diagram

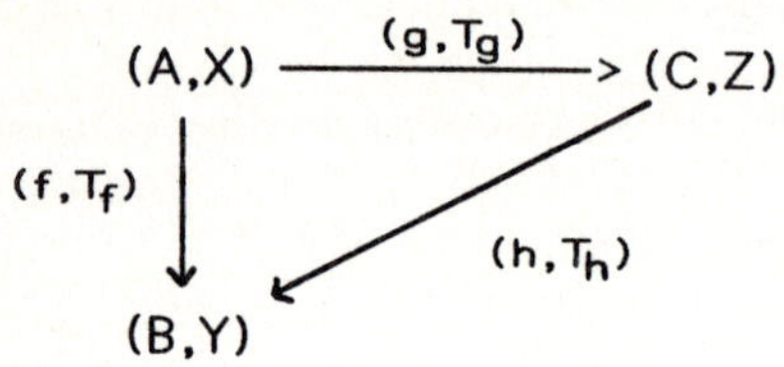

is commutative , i.e. (B,Y) is a morphic extension of (C,Z) via the map h and $h \circ g = f$.

<u>Remarks.</u> i) If (B,Y) is a morphic extension of (A,X) then f is necessarily injective, since A separates the points of X. So we can interpret the elements of B as extended functions of A. Part ii) guarantees that Y is the largest space to which all elements of A can be extended.

ii) If (B,Y) is a morphic extension of (A,X) via $f\colon X \longrightarrow Y$ and (C,Z) is a morphic extension of (B,Y) via $g\colon Y \longrightarrow Z$, then (C,Z) is a morphic extension via $g \circ f\colon X \longrightarrow Z$, since

$$T_{g\circ f}(c) = T_f \circ T_g(c) \text{ for all } c \in C.$$

iii) The structure map $j\colon X \longrightarrow M(A)$ is continuous and

$$T_j\colon \hat{A} \longrightarrow A,\ \hat{f} \longrightarrow \hat{f} \circ j = f$$

is just the inverse of the mapping Γ (cf. 3.2.12) and hence a topological algebra isomorphism by (4.1.3)), so $(\hat{A},M(A))$ is a morphic extension of (A,X).

Before we show that $(\hat{A},M(A))$ is the morphic hull of (A,X) we prove two lemmata. Recall from (3.2.5)iv) that

$$T_f^*: M(A) \longrightarrow M(B),\ \varphi \longrightarrow \varphi \circ T_f,$$

is a homeomorphism, if B is a morphic extension of A via f.

(4.2.3) Lemma. If (B,Y) is a morphic extension of (A,X) via the continuous map f, then the diagram

$$\begin{array}{ccc} X & \xrightarrow{\quad f \quad} & Y \\ {\scriptstyle j_X}\downarrow & & \downarrow{\scriptstyle j_Y} \\ M(A) & \xleftarrow{\ (T_f^*)^{-1}\ } & M(B) \end{array}$$

is commutative, where j_X (resp. j_Y) denote the structure maps.

Proof. Since T_f^* is a homeomorphism it suffices to show that

$$T_f^* \circ j_X = j_Y \circ f.$$

Let $z \in X$, and let $b \in B$ be arbitrary elements, then

$$\hat{b}(T_f^* \circ j_X(z)) = \hat{b}(\varphi_z \circ T_f) = \varphi_z(T_f(b)) = \varphi_z(b \circ f) =$$
$$= b \circ f(z) = \hat{b}(j_Y \circ f(z)).$$

Our assertion follows, since $\hat{B}$ separates the points of M(B) □

(4.2.4) Lemma. Let A,B be two uF-algebras, and let $T: B \longrightarrow A$ be a topological algebra isomorphism, then $(\hat{B}, M(B))$ is a morphic extension of $(\hat{A}, M(A))$ via the adjoint spectral map T^*.

Proof. The map

$$s: \hat{B} \longrightarrow \hat{A},\ \hat{b} \longrightarrow \hat{b} \circ T^*,$$

is a well-defined injective algebra homomorphism. It is continuous, since T^* is a homeomorphism. Let $\hat{a} \in \hat{A}$ be an arbitrary element. Let $b \in B$ such that $T(b) = a$, then

$$\hat{b} \circ T^*(\varphi) = \varphi(T(b)) = \hat{a}(\varphi) \text{ for all } \varphi \in M(A),$$

hence s is a topological isomorphism by the open mapping theorem. □

<u>(4.2.5) Theorem .</u> $(\hat{A}, M(A))$ is the morphic hull of (A,X).

<u>Proof.</u> We have already remarked that $(\hat{A}, M(A))$ is a morphic extension of (A,X).

Let (B,Y) be another morphic extension with respect to the map $f: X \longrightarrow Y$.

Define

$$h: Y \longrightarrow M(A),\ y \longrightarrow (T_f^*)^{-1} \circ j_Y, \text{ then the diagram}$$

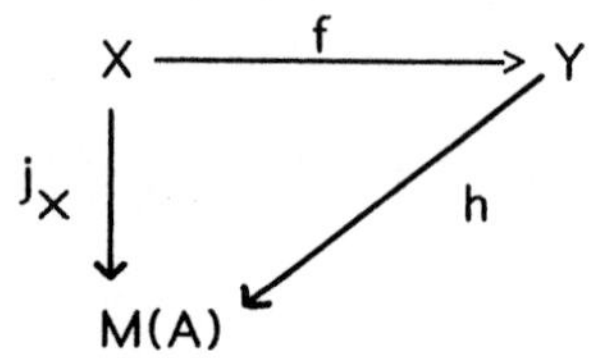

is commutative by (4.2.3) and $(\hat{A}, M(A))$ is a morphic extension of (B,Y) via h by lemma (4.2.4) and remark (4.2.2)ii)-iii), since

$$(T_f^*)^{-1} = (T_f^{-1})^*$$

by the proof of lemma (3.2.5)iv).

Let $h_1, h_2 : Y \longrightarrow M(A)$ be two maps such that the diagram

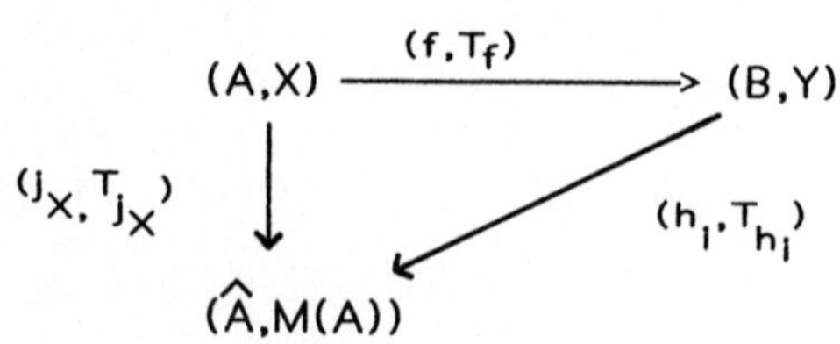

is commutative for i=1,2. Then

$$T_f(\hat{a} \circ h_i) = \hat{a} \circ h_i \circ f = \hat{a} \circ j_X = a \quad \text{for all } a \in A,\ i=1,2,$$

hence

$$\hat{a} \circ h_i = T_f^{-1}(a) \quad \text{for all } a \in A,\ i=1,2,$$

and we have $h_1 = h_2$ since $\hat{A}$ separates the points of $M(A)$. □

(4.3) Convexity for uF-algebras

Recall from (2.2.6) that a domain $G \subset \mathbb{C}^n$ (or more general a reduced complex space X) is said to be holomorphically convex if the holomorphically convex hull of each compact subset $K \subset G$ is again a compact subset of G. As noted in (2.2.7) G is holomorphically convex iff $M(\mathrm{Hol}(G)) = G$.

In this section we introduce convexity for uF-algebras in an analogous way.

(4.3.1) Definition. Let (A,X) be a uF-algebra, and let $K \subset X$ be a compact set. We call the set

$$\hat{K}_{(A,X)} = \{x \in X: |f(x)| \le \|f\|_K, \text{ for all } f \in A\}$$

the A-convex hull of K in X. We say that a subset $S \subset X$ is A-convex if $\hat{K}_{(A,X)} \subset S$ for each compact set $K \subset S$. If $K \subset M(A)$ is compact we use the abbreviation $\hat{K}_A = \hat{K}_{(\hat{A},M(A))}$.

(4.3.2) Example. If $K \subset X$ is compact then in general $\hat{K}_{(A,X)}$ is not compact. For example let $\mathbb{C}^* = \mathbb{C}\backslash\{0\}$ and set $A = \mathrm{Hol}(\mathbb{C})|_{\mathbb{C}^*}$. It is easy to show that A is a complete subalgebra of $C(\mathbb{C}^*)$, hence $(A,\mathbb{C}^*)$ is a uF-algebra by (4.1.3). Set $\partial D = \{x \in \mathbb{C}^*: |x| = 1\}$. Then

$$\widehat{\partial D}_{(A,\mathbb{C}^*)} = \{ x \in \mathbb{C}^*: |x| \le 1\},$$

since each $f \in (A,\mathbb{C}^*)$ is the restriction of an element $\tilde{f} \in \mathrm{Hol}(\mathbb{C})$.

Definition. Let (A,X) be a uF-algebra. We say that X is A-morphically convex if $\hat{K}_{(A,X)}$ is compact for every compact set $K \subset X$.

<u>Remark.</u> As noted above a domain $G \subset \mathbb{C}^n$ is holomorphically convex iff $M(\mathrm{Hol}(G)) = G$. Of course this result is not valid in our general setting. Take for example $(P(\partial D),\partial D)$, then ∂D is $P(\partial D)$-morphically convex, but $M(P(\partial D)) = \{x \in \mathbb{C}: |x| \le 1\}$ (cf. (1.2.4)ii)).
However we can prove that $M(A)$ is $\hat{A}$-morphically convex.

<u>(4.3.3) Theorem.</u> Let A be a uF-algebra, and let $K \subset M(A)$ be a compact set. Then $M(\hat{A}_K)$ can be identified with $\hat{K}_A$ and $\hat{K}_A$ is a compact set. In particular $M(A)$ is $\hat{A}$-morphically convex.

<u>Proof.</u> Since A is a uF-algebra $\Gamma: A \longrightarrow \hat{A},\ f \longrightarrow \hat{f}$, is a topological algebra isomorphism, cf. (4.1.3), hence we can naturally identify $M(A)$ and $M(\hat{A})$. Since the restriction mapping

$$r: \hat{A} \longrightarrow \hat{A}_K,\ \hat{f} \longrightarrow \hat{f}|_K,$$

has dense range the adjoint spectral map

$$r^*: M(\hat{A}_K) \longrightarrow M(\hat{A}) = M(A)$$

is continuous and injective (cf. (3.2.5)ii)). $M(\hat{A})_K$ - as the spectrum of a uB-algebra - is compact, hence

$$r^*: M(\hat{A}_K) \longrightarrow r^*(M(\hat{A}_K))$$

is a homeomorphism. Our assertions will follow if we can show that $r^*(M(\hat{A}_K)) = \hat{K}_A$.
Let $\varphi \in M(\hat{A}_K)$ and $\hat{f} \in \hat{A}$ be arbitrary elements, then

$$|\hat{f}(r^*(\varphi))| = |\varphi(r(\hat{f}))| \le \|\hat{f}\|_K,$$

for the inequality cf. (3.2.2)ii).
On the other hand, if $\varphi \in \hat{K}_A$ then

$$\tilde{\varphi}: \hat{A}|_K \longrightarrow \mathbb{C},\ \hat{f}|_K \longrightarrow \hat{f}(\varphi),$$

is a well-defined mapping, since $\hat{f}|_K \equiv 0$ implies that $\hat{f}|_{\hat{K}_A} \equiv 0$ and hence in particular $\hat{f}(\varphi) = 0$. Clearly $\tilde{\varphi}$ is an algebra homomorphism.

It is continuous since

$$|\tilde{\varphi}(\hat{f}|_K)| = |\hat{f}(\varphi)| \le \|\hat{f}\|_{\hat{K}_A} = \|\hat{f}\|_K \quad \text{for all } f \in A.$$

Hence $\tilde{\varphi}$ can be extended to an element of $M(\hat{A}_K)$ and $r^*(\tilde{\varphi}) = \varphi$. □

<u>(4.3.4) Remarks.</u> i) It follows from the theorem that we can always choose an admissible exhaustion $(K_n)_n$ of $M(A)$, which is moreover $\hat{A}$-convex, i.e. $K_n = (\hat{K}_n)_A$ for all $n \in \mathbb{N}$.

ii) If (A,X) is a uF-algebra and $K \subset X$ is a compact subset, then it is easy to show that A_K is topologically and algebraically isomorphic to $\hat{A}_{j(K)}$ ($j: X \longrightarrow M(A)$ the structure map, cf. (4.2.1)). Hence we can identify $M(A_K)$ with $j(K)\hat{}_A$, the $\hat{A}$-convex hull of $j(K)$.

iii) Let $L \subset X$ be a hemicompact set, and let $(K_n)_n$ be an admissible exhaustion of L. Recall the definition of A_L from (4.1.5). Then

$$M(A_L) = \cup\ j(K_n)\hat{}_A,$$

and $(j(K_n)\hat{}_A)_n$ is an admissible exhaustion of $M(A_L)$ by theorem (4.1.6) and ii).

In particular if $L \subset M(A)$ is a hemicompact and $\hat{A}$-convex set we can identify $M(\hat{A}_L)$ and L as sets.

Let $L \subset M(A)$ be a hemicompact and $\hat{A}$-convex subset, and let $K \subset L$ be a compact subset. Identify L and $M(\hat{A}_L)$ as sets. We claim that on K the relative topology (inherited from $M(\hat{A}_L)$) and the original topology (inherited from $M(A)$) coincide.

We interpret the elements of $\hat{A}_L$ as complex-valued functions on L which can be approximated uniformly on each compact set of L by elements of $\hat{A}$, cf. (4.1.5). Since $\hat{f}|_L \in \hat{A}_L$ for each $\hat{f} \in \hat{A}$, we see that the original topology on K is coarser. To prove that both topologies coincide we have only to show by (3.2.2)i) that sets of the form

$$U_{f_1,\dots,f_r,\varphi} = \{\psi \in K: |\hat{f}_i(\varphi)-\hat{f}_i(\psi)| < 1,\ i=1,\dots,r\}$$

$\hat{f}_1,\dots,\hat{f}_r \in \hat{A}_L, \varphi \in K$, are a neighbourhood of φ with respect to the original topology. To see this choose $\hat{g}_1,\dots,\hat{g}_r \in \hat{A}$ so that

$$\|\hat{f}_i-\hat{g}_i\|_K < 1/4 \quad , i=1,\dots,r,$$

then

$$\{\psi \in K: |\hat{g}_i(\psi)-\hat{g}_i(\varphi)| < 1/4,\ i=1,\dots,r\} \subset U_{f_1,\dots,f_r,\varphi}.$$

In particular we get the following result:
Let $U \subset M(A)$ be a hemicompact, $\hat{A}$-convex and locally compact subset. Then $M(\hat{A}_U)$ can be identified with U as a topological space, and we have $\hat{A}_U \subset C(U)$.

<u>(4.3.5)</u> We next deal with the question when X is A-morphically convex.

<u>Lemma.</u> Let (A,X) be a uF-algebra. Then X is A-morphically convex iff the structure map j is a proper mapping, i.e. $j^{-1}(K)$ is compact for each compact set $K \subset M(A)$.

<u>Proof.</u> i) If j is a proper mapping and $K \subset X$ is compact, then by theorem (4.3.3) $j(K)^{\wedge}_A$ and hence

$$\hat{K}_{(A,X)} = j^{-1}(j(K)^{\wedge}_A)$$

is compact.
ii) Let $K \subset M(A)$ be a compact subset. Choose an admissible exhaustion $(K_n)_n$ of X, then $(j(K_n)^{\wedge}_A)_n$ is an admissible exhaustion of M(A) by (4.3.4)iii). Hence there is $l \in \mathbb{N}$ such that

$$K \subset j(K_l)^{\wedge}_A,$$

so $j^{-1}(K)$, as a closed subset of the compact set

$$(K_l)_{(A,X)} = j^{-1}(j(K_l)^{\wedge}_A),$$

is compact. □

(4.3.6) The next theorem is again motivated by a well known result in the theory of several complex variables (cf. (2.2.7)). The proof of this theorem can be extended to the case of uF-algebras.

Theorem. Let (A,X) be a uF-algebra, then X is A-morphically convex iff for every sequence $(x_n)_n \subset X$ which has no accumulation point in X there exists a function $f \in A$ such that $\sup\{|f(x_n)|,\ n \in \mathbb{N}\} = \infty$.

Proof. i) Assume that X is A-morphically convex, and let $(x_n)_n$ be a sequence which has no accumulation point in X.
Choose an A-convex admissible exhaustion $(K_n)_n$ of X, w.l.o.g. we can assume that

$$x_n \in K_{n+1} \backslash K_n \quad \text{for all } n \in \mathbb{N}.$$

Now we choose inductively functions $f_n \in A$ so that

a) $\|f_n\|_{K_n} \le 2^{-n}$ and

b) $\left|f_n(x_n)\right| \ge n + \left|\sum_{k=1}^{n-1} f_k(x_n)\right|$.

This is always possible, because there exist functions $g_n \in A$ so that $\|g_n\|_{K_n} < 1 < |g_n(x_n)|$, since K_n is A-convex.
Then $f_n = (g_n)^{l(n)}$ fulfills a) and b) for sufficiently large $l(n) \in \mathbb{N}$.
The series $\sum f_n$ defines an element f in A by a) and we have

$$\left|f(x_n)\right| \ge \left|f_n(x_n)\right| - \left|\sum_{k=1}^{n-1} f_k(x_n)\right| - \left|\sum_{k=n+1}^{\infty} f_k(x_n)\right|$$

$$\ge n - \sum_{k=n+1}^{\infty} 2^{-k} > n-1.$$

ii) We argue indirectly. Assume there is a compact set $K \subset X$ such that $\hat{K}_{(A,X)}$ is not compact.
X is a Lindelöf space, i.e. each open cover of X has a countable sub-

cover, since X is hemicompact. Hence $\hat{K}_{(A,X)}$ - as a closed subset of X - is a Lindelöf space. Since $\hat{K}_{(A,X)}$ is not compact there exists a sequence $(x_n)_n \subset \hat{K}_{(A,X)}$, which has no accumulation point in $\hat{K}_{(A,X)}$ (cf. [KEL], p. 137). Hence $(x_n)_n$ has no accumulation point in X. So there is a function $f \in A$ such that $\sup\{|f(x_n)|, n \in \mathbb{N}\} = \infty$. But this is a contradiction to

$$|f(x_n)| \le \|f\|_K \text{ for all } n \in \mathbb{N}.$$

□

<u>(4.3.7)</u> We close our remarks about convexity with some further results, which will be needed later.
Let (A,X) be a uF-algebra. For a subfamily of functions $B \subset A$ we can analogously define the concept of B-convexity, for instance

$$\hat{K}_{(B,X)} = \{x \in X: |f(x)| \le \|f\|_K \text{ for all } f \in B\}$$

is called the B-convex hull of a compact set $K \subset X$.
If B is a dense subset of A, then a set $M \subset X$ is A-convex iff it is B-convex. So $M \subset \mathbb{C}^n$ is polynomially convex, i.e. the polynomially convex hull $\hat{K}$ of each compact set $K \subset M$ is contained in M, iff it is $Hol(\mathbb{C}^n)$-convex.

<u>Proposition.</u>i) Let A be a uF-algebra, and let $\varphi \in M(A)$. Then there exists a basis for the neighbourhood system of φ consisting of $\hat{A}$-convex open (resp. closed) subsets.
ii)If φ has a compact neighbourhood, then the elements of this basis can be chosen so that they are moreover hemicompact and relatively compact.

<u>Proof.</u> i) By (3.2.2)i) we only have to show that sets of the form

$$U_{f_1,\dots,f_r} = \{\psi \in M(A): |\hat{f}_i(\psi)-\hat{f}_i(\varphi)| < 1,\ i=1,\dots,r\},$$

$f_1,\dots,f_r \in A$, are $\hat{A}$-convex.

Let $K \subset U_{f_1,\dots,f_r}$ be a compact set, and let $x \in M(A)\backslash U_{f_1,\dots,f_r}$. W.l.o.g. $|\hat{f}_1(x)-\hat{f}_1(\varphi)| \geq 1$, hence $|\hat{g}(x)| > \|\hat{g}\|_K$ for $g = f_1-\hat{f}_1(\varphi)$.

ii) Let K be a compact neighbourhood of φ. Choose $f_1,\dots,f_r \in A$ by (3.2.2)i) so that

$$U_{f_1,\dots,f_r,\varphi} = \{\psi \in M(A)\colon |\hat{f}_i(\psi)-\hat{f}_i(\varphi)| < 1\} \subset K,$$

Then $U_{f_1,\dots,f_r,\varphi}$ is $\hat{A}$-convex and relatively compact. It is moreover hemicompact, since

$$U_n = \{\psi \in M(A)\colon |\hat{f}_i(\psi)-\hat{f}_i(\varphi)| \leq 1-(1/n)\},\ n \in \mathbb{N},$$

is an admissible exhaustion. □

Example. The closure of an A-convex set need not be A-convex. For instance consider the uF-algebra $(\mathrm{Hol}(\mathbb{C}),\mathbb{C})$ and define

$$G = \{z \in \mathbb{C}\colon |z+1| < 2 \text{ and } |z| > 1\}.$$

Then G is a simply connected domain, hence G is polynomially convex ([NAR 1], p. 151), hence - by the remark above - G is also $\mathrm{Hol}(\mathbb{C})$-convex, but $\overline{G}$ - the closure of G - is not polynomially convex, since it contains $\{z \in \mathbb{C}\colon |z+1| = 2\}$.

(4.4) Uniform Fréchet algebras with locally compact spectrum

Many of our later results will be valid only for uF-algebras with locally compact spectrum. Although the hypothesis "locally compact spectrum" is quite elementary, it would be important to find criterions for A which imply this hypothesis. No such criterions seem to be known, besides very special situations, cf. for instance (18.2.8).

In [H/V 1] Hayes and Vigue constructed even a reduced complex space X with countable topology such that M(Hol(X)) is not locally compact at any point.

(4.4.1) We first state a sufficient topological condition.

Proposition. If a hemicompact space X satisfies the first axiom of countability then X is locally compact.

Proof. Let $(K_n)_n$ be an admissible exhaustion of X. Assume there is a point $x \in X$ without compact neighbourhood. Let $(U_n)_n$ be a countable basis for the neighbourhood system of x. Choose for every $n \in \mathbb{N}$ a point $x_n \in U_n \backslash K_n$, then

$$K = \{x_n : n \in \mathbb{N}\} \cup \{x\}$$

is compact, but there is no $n \in \mathbb{N}$ such that $K \subset K_n$. This is a contradiction.

□

With regard to the problem described above it would be enough to find criterions for A which imply the axiom of first countability for the spectrum M(A). Unfortunately in the next theorem we have to assume a priori that M(A) is locally compact.

(4.4.2) Theorem [KRA 1]. Let A be a uF-algebra with locally compact spectrum M(A), and let $\varphi \in M(A)$. Suppose there is a sequence $(f_n)_n$ in A so that the ideal $(f_1, f_2, \ldots)$ lies dense in the maximal ideal $\ker \varphi = \{f \in A : \varphi(f) = 0\}$, then there exists a countable basis for the neighbourhood system of φ.

For the proof we use a proposition.

Proposition. Let X be a locally compact Hausdorff space. Suppose that $x \in X$ is a G_δ-point, i.e. there exists a sequence of open neighbourhoods $(U_n)_n$ such that $\{x\} = \bigcap_{n \in \mathbb{N}} U_n$.

Then x has a countable neighbourhoodbasis.

Proof. Let K be a compact neighbourhood of x. Set

$$L_1 = (\text{int } K) \cap U_1.$$

Choose a compact neighbourhood Q_1 of x such that $Q_1 \subset L_1$, and set

$$L_2 = (\text{int } Q_1) \cap U_2.$$

Continuing in this manner we get a sequence $(L_n)_n$ of open neighbourhoods such that

$$\ldots \supset L_n \supset \overline{L}_{n+1} \supset L_{n+1} \supset \overline{L}_{n+2} \supset \ldots$$

where $\overline{L}_n$ denotes the closure of L_n. By our assumption we have

$$(*) \qquad \bigcap_{n \in \mathbb{N}} \overline{L}_n = \{x\}.$$

Now let W be an arbitrary open neighbourhood of x. Let $y \in K \setminus W$ be an arbitrary point. By (*) there is $n \in \mathbb{N}$ such that $y \in X \setminus \overline{L}_n$. By the compactness of $K \setminus W$ there is $m \in \mathbb{N}$ so that $L_m \cap (K \setminus W) = \emptyset$, i.e. $L_m \subset W$, and we are done. □

Proof of (4.4.2). For $n \in \mathbb{N}$ set

$$U_n = \{\psi \in M(A) : |\hat{f}_i(\psi)| < 1/n,\ i=1,\ldots,n\}.$$

Then $\bigcap U_n = \{\varphi\}$. □

(4.4.3) We say that a uF-algebra is topologically countably generated if there is a sequence $(f_n)_n$ in A such that $\mathbb{C}[f_1, f_2, \ldots]$, the ring of polynomials in $f_1, f_2, \ldots$ lies dense in A. Note that in this case A is separable.

Corollary. Let A be a separable uF-algebra with locally compact spectrum M(A). Then M(A) satisfies the first axiom of countability.

Proof. Let $(g_n)_n$ be a countable dense subset of A, and let $\varphi \in M(A)$

be an arbitrary point, then the ideal $(g_1-\hat{g}_1(\varphi), g_2-\hat{g}_2(\varphi), \ldots)$ lies dense in ker φ.

□

Remarks. i) Both results - even for finitely generated uF-algebras - are in general not valid if we drop the assumption "locally compact spectrum". Otherwise the spectrum of every finitely generated uF-algebra would be automatically locally compact by (4.4.1). This is false as we shall see in (5.1.12).

ii) On the other hand let K be a compact Hausdorff space which does not satisfy the first axiom of countability. Then (C(K),K) is an example of a uB-algebra with (locally) compact spectrum, which does not satisfy the first axiom of countability.

iii) In (5.1.12) we shall give a simple example of a uF-algebra (A,X) on a locally compact space X such that M(A) is not locally compact (see also the example of Hayes and Vigue, mentioned above).

CHAPTER 5

FINITELY GENERATED F-ALGEBRAS

In this chapter we deal with a class of F-algebras which admit a finite set of topological generators. In this case the problem of determining the spectrum is equivalent to the problem of determining the polynomially convex hull of compact subsets of $\mathbb{C}^n$.

(5.1.1) Definition. We call an F-algebra A (topologically) n-generated provided there exist elements $f_1,\dots,f_n$ in A such that A is the closure of the polynomials in $f_1,\dots,f_n$.

(5.1.2) Examples. i) Let $G \subset \mathbb{C}^m$ be a polynomially convex open set, i.e. there exists an admissible exhaustion $(K_n)_n$ of G consisting of polynomially convex compact subsets. Let $f \in \mathrm{Hol}(G)$, then f can be approximated uniformly on each K_n by polynomials (cf. (1.4.5)). So Hol(G) is n-generated by the coordinate functions $z_1,\dots,z_n$.
ii) Using the theorem of Stone-Weierstraß we see that $C(\mathbb{R})$ is singly generated by the function $z: \mathbb{R} \longrightarrow \mathbb{C}$, $x \longrightarrow x$.
iii) Let $f \in C^\infty([0,1])$ be an arbitrary element, and let $n \in \mathbb{N}$. By the theorem of Stone-Weierstraß we can choose a sequence of polynomials $(p_k)_k$ such that

$$\|f^{(n)}-p_k\|_{[0,1]} \longrightarrow 0, \text{ as } k \longrightarrow \infty,$$

where $f^{(n)}$ denotes the n-th derivative of f. Consider on $[0,1]$ the polynomials

$$\tilde{p}_k(x) = \int_0^x p_k(t)\, dt - f^{(n-1)}(0).$$

Then

$$|f^{(n-1)}(x)-\tilde{p}_k(x)| = |\int_0^x f^{(n)}(t)-p_k(t)\, dt\, |$$

$$\leq \|f^{(n)}-p_k\|_{[0,1]} \longrightarrow 0, \text{ as } k \longrightarrow \infty.$$

Continuing in this manner we get a sequence of polynomials $(q_k)_k$ so that

$$\|f^{(i)}-q_k^{(i)}\|_{[0,1]} \longrightarrow 0, \text{ as } k \longrightarrow \infty, \text{ for } i=0,\dots,n.$$

It follows that $C^\infty([0,1])$ is singly generated by the function

$$z: \mathbb{R} \longrightarrow \mathbb{C},\ x \longrightarrow x.$$

For more information about the algebra $C^\infty(M)$, M a differentiable manifold, we refer the reader to the book of Mallios [MAL].

<u>(5.1.3) Remarks.</u> i) Generating elements are not uniquely determined. For example every real-valued function $f \in C(\mathbb{R})$, which separates the points of $\mathbb{R}$, generates $C(\mathbb{R})$ by the theorem of Stone-Weierstraß. ii) We do not assume that the number n is minimally chosen, so if A is m-generated, then A is also n-generated for $n \geq m$.

Next we want to reformulate our main theorems for finitely-generated uF-algebras. First we consider the joint spectrum of the generating elements.

<u>(5.1.4)</u> Let A be an n-generated F-algebra with generating elements $f_1,\dots,f_n$. We denote by $\hat{F}$ the mapping

$$\hat{F}: M(A) \longrightarrow \mathbb{C}^n,\ \varphi \longrightarrow (\hat{f}_1(\varphi),\dots,\hat{f}_n(\varphi)).$$

Clearly $\hat{F}$ is continuous. If $\hat{F}(\varphi) = \hat{F}(\psi)$, then

$$\hat{p}(f_1,\dots,f_n)(\varphi) = \hat{p}(f_1,\dots,f_n)(\psi)$$

for all polynomials p, where $\hat{p}(f_1,\ldots,f_n)$ denotes the Gelfand transform of $p(f_1,\ldots,f_n)$. Hence $\hat{g}(\varphi) = \hat{g}(\psi)$ for all $g \in A$, and we have $\varphi = \psi$, i.e. $\hat{F}$ is injective.

(5.1.5) Proposition. Let A be a uF-algebra with generating elements $f_1,\ldots,f_n$, and let $K \subset M(A)$ be an $\hat{A}$-convex compact set. Then $\hat{F}(K)$ is a polynomially convex compact set and $\hat{A}_K$ is topologically and algebraically isomorphic to $P(\hat{F}(K))$.

Proof. By our assumptions $\hat{A}_K$ is n-generated by $\hat{f}_1|_K,\ldots,\hat{f}_n|_K$. Since $M(\hat{A}_K) = K$ by (4.3.3), the result follows from (1.2.13). □

(5.1.6) Let $(K_n)_n$ be a sequence of compact sets in $\mathbb{C}^m$ such that $K_n \subset K_{n+1}$ for all $n \in \mathbb{N}$. Consider the dense projective system of uB-algebras

$$\ldots \longrightarrow P(K_{n+1}) \xrightarrow{r_n} P(K_n) \longrightarrow \ldots$$

with the restriction mappings

$$r_n : P(K_{n+1}) \longrightarrow P(K_n),\; f \longrightarrow f|_{K_n} .$$

Then $A = \varprojlim P(K_n)$ is a uF-algebra by (4.1.3), the spectrum of A can be identified as a set with $\cup \hat{K}_n$, where $\hat{K}_n$ denotes the polynomially convex hull of K_n, and $(\hat{K}_n)_n$ is an admissible exhaustion with respect to the Gelfand topology (cf. theorem (4.1.6) and (1.2.4)ii)).

Clearly A is m-generated by the coordinate functions $z_1,\ldots,z_m$.

(5.1.7) Remarks. i) Let A be as above. Then we can interpret each element $\hat{f} \in \hat{A}$ as a complex-valued function on $\cup \hat{K}_n$, which can be approximated uniformly on each $\hat{K}_n$ by polynomials.

ii) On every $\hat{K}_n$ the induced Gelfand topology and the induced euclidean topology are equivalent (cf. (1.2.4)ii)).

In general the Gelfand topology on $\cup \hat{K}_n$ is finer than the induced euclidean topology, since the coordinate functions $z_1,\dots,z_m$ define elements of A by i) and for each $x \in \cup \hat{K}_n$, $\varepsilon>0$ the set

$$\{y \in \cup \hat{K}_n : |z_i(x)-z_i(y)| < \varepsilon,\ i=1,\dots,m\}$$
$$= \{y \in \cup \hat{K}_n : |x_i-y_i| < \varepsilon,\ i=1,\dots,m\}$$

is Gelfand open (cf. (3.2.2)i)).
Both topologies are equivalent iff each element $\hat{f} \in \hat{A}$ is continuous with respect to the euclidean topology on $\cup \hat{K}_n$.
Suppose that $(\hat{K}_n)_n$ is an admissible exhaustion of $\cup \hat{K}_n$ with respect to the euclidean topology. Let $\hat{f} \in \hat{A}$ be an arbitrary element, and let $(x_n)_n$ be a sequence in $\cup \hat{K}_n$ which converges to $x_o \in \cup \hat{K}_n$ (euclidean topology). Then $K = \{x_n : n\geq 0\}$ is a compact set (euclidean topology), hence there exists $l \in \mathbb{N}$ such that $K \subset \hat{K}_l$. Since on $\hat{K}_l$ both topologies coincide, we see that $(\hat{f}(x_n))_n$ converges to $\hat{f}(x_o)$ and it follows that $\hat{f}$ is continuous with respect to the euclidean topology. Hence both topologies coincide iff $(\hat{K}_n)_n$ is an admissible exhaustion of $\cup \hat{K}_n$ with respect to the euclidean topology.
We give a simple example, where both topologies do not coincide. For $n \in \mathbb{N}$ set

$$K_n = \{\exp(it) : 2\pi -(1/n) \geq t \geq 0\}.$$

Then each K_n is polynomially convex, since $\mathbb{C}\backslash K_n$ is connected (cf. (1.2.4)ii)) and $\cup K_n = \{z \in \mathbb{C} : |z| = 1\} = \partial D$. Clearly ∂D is a compact set (euclidean topology) but there is no $l \in \mathbb{N}$ so that $\partial D \subset K_l$.

<u>Example.</u> Let $G \subset \mathbb{C}^n$ be a polynomially convex domain, and let $(K_n)_n$ be an admissible polynomially convex exhaustion of G. Then

$$T: \mathrm{Hol}(G) \longrightarrow \varprojlim P(K_n),\ f \longrightarrow (f|_{K_n})_n,$$

defines a topological and algebraical isomorphism by (1.4.5). Hence

$$M(Hol(G)) = \cup K_n = G \quad \text{as sets.}$$

Since $(K_n)_n$ is an admissible exhaustion of G with respect to the euclidean topology we see that M(Hol(G)) = G as topological spaces by (5.1.7)ii). (We could have also derived this result from (2.2.7), since G is holomorphically convex.)

(5.1.8) Structure theorem for finitely generated uF-algebras.

Let

$$\cdots \longrightarrow A_{n+1} \longrightarrow A_n \longrightarrow \cdots$$

be a dense projective system of uB-algebras. Suppose that $A = \varprojlim A_n$ is n-generated by $f_1,\ldots,f_n$. Then

i) A is topologically and algebraically isomorphic to $\varprojlim P(\hat{F}(M(A_k)))$;

ii) the spectrum M(A) can be identified as a set with $\hat{F}(M(A))$ and $(\hat{F}(M(A_k)))_k$ is an admissible exhaustion of $\hat{F}(M(A))$ with respect to the Gelfand topology, consisting of polynomially convex compact subsets, where $\hat{F}$ is defined as in (5.1.4).

Proof. Denote by p_k the norm of A_k.
By (3.3.4) $M(A) = \cup M(A_k)$ and $(M(A_k))_k$ is an admissible exhaustion of M(A). By (4.1.3) and (4.1.6) A is topologically and algebraically isomorphic to $\hat{A} = \varprojlim \hat{A}_{M(A_k)}$. Note that $\hat{A}$ is n-generated by $\hat{f}_1,\ldots,\hat{f}_n$.
Since $p_k(f) = \|\hat{f}\|_{M(A_k)}$ for all $f \in A_k$ by (1.2.10), we have by (3.2.6) (*)

$$M(A_k) = \{\varphi \in M(A)\colon |\hat{f}(\varphi)| \le \|\hat{f}\|_{M(A_k)} \text{ for all } \hat{f} \in \hat{A}\},$$

i.e. $M(A_k)$ is $\hat{A}$-convex. So $\hat{A}_{M(A_k)}$ is isomorphic to $P(\hat{F}(M(A_k)))$ by (5.1.5). For each $k \in \mathbb{N}$ the diagram

$$\begin{array}{ccc} \hat{A}_{M(A_{k+1})} & \longrightarrow & \hat{A}_{M(A_k)} \\ \downarrow & & \downarrow \\ P(\hat{F}(M(A_{k+1}))) & \longrightarrow & P(\hat{F}(M(A_k))) \end{array}$$

is commutative with respect to the natural restriction mappings and the isomorphism given in (5.1.5).

Hence $A = \varprojlim A_k$ is topologically and algebraically isomorphic to $\varprojlim P(\hat{F}(M(A_k)))$ and our assertion follows from the considerations in (5.1.6) and (5.1.5).

□

<u>(5.1.9)</u> We have seen that $\hat{F}$ is an injective and continuous map. We shall now discuss the question, whether $\hat{F}$ is a homeomorphism. We start with an obvious result.

<u>Proposition.</u> Let A and $\hat{F}$ be as above. If $\hat{F}: M(A) \longrightarrow \hat{F}(M(A))$ is a homeomorphism then M(A) is metrizable.

<u>(5.1.10)</u> Recall that for finitely generated uB-algebras B the map $\hat{F}$ is always a homeomorphism and $\hat{F}(M(B))$ is polynomially convex (cf. (1.2.13)). The next examples - due to Brooks [BRO 3] - show that these results are not valid for F-algebras.

<u>Example.</u> We construct a finitely generated uF-algebra A, such that for no generating elements $f_1,\dots,f_r$ the map $\hat{F}$ is a homeomorphism.

Let $(t_n)_n$ be a decreasing sequence of positive numbers such that $t_1 < \pi/2$ and $\lim t_n = 0$. Let L_n denote the segment $[0,\exp(it_n)]$ in the plane and set

$$K_n = \bigcup_{i=1}^{n} L_i .$$

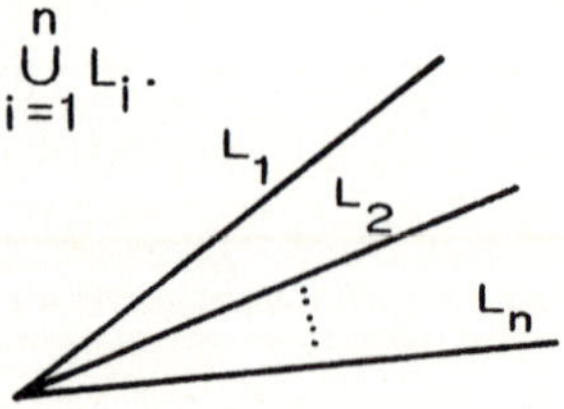

Each K_n is polynomially convex (cf. (1.2.4)ii)). By (5.1.6) the spectrum of the singly generated uF-algebra $A = \varprojlim P(K_n)$ can be identified with $\cup K_n$ and $(K_n)_n$ is an admissible exhaustion with respect to the Gelfand topology on $\cup K_n$.
We shall show that M(A) is not metrizable, which yields our result by proposition (5.1.9).
Suppose there is a metric d on $\cup K_n$ which defines the Gelfand topology. Since Gelfand topology and euclidean topology coincide on each K_n by (5.1.7)ii), we can choose for every $n \in \mathbb{N}$ an element $x_n \in L_n \setminus \{0\} \subset K_n$ so that $d(x_n,0) < 2^{-n}$.
Hence $K = \{0\} \cup \{x_n : n \in \mathbb{N}\}$ is a compact set with respect to the topology defined by d on $\cup K_n$, but K is not contained in any K_n, therefore K cannot be Gelfand compact. This is a contradiction.

<u>(5.1.11) Example.</u> We construct a singly generated uF-algebra such that $\hat{F}(M(A))$ is not polynomially convex.
Recall that a set $M \subset \mathbb{C}^n$ is called polynomially convex provided that the polynomially convex hull $\hat{K}$ of each compact set $K \subset M$ is contained in M.
Let $K_n = [0,2\pi-(1/n)]$, then K_n is polynomially convex (cf. (1.2.4)ii)), and let $A = \varprojlim P(K_n)$. Then $M(A) = [0,2\pi)$ and Gelfand topology and euclidean topology are equivalent on $[0,2\pi)$ by (5.1.7)ii).
We have $P(K_n) = C(K_n)$ by the theorem of Stone-Weierstraß, thus $A = C([0,2\pi))$.
Equally the function

$$g: [0,2\pi) \longrightarrow \mathbb{C},\ t \longrightarrow \exp(it)$$

generates A. But $\hat{g}([0,2\pi)) = \{z \in \mathbb{C} : |z|=1\}$ is not polynomially convex. Moreover $\hat{g}$ is not a homeomorphism, while $\hat{f}$ is a homeomorphism for the generating element

$$\hat{f}: [0,2\pi) \longrightarrow \mathbb{C},\ t \longrightarrow t.$$

So we have seen that there are singly generated uF-algebras such that $\hat{f}$ is a homeomorphism for one generating element, while $\hat{f}$ fails to be a homeomorphism for another generating element.

<u>(5.1.12) Example.</u> We construct a doubly generated uF-algebra (A,X) on a locally compact space X such that M(A) is not locally compact (cf. remark (4.4.3)iii)).
For $n \in \mathbb{N}$ define the following subsets of $\mathbb{C}^2$

$$K_n = \{(z,0): |z|\le 1\} \cup \{(0,w): w \in [0,2]\} \cup \{(z,w): |z|=1, |w|\le 1-(1/n)\}$$

Then K_n may be illustrated in the following way

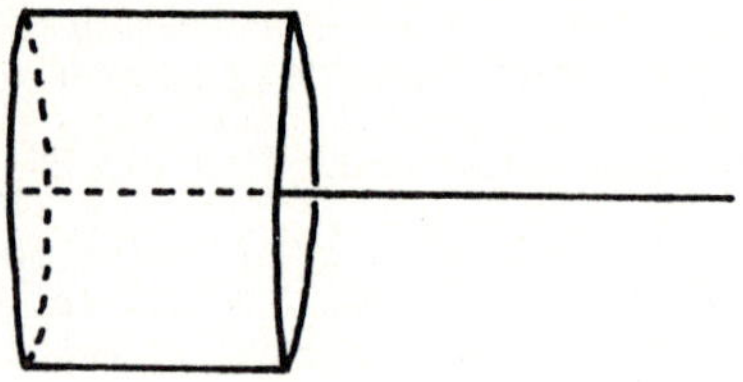

i) Let

$$L_n = \{(z,w): |z|\le 1,\ |w|\le 1-(1/n)\} \cup \{(0,w): w \in [0,2]\}$$

Claim. L_n is the polynomially convex hull of K_n.
We first show that L_n is polynomially convex.
Let (x,y) be an arbitrary point of $\mathbb{C}^2\backslash L_n$.
If $|x|>1$, then $|z_1(x,y)| > \|z_1\|_{L_n}$, where z_1 denotes the first coordinate function.
If $0<|x|\le 1$, then $|y|>1-(1/n)$, hence

$$|p(x,y)| > \|p\|_{L_n}$$

for the polynomial $p(z_1,z_2) = z_1(z_2/(1-(1/n)))^k$, k sufficiently large.
If x=0, then

$$y \in \mathbb{C}\backslash M \text{ with } M = \{w \in \mathbb{C}: |w| \le 1-(1/n)\} \cup [0,2].$$

Since M is a polynomially convex compact subset of the plane (cf. (1.2.4)ii)), we find a polynomial p so that

$$|p(y)| > \|p\|_M$$

and so

$$|\tilde{p}(x,y)| > \|\tilde{p}\|_{L_n}$$

for the polynomial $\tilde{p}(z_1,z_2) = p(z_2)$, hence L_n is polynomially convex and the polynomially convex hull $\hat{K}_n$ of K_n is contained in L_n.
Now let p be an arbitrary polynomial, and let $(x,y) \in L_n \setminus K_n$. Then - applying the maximum principle to the polynomial $q(z) = p(z,y)$ - we get

$$|p(x,y)| = |q(x)| \leq \|q\|_{\{|z|=1\}} = \|p\|_{\{(z,y)\,:\,|z|=1\}} \leq \|p\|_{K_n},$$

so $L_n \subset \hat{K}_n$.
ii) Set $X = \cup K_n$. Then X, endowed with the induced euclidean topology, is a locally compact space and $(K_n)_n$ is an admissible exhaustion of X. Set

$$A = \{f \in C(X)\colon f \text{ can be approximated uniformly on each } K_n \text{ by polynomials}\}.$$

Then (A,X) is a uF-algebra by theorem (4.1.3)iii) and (A,X) is doubly generated by the coordinate functions. We have $A = \varprojlim P(K_n)$ by theorem (4.1.6) and

$$M(A) = \cup \hat{K}_n = \cup L_n$$

as sets and $(L_n)_n$ is an admissible exhaustion of M(A) with respect to the Gelfand topology by (5.1.6).
iii) Assertion. The point $(0,1) \in M(A)$ has no compact neighbourhood (Gelfand topology).
We assume the contrary. Then there is $n \in \mathbb{N}$ so that L_n is a compact neighbourhood of (0,1). By (3.2.2)i) there exist $f_1,\dots,f_r \in A$ such that

$$(*) \qquad \{(y_1,y_2) \in M(A)\colon |\hat{f}_i(y_1,y_2) - \hat{f}_i(0,1)| < 1,\ i=1,\dots,r\} \subset L_n.$$

The functions $\hat{f}_1,\dots,\hat{f}_r$ are continuous on every L_n (euclidean topology) by (5.1.7)i). So we can choose $m \in \mathbb{N}$ and $w \in \mathbb{R}$ such that $1-(1/m) > w > 1-(1/n)$ and

$$|\hat{f}_i(0,w) - \hat{f}_i(0,1)| < 1/4 \quad \text{for } i=1,\dots,r.$$

Again because of the continuity of the $\hat{f}_i$'s we can choose $u = (u_1,u_2) \in L_m \backslash L_n$ such that

$$|\hat{f}_i(0,w)-\hat{f}_i(u_1,u_2)| < 1/4 \quad \text{for } i=1,\dots,r,$$

hence

$$|\hat{f}_i(0,1)-\hat{f}_i(u_1,u_2)| < 1 \quad \text{for } i=1,\dots,r.$$

But this is a contradiction to (*).

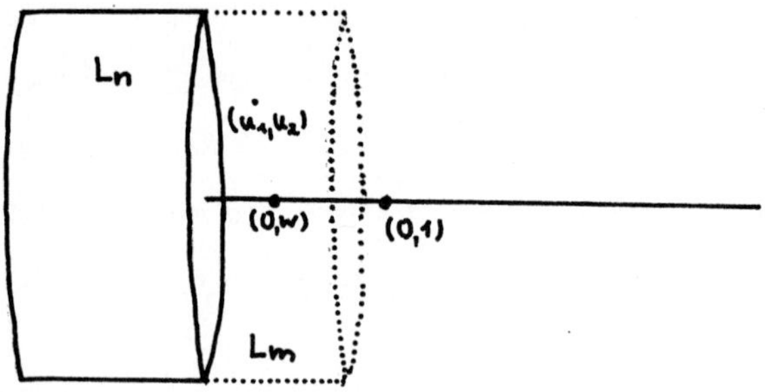

(5.2) Rationally finitely generated F-algebras

(5.2.1) Definition. We say that an F-algebra A is rationally n-generated provided there exist elements $f_1,\dots,f_n$ in A such that all functions of the form $p\cdot q^{-1}$, p and q polynomials in $f_1,\dots,f_n$, q invertible in A, lie dense in A.

(5.2.2) Example. Let $G \subset \mathbb{C}$ be a domain, and let $(K_n)_n$ be an admissible exhaustion of G such that for all $n \in \mathbb{N}$ there is no bounded component of $\mathbb{C}\backslash K_n$, which is relatively compact in G.
Fix $n \in \mathbb{N}$. Let $V_1,V_2,\dots$ be the bounded components of $\mathbb{C}\backslash K_n$. For each k we choose a point $x_k \in V_k\backslash G$. Then $R(K_n)$ is doubly generated by the functions

$$z:K_n \longrightarrow \mathbb{C},\ x \longrightarrow x, \text{ and } f:K_n \longrightarrow \mathbb{C},\ x \longrightarrow \sum_k \varepsilon_k/(x-x_k)$$

where ε_k are suitable chosen constants (cf. the proof of (24.4) in [STO]). $R(K_n)$ is singly generated by z if there is no bounded component. Let $g \in Hol(G)$ be an arbitrary element. Then $g|_{K_n} \in R(K_n)$ by (1.4.5). Hence g can be approximated uniformly on K_n by polynomials in f and z. Since all $x_k \in \mathbb{C}\setminus G$, the partial sums of f are of the form p/q, p,q polynomials, q invertible in Hol(G). It follows from this that Hol(G) is singly rationally generated by

$$z: G \longrightarrow \mathbb{C},\ x \longrightarrow x.$$

<u>(5.2.3)</u> We note that n-generated F-algebras are in particular rationally n-generated.
In the same way as for n-generated F-algebras we can show that the map

$$\hat{F}: M(A) \longrightarrow \mathbb{C}^n,\ \varphi \longrightarrow (\hat{f}_1(\varphi),\dots,\hat{f}_n(\varphi))$$

is injective and continuous, if A is rationally n-generated by $f_1,\dots,f_n$.

<u>(5.2.4)</u> Recall that a compact set $K \subset \mathbb{C}^n$ is called rationally convex, if $K = M(R(K))$.

<u>Proposition.</u> i) Let A be a rationally n-generated uF-algebra with generating elements $f_1,\dots,f_n \in A$ and let $K \subset M(A)$ be an $\hat{A}$-convex compact subset, then $\hat{F}(K)$ is rationally convex and $\hat{A}_K$ is topologically and algebraically isomorphic to $R(\hat{F}(K))$, where $\hat{F}$ is defined as in (5.2.3).
ii) If $(K_n)_n$ is an admissible $\hat{A}$-convex exhaustion of M(A), then A is topologically and algebraically isomorphic to $\varprojlim R(\hat{F}(K_n))$.

<u>Proof.</u> $\hat{F}|_K: K \longrightarrow \hat{F}(K)$

is a homeomorphism by (5.2.3) and it is easy to show that

$$T: \hat{A}_K \longrightarrow R(\hat{F}(K)),\ f \longrightarrow f \circ \hat{F}^{-1}$$

is a well-defined algebra homomorphism. Since

$$\|T(g)\|_{\hat{F}(K)} = \|g\|_K \text{ for all } g \in \hat{A}_K,$$

$\hat{A}_K$ is isomorphic to a closed subalgebra of $R(\hat{F}(K))$. Let $g \in R(\hat{F}(K))$ be an arbitrary element. Using (1.4.2) we can show that there is $h \in \hat{A}_K$ such that $h = g \circ \hat{F}$ (note that $M(\hat{A}(K)) = K$ by (4.3.3)), i.e. $T(h) = g$ and T is surjective, hence $\hat{A}_K$ is isomorphic to $R(\hat{F}(K))$. The adjoint spectral map

$$(T^{-1})^*\colon M(\hat{A}_K) \longrightarrow M(R(\hat{F}(K)))$$

is a homeomorphism by (3.2.5)iv). Let $\varphi \in K = M(\hat{A}_K)$ and $f \in R(\hat{F}(K))$ be arbitrary elements, then

$$(T^{-1})^*(\varphi)(f) = \varphi(T^{-1}(f)) = \varphi(f \circ \hat{F}) = f(\hat{F}(\varphi)),$$

so

$$(T^{-1})^*(\varphi) = \hat{F}(\varphi),$$

and

$$(T^{-1})^*(K) = \hat{F}(K),$$

hence

$$\hat{F}(K) = M(R(\hat{F}(K))).$$

ii) For all $n \in \mathbb{N}$ the diagram

$$\begin{array}{ccc} \hat{A}_{K_{n+1}} & \xrightarrow{\quad r'_n \quad} & \hat{A}_{K_n} \\ T_{n+1} \downarrow & & \downarrow T_n \\ R(\hat{F}(K_{n+1})) & \xrightarrow{\quad r_n \quad} & R(\hat{F}(K_n)) \end{array}$$

commutes, where r_n resp. r'_n denote the natural restriction mappings and T_{n+1} resp. T_n denote the isomorphisms of i).
Hence

$$\ldots \longrightarrow R(\hat{F}(K_{n+1})) \xrightarrow{\ r_n\ } R(\hat{F}(K_n)) \longrightarrow \ldots$$

constitutes a dense projective system of uB-algebras and A is topologically and algebraically isomorphic to $\varprojlim R(\hat{F}(K_n))$. □

In the case of ii) of the theorem we can identify $M(\varprojlim R(\hat{F}(K_n)))$ with $\cup\, \hat{F}(K_n) = \hat{F}(M(A))$ and $(\hat{F}(K_n))_n$ is an admissible exhaustion of

$\cup \hat{F}(K_n)$ with respect to the Gelfand topology (cf. (4.1.6)).

(5.2.5) We close this chapter considering the projective system

$$(*) \qquad \ldots \longrightarrow R(K_{n+1}) \xrightarrow{r_n} R(K_n) \longrightarrow \ldots ,$$

where $\ldots \subset K_n \subset K_{n+1} \subset \ldots$ is a sequence of compact subsets of $\mathbb{C}^m$ and r_n denotes the restriction mapping.
We first remark that in general (*) is not a dense projective system, hence we can not apply our previous results in this case. We give an example.
For $x \in \mathbb{C}$, $r>0$ set $D_{x,r} = \{z \in \mathbb{C}: |z-x| < r\}$ and $\overline{D} = \{|z|\leq 1\}$. For $n \in \mathbb{N}$ set

$$K_n = \overline{D} \setminus \bigcup_{k>n} D_{1/k,((1/k)-(1/k+1))/2} .$$

Then the restriction mappings

$$r_{l,n}: R(K_l) \longrightarrow R(K_n),\ l > n,$$

don't have dense range, since for example $1/(z-(1/n))$ cannot be approximated uniformly on K_n by elements of $R(K_l)$.
Each element $f \in \varprojlim R(K_n)$ can be interpreted as complex-valued function on $\cup K_n$ such that $f|_{K_n} \in R(K_n)$ for all $n \in \mathbb{N}$. The topology of $\varprojlim R(K_n)$ is generated by the seminormsystem $\|\cdot\|_{K_n}$ and it is easy to see that $\varprojlim R(K_n)$ is in fact a uF-algebra.

(5.2.6) Even if (*) in (5.2.5) constitutes a dense projective system it is in general not true that $\varprojlim R(K_n)$ is rationally m-generated, while $\varprojlim P(K_n)$ is clearly m-generated by the coordinate functions.
For $n \in \mathbb{N}$ set

$$K_n = \{z \in \mathbb{C}: 1/n \leq |z| \leq 1-(1/n)\} \cup [0,1/2].$$

Suppose that $A = \varprojlim R(K_n)$ is rationally singly generated by $f \in A$.

By (3.3.4) and (1.2.4)iii) we have

$$M(A) = \cup M(R(K_n)) = \cup K_n = \{z: |z|<1\} = D,$$

and $(K_n)_n$ is an admissible exhaustion of M(A). As above we interpret each element $\hat{g} \in \hat{A}$ as complex-valued function on D, such that $\hat{g}|_{K_n} \in R(K_n)$ for all $n \in \mathbb{N}$. Hence $\hat{f}|_{D\setminus 0} \in Hol(D\setminus 0)$ and $\hat{f}|_{[0,1/2]}$ is continuous, hence $\hat{f}$ has no pole in 0. By (5.2.3) $\hat{f}$ is injective on D, hence $\hat{f}$ has no essential singularity in 0. It follows that $\hat{f} \in Hol(D)$. The mapping

$$T: Hol(D) \longrightarrow \hat{A},\ g \longrightarrow g,$$

defines a continuous and injective algebra homomorphism. We claim that T is surjective, too.

Let $h \in A$ be an arbitrary element. Choose polynomials q_n, p_n so that

$$(*) \qquad \|\hat{p}_n(f)/\hat{q}_n(f)-\hat{h}\|_{K_n} < 1/n,$$

where $\hat{p}_n(f)$ denotes the Gelfand transform of $p_n(f)$. Since $\hat{q}_n(f)$ is invertible in $\hat{A}$, $\hat{q}_n(f)$ has no zero on D, hence $\hat{p}_n(f)/\hat{q}_n(f) \in Hol(D)$ for all $n \in \mathbb{N}$. Set $D' = \{z: |z| \le 1/2\}$. It follows by (*) and the maximum principle that

$$(\hat{p}_n(f)/\hat{q}_n(f)|_{D'})_n$$

is a Cauchy sequence with respect to the norm $\|\cdot\|_{D'}$, hence this sequence converges on $\{|z|<1/2\}$ to a holomorphic function $\tilde{h}$. Since on the other hand this sequence converges pointwise on $\{|z|<1/2\}$ to $\hat{h}$ by (*), we get

$$\hat{h}|_{\{|z|<1/2\}} = \tilde{h},$$

and our claim is proved.

By the open mapping theorem for Fréchet spaces Hol(D) and $\hat{A}$ are topologically and algebraically isomorphic. So the adjoint spectral map

$$T^*: D = M(A) = M(\hat{A}) \longrightarrow D = M(Hol(D)),\ x \longrightarrow x,$$

is a homeomorphism by (3.2.5)iv). But this is false, since $(K_n)_n$ is not an admissible exhaustion of D with respect to the Gelfand topology.

CHAPTER 6

APPLICATIONS OF THE PROJECTIVE LIMIT REPRESENTATION

The projective limit representation (3.3) enables us to prove some results for F-algebras, which we have already stated for B-algebras (cf. chapter 1). In this chapter we extend among others the holomorphic functional calculus and Shilov's idempotent theorem. We are also concerned with an extension of the theorem of Arens-Royden. Here however we are able to prove a perfect analogue to the result in the B-algebra case only in special cases. Furthermore we shall discuss the problems which appear in connection with this result.

(6.1) The holomorphic functional calculus

(6.1.1) Definition. Let A be an F-algebra, and let $f_1,\dots,f_n \in A$. We denote by $\hat{F}$ the mapping

$$\hat{F}: M(A) \longrightarrow \mathbb{C}^n,\ \varphi \longrightarrow (\hat{f}_1(\varphi),\dots,\hat{f}_n(\varphi)).$$

The set $\hat{F}(M(A))$ is called the joint spectrum of $f_1,\dots,f_n$.

(6.1.2) Theorem [ZAM]. Let A be an F-algebra, and let $f_1,\dots,f_n \in A$. Let $U \subset \mathbb{C}^n$ be an open set which contains the joint spectrum of $f_1,\dots,f_n$. Then there is a unique continuous homomorphism

$$\vartheta_F: Hol(U) \longrightarrow A$$

such that

i) $\vartheta_F(z_i) = f_i$, $i=1,\dots,n$, and

ii) $\vartheta_F(g)\hat{} = g \circ \hat{F}$ for all $g \in Hol(U)$, where $z_1,\dots,z_n$ denote the coordinate functions and $\vartheta_F(g)^\wedge$ denotes the Gelfand transform of $\vartheta_F(g)$.

Proof. Represent A as the projective limit of a dense projective system of B-algebras

$$\dots \longrightarrow A_{k+1} \longrightarrow A_k \longrightarrow \dots$$

(cf. (3.3.7)). Denote by $\pi_k: A \longrightarrow A_k$, resp. $\pi_{m,k}: A_m \longrightarrow A_k$ ($m \geq k$), the canonical maps (cf. (3.3.6)). As usual we identify $M(A_k)$ and $\pi_k^*(M(A_k))$ (cf. (3.2.7)).
We have

$$\sigma(\pi_m(f_1),\dots,\pi_m(f_n)) \subset \sigma(\pi_s(f_1),\dots,\pi_s(f_n)) \subset \hat{F}(M(A)) \subset U,$$

for all m,s ($s \geq m$), where $\sigma(\pi_s(f_1),\dots,\pi_s(f_n))$ denotes the joint spectrum of the elements $\pi_s(f_1),\dots,\pi_s(f_n) \in A_s$.
By (1.4.2) there exists for each $m \in \mathbb{N}$ a unique continuous algebra homomorphism

$$\vartheta^m: \mathrm{Hol}(U) \longrightarrow A_m,$$

such that

(*) i) $\vartheta^m(z_i) = \pi_m(f_i)$, $i=1,\dots,n$, and

ii) $\vartheta^m(g)^\wedge = g \circ \hat{F}|_{M(A_m)}$ for all $g \in \mathrm{Hol}(U)$.

(Note that $\varphi(\pi_m(f_i)) = \pi_m^*(\varphi)(f_i)$, for $i=1,\dots,n$ and for all $\varphi \in M(A_m)$.)
For $s \geq m$ the homomorphism

$$\pi_{s,m} \circ \vartheta^s: \mathrm{Hol}(U) \longrightarrow A_m$$

is continuous and satisfies i) and ii) of (*). Hence

$$\pi_{s,m} \circ \vartheta^s = \vartheta^m \text{ for } s \geq m$$

by the uniqueness assertion of (1.4.2). It follows that

$$\vartheta_F: \mathrm{Hol}(U) \longrightarrow A = \varprojlim A_m, \quad g \longrightarrow (\vartheta^m(g))_m$$

is a well-defined continuous homomorphism which has the required properties i) and ii).

Let τ: Hol(U) $\longrightarrow$ A be another continuous homomorphism which fulfills i) and ii). Then $\pi_m \circ \tau = \vartheta^m$ for all $m \in \mathbb{N}$, by the uniqueness of the homomorphisms ϑ^m and it follows that $\tau = \vartheta_F$. □

(6.1.3) Let $S \subset \mathbb{C}^n$ be an arbitrary set. Recall from (1.4.1) the definition of the algebra H(S).

Corollary. Let A be an F-algebra, and let $f_1,\dots,f_n \in A$. There exists a unique continuous homomorphism

$$\vartheta_F: H(\hat{F}(M(A))) \longrightarrow A$$

such that

i) $\vartheta_F(\tilde{z}_i) = f_i$, $i=1,\dots,n$ and

ii) $\vartheta_F(\tilde{g})\hat{\ } = \tilde{g}\circ\hat{F}$ for all $\tilde{g} \in H(\hat{F}(M(A)))$,

where $\tilde{z}_1,\dots,\tilde{z}_n$ denote the germs of the coordinate functions.

Proof. Let $U, W \subset \mathbb{C}^n$ be open sets so that $\hat{F}(M(A)) \subset U \subset W$. Denote by $r_{U,W}$ the natural restriction mapping

$$\text{Hol}(W) \longrightarrow \text{Hol}(U),\ f \longrightarrow f|_U$$

and denote by ϑ_F^U: Hol(U) $\longrightarrow$ A the unique homomorphism of (6.1.2). Then

$$(*)\qquad \vartheta_F^U \circ r_{U,W} = \vartheta_F^W$$

by the uniqueness assertion of (6.1.2).

Now let $\tilde{g} \in H(\hat{F}(M(A)))$ be an arbitrary element, and let $g \in$ Hol(U) be a representative of $\tilde{g}$ in an open neighbourhood U of $\hat{F}(M(A))$. Set

$$\vartheta_F(\tilde{g}) = \vartheta_F^U(g).$$

It follows from (*) that $\vartheta_F(\tilde{g})$ is a well-defined mapping and it is easy to check that ϑ_F is in fact a homomorphism, which has the required properties i) and ii). (Recall that $\tilde{g}(x)$ is well-defined for $\tilde{g} \in H(\hat{F}(M(A)))$ and $x \in \hat{F}(M(A))$.)

To see that ϑ_F is continuous let U be an arbitrary open neighbourhood of $\hat{F}(M(A))$ and denote by τ_U the natural mapping

$$\tau_U : \mathrm{Hol}(U) \longrightarrow H(\hat{F}(M(A))),\ g \longrightarrow \tilde{g}.$$

Then $\vartheta_F \circ \tau_U = \vartheta_F^U$ by the definition of ϑ_F, i.e. $\vartheta_F \circ \tau_U$ is continuous for each open neighbourhood U, so ϑ_F is continuous by (1.4.1). Let $\rho : H(\hat{F}(M(A))) \longrightarrow A$ be another continuous homomorphism with the required properties, then $\rho \circ \tau_U = \vartheta_F^U$ for all open neighbourhoods U of $\hat{F}(M(A))$ by (6.1.2), hence $\rho = \vartheta_F$.

□

Remark. A weaker form of the holomorphic functional calculus for F-algebras was first proved by Arens [ARE 2] and Waelbroeck [WAE] , cf. also Rosenfeld [ROS].

(6.1.4) As an immediate consequence of the holomorphic functional calculus we can describe the invertible elements of an F-algebra.

Corollary. If A is an F-algebra, then

$$A^{-1} = \{f \in A : \varphi(f) \neq 0, \text{ for all } \varphi \in M(A)\}.$$

Proof. i) If $f \in A^{-1}$ there exists $g \in A$ so that $fg = 1$, hence

$$1 = \varphi(f)\varphi(g) \quad \text{for all } \varphi \in M(A).$$

ii) If $\varphi(f) \neq 0$ for all $\varphi \in M(A)$, then $\hat{f}(M(A)) \subset \mathbb{C}^* = \mathbb{C}\backslash\{0\}$. Let $\vartheta_f : \mathrm{Hol}(\mathbb{C}^*) \longrightarrow A$ be as in (6.1.2), then

$$1 = \vartheta_f(z\cdot(1/z)) = \vartheta_f(z)\vartheta_f(1/z) = f\cdot\vartheta_f(1/z),$$

where $z : \mathbb{C}^* \longrightarrow \mathbb{C},\ x \longrightarrow x$.

□

(6.1.5) Shilov's idempotent theorem. Let A be an F-algebra and let $E \subset M(A)$ be an open and closed set. Then there is a unique idempotent $f \in A$ such that $\hat{f}$ is the characteristic function of E.

Proof. Let $A = \varprojlim A_n$ (3.3.7). Then

$$\ldots \subset M(A_n) \subset M(A_{n+1}) \subset \ldots$$

is an admissible exhaustion of M(A) by (3.3.4) and $M(A_n) \cap E$ is a closed and open subset of $M(A_n)$ for each $n \in \mathbb{N}$.
By (1.4.6) there is a unique idempotent $f_n \in A_n$, so that $\hat{f}_n$ is the characteristic function of $M(A_n) \cap E$ for each $n \in \mathbb{N}$. Denote by $\pi_{m,n}$ $(m \geq n)$ the canonical maps $A_m \longrightarrow A_n$ (cf. (3.3.6)), then $\pi_{m,n}(f_m)$ is an idempotent element of A_n and $\hat{\pi}_{m,n}(f_m) = \hat{f}_n$, where $\hat{\pi}_{m,n}(f_m)$ denotes the Gelfand transform of $\pi_{m,n}(f_m)$.
It follows from the uniqueness assertion of (1.4.6) that

$$\pi_{m,n}(f_m) = f_n.$$

Hence $f = (f_n)_n \in \varprojlim A_n$ has the desired properties. □

(6.1.6) Corollary [ROS]. Let A be an F-algebra, and let $\varphi \in M(A)$. If $\varphi \in M(A)$ is an isolated point of each compact set $K \subset M(A)$ which contains it, then φ is isolated in M(A).

Proof. Let $A = \varprojlim A_n$ be as in the proof of (6.1.5). By our assumption there exists an idempotent $f_m \in A_m$ such that

$$\hat{f}_m(\psi) = \begin{cases} 0 \text{ if } \psi \in M(A_m)\backslash\{\varphi\} \\ 1 \text{ if } \psi = \varphi \end{cases} \quad \text{for each } m \in \mathbb{N} \text{ (cf. (1.4.6)).}$$

Using the uniqueness assertion of (1.4.6) we can show as in the proof of (6.1.5) that $f = (f_m)_m \in \varprojlim A_m$.
Since $\hat{f}$ is continuous on M(A) and

$$\hat{f}(\psi) = \begin{cases} 0 \text{ if } \psi \neq \varphi \\ 1 \text{ if } \psi = \varphi \end{cases}$$

φ is an isolated point of M(A). □

Remark. Arens constructed a Lindelöf space X such that there is a point $x \in X$ so that x is an isolated point of each compact set which contains it, but x is not an isolated point of X (cf. [KEL],p. 77).

(6.1.7) Corollary. Let (A,X) be a uF-algebra, and let X be connected. Then $M(A)$ is connected, too.

Proof. Suppose that

$$M(A) = U \cup W, \; U \cap W = \emptyset,$$

with nonempty closed and open subsets U,W. Let $(K_n)_n$ be an admissible exhaustion of X. Since the structure mapping

$$j: X \longrightarrow M(A), \; x \longrightarrow \varphi_x,$$

is continuous by (4.2.1) and since X is connected, we can w.l.o.g. assume that $j(X) \subset U$, in particular $j(K_n) \subset U$ for all $n \in \mathbb{N}$.
The characteristic function χ_W is contained in $\widehat{A}$ by Shilov's theorem, thus

$$j(K_n)^{\wedge}_{A} \cap W = \emptyset \quad \text{for all } n \in \mathbb{N},$$

where $j(K_n)^{\wedge}_{A}$ denotes the $\widehat{A}$-convex hull of $j(K_n)$. This is a contradiction to

$$\bigcup j(K_n)^{\wedge}_{A} = M(A) \quad \text{(cf. (4.3.4)iii)).}$$

□

(6.1.8) The next proposition will be needed later.
Proposition. Let A be a uF-algebra, and let $U \subset M(A)$ be a nonempty closed and open set. Then $\widehat{A}_U$ is a uF-algebra, $M(\widehat{A}_U) = U$ and $\widehat{A}_U$ can be considered as a closed subalgebra of $\widehat{A}$ via the map

$$\text{incl}: \widehat{A}_U \longrightarrow \widehat{A}, \; f \longrightarrow \tilde{f},$$

where

$$\tilde{f}(x) = \begin{cases} f(x) & \text{if } x \in U \\ 0 & \text{if } x \in M(A)\setminus U \end{cases}.$$

Proof. By our hypothesis U is a hemicompact space, hence $\hat{A}_U$ is a uF-algebra by (4.1.5).

We interpret each element of $\hat{A}_U$ as complex-valued function on U, which can be approximated uniformly on each compact subset of U by elements of $\hat{A}$.

By Shilov's theorem χ_U is contained in $\hat{A}$. Let $f \in \hat{A}_U$ be an arbitrary element. Choose a sequence $(\hat{f}_n)_n$ in $\hat{A}$ such that $\hat{f}_n|_U$ converges to f in $\hat{A}_U$. Then $(\hat{f}_n \cdot \chi_U)_n$ converges to $\tilde{f}$ with respect to the compact open topology, hence $\tilde{f} \in \hat{A}$ and the map incl is a well-defined injective and continuous homomorphism. We have

$$(*) \qquad \mathrm{incl}(\hat{A}_U) = \hat{A} \cdot \chi_U,$$

so $\mathrm{incl}(\hat{A}_U)$ is a closed subalgebra of $\hat{A}$, which is isomorphic to $\hat{A}_U$. Let $K \subset U$ be a compact subset, then we can show as in the proof of (6.1.7) that $\hat{K}_A \subset U$, hence

$$M(\hat{A}_U) = U$$

by (4.3.4)iii) and it follows from (*) and (3.2.2)i) that the Gelfand topology on $M(\hat{A}_U)$ coincides with the original topology on U. □

Remark. Note that we have proved that $\hat{A}_U \subset C(U)$, i.e. $(\hat{A}_U, U)$ is a uF-algebra.

(6.1.9) In (1.2.6) we established a one-to-one correspondence between the maximal ideals and the elements of M(A) for B-algebras A. As an easy consequence we proved that the ideal, which is generated by $f_1,\dots,f_r \in A$, is proper iff

$$\{\varphi \in M(A) : \varphi(f_i) = 0,\ i=1,\dots,r\} \neq \emptyset \qquad (\text{cf.}(1.2.9)).$$

The analogous result for F-algebras is true (Corollary (6.1.10) below) but requires much more effort. It was first proved by Arens [ARE 1]. We give a generalization of this result due to Brooks [BRO 2].

Theorem [BRO 2]. Let A be an F-algebra and let $(a_n)_n$ be a sequence in A such that

$$\{\varphi \in M(A): \varphi(a_i) = 0,\ i=1,2,\ldots\} = \emptyset.$$

Then there exists a sequence $(b_n)_n$ in A such that

$$\sum_{i=1}^{\infty} a_i b_i = 1,$$

i.e. the sequence of the partial sums converges to 1.

Proof. Let $(p_n)_n$ be a defining sequence of seminorms for A (cf. (3.1.7) and recall in particular that $p_n \le p_{n+1}$ for all n). Recall from (3.2.6) the definitions of the algebras A_n and of the norms p'_n, resp. from (3.3.6) the definitions of the maps $\pi_{n,k}$ $(n \ge k)$. As usual (cf. (3.2.7)) we identify $M(A_n)$ and $\pi_n^*(M(A_n))$. We divide the proof into several steps.

i) Set $i_o = 1$ and choose for each $n \in \mathbb{N}$ an integer i_n so that

$$(*) \qquad M(A_n) \cap \{\varphi \in M(A): \varphi(a_1) = \varphi(a_2) = \ldots = \varphi(a_{i_n}) = 0\} = \emptyset$$

and $i_n > i_{n-1}$, hence $i_n > n$ for all $n \in \mathbb{N}$.

Suppose this choice is not possible. Then, for each $k \in \mathbb{N}$, there is $\varphi_k \in M(A_n)$ such that

$$\varphi_k(a_1) = \varphi_k(a_2) = \ldots = \varphi_k(a_k) = 0.$$

Since $M(A_n)$ is compact, a subsequence of $(\varphi_k)_k$ converges to an element $\varphi \in M(A_n)$. Hence by continuity $\varphi(a_i) = 0$ for all $i \in \mathbb{N}$, and we get a contradiction.

ii) Let $(\delta_n)_n$ be a sequence of positive numbers such that $\sum \delta_n < \infty$. We first prove by induction that for every $n \in \mathbb{N}$ there exist elements $x_1^{(n)}, \ldots, x_{i_n}^{(n)} \in A_n$ so that

$$\alpha) \ \sum_{i=1}^{i_n} \pi_n(a_i) x_i^{(n)} = 1,$$

$$\beta) \ p'_{n-1}(\pi_{n,n-1}(x_i^{(n)}) - x_i^{(n-1)}) < \delta_n, \text{ if } 1 \le i \le i_{n-1},$$

γ) $p'_{n-1}(\pi_{n,n-1}(x_i^{(n)})) < \delta_n$, if $i_{n-1}<i\le i_n$.

Note that (*) implies that

$$\{\varphi \in M(A_n)\colon \varphi(\pi_n(a_1)) = \ldots = \varphi(\pi_n(a_{i_n})) = 0\} = \emptyset.$$

Since each A_n is a B-algebra, we always find $z_1,\ldots,z_{i_n} \in A_n$ so that α) is fulfilled (cf. (1.2.9)).

For n=1 the conditions β) and γ) are vacuous.

Now let n>1 and suppose that $x_1^{(n-1)},\ldots,x_{i_{n-1}}^{(n-1)} \in A_{n-1}$ have already been chosen. Let $z_1,\ldots,z_{i_n} \in A_n$ so that α) is fulfilled. Choose arbitrary $t'_1,\ldots,t'_{i_{n-1}} \in A_n$ and set $t'_i=0$ for $i_{n-1}<i\le i_n$. Furthermore define

$$t_i = t'_i + z_i\Big(1 - \sum_{j=1}^{i_{n-1}} \pi_n(a_j)t'_j\Big), \quad 1\le i\le i_n.$$

One varifies easily that $t_1,\ldots,t_{i_n}$ fulfill α). We have for $1 \le i\le i_n$

$$c_i = p'_{n-1}\Big(\pi_{n,n-1}\Big(z_i\Big(1 - \sum_{j=1}^{i_{n-1}} \pi_n(a_j)t'_j\Big)\Big)\Big) =$$

$$= p'_{n-1}\Big(\pi_{n,n-1}(z_i)\Big(1 - \sum_{j=1}^{i_{n-1}} \pi_{n-1}(a_j)\pi_{n,n-1}(t'_j)\Big)\Big) =$$

$$= p'_{n-1}\Big(\pi_{n,n-1}(z_i)\Big(\sum_{j=1}^{i_{n-1}} \pi_{n-1}(a_j)(x_j^{(n-1)} - \pi_{n,n-1}(t'_j))\Big)\Big) \le$$

$$\le p'_{n-1}(\pi_{n,n-1}(z_i))\Big(\sum_{j=1}^{i_{n-1}} p'_{n-1}(\pi_{n-1}(a_j))p'_{n-1}(x_j^{(n-1)} - \pi_{n,n-1}(t'_j))\Big).$$

Since $\pi_{n,n-1}$ has dense range, we can choose $t'_1,\ldots,t'_{i_{n-1}}$ so that

$$p'_{n-1}(\pi_{n,n-1}(t'_i) - x_i^{(n-1)}) < \delta_n/2, \text{ for } i=1,\ldots,i_{n-1},$$

and $c_i < \delta_n/2$, for $i=1,\dots,i_n$. Set $x_i = t_i$ for $i=1,\dots,i_n$ and claim ii) is proved.

Next we fix $n \in \mathbb{N}$. For $i \le i_n$ and $k \ge n$ set

$$y_i^{(k,n)} = \pi_{k,n}(x_i^{(k)}).$$

Then for $k \ge l \ge n$

$$p'_n(y_i^{(k,n)} - y_i^{(l,n)}) =$$

$$p'_n(y_i^{(k,n)} - y_i^{(k-1,n)} + y_i^{(k-1,n)} - \dots + y_i^{(l+1,n)} - y_i^{(l,n)}) \le$$

$$p'_n(\pi_{k,n}(x_i^{(k)}) - \pi_{k-1,n}(x_i^{(k-1)})) + \dots + p'_n(\pi_{l+1,n}(x_i^{(l+1)}) - \pi_{l,n}(x_i^{(l)}))$$

$$= p'_n(\pi_{k-1,n}(\pi_{k,k-1}(x_i^{(k)}) - x_i^{(k-1)})) + \dots$$

$$\le p'_{k-1}(\pi_{k,k-1}(x_i^{(k)}) - x_i^{(k-1)})) + \dots$$

$$\le \delta_k + \dots + \delta_{l+1} \quad \text{by } \beta) \text{ of ii)}.$$

(**) The next to last inequality follows from

$$p'_n(\pi_{k,n}(f)) \le p'_k(f), \text{ for all } f \in A_k.$$

(Note that $p_n \le p_{n+1}$.) Hence we have for $k \ge l \ge n$, $i \le i_n$

$$\text{iii) } p'_n(y_i^{(k,n)} - y_i^{(l,n)}) \le \sum_{j=l+1}^{k} \delta_j.$$

It follows that $(y_i^{(k,n)})_{k \ge n}$ is a Cauchy sequence in A_n. We denote the limit by $y_i^{(n)}$. Since

$$\pi_{n+1,n}(y_i^{(k,n+1)}) = y_i^{(k,n)},$$

we see that

$$\text{iv) } \pi_{k,n}(y_i^{(k)}) = y_i^{(n)}, \text{ if } k \ge n,\ i \le i_n.$$

For given $i \in \mathbb{N}$ let $n(i)$ be the smallest integer so that $y_i^{(k)}$ is defined for $k \geq n(i)$. For $k < n(i)$ define

$$y_i^{(k)} = \pi_{n(i),k}(y_i^{(n(i))}).$$

Together with iv) we see that $y_i = (y_i^{(n)})_n$ defines an element of $\varprojlim A_n = A$.

We claim that $\sum a_i y_i = 1$.

To prove this we establish two inequalities. We claim that for each $n \in \mathbb{N}$

$$\text{v)}\quad p_n'(\pi_n(y_i) - x_i^{(n)}) \leq \sum_{j=n+1}^{\infty} \delta_j, \text{ for } i \leq i_n,$$

$$\text{vi)}\quad p_n'(\pi_n(y_i)) \leq \sum_{j=n+1}^{\infty} \delta_j, \text{ for } i_n < i \leq i_{n+1}.$$

If $i \leq i_n$, then

$$p_n'(\pi_n(y_i) - x_i^{(n)}) = p_n'(y_i^{(n)} - x_i^{(n)}) = p_n'(y_i^{(n)} - y_i^{(n,n)})$$

$$= \lim_{k \to \infty} p_n'(y_i^{(k,n)} - y_i^{(n,n)}) \leq \sum_{j=n+1}^{\infty} \delta_j \quad \text{by iii).}$$

If $i_n < i \leq i_{n+1}$, then

$$p_n'(\pi_n(y_i)) = p_n'(\pi_{n+1,n}(y_i^{(n+1)})) = \lim_{k \to \infty} p_n'(\pi_{n+1,n}(y_i^{(k,n+1)}))$$

$$\leq \lim_{k \to \infty} p_n'(\pi_{n+1,n}(y_i^{(k,n+1)} - y_i^{(n+1,n+1)})) + p_n'(\pi_{n+1,n}(y_i^{(n+1,n+1)}))$$

$$\leq \lim_{k \to \infty} p_{n+1}'(y_i^{(k,n+1)} - y_i^{(n+1,n+1)}) + \delta_{n+1}$$

$$\leq \sum_{j=n+2}^{\infty} \delta_j + \delta_{n+1} \quad \text{by iii).}$$

The next to last inequality follows from $y_i^{(n+1,n+1)} = x_i^{(n+1)}$, γ) of ii) and by the argument used in (**).

For $k \in \mathbb{N}$ set

$$s_k = \sum_{i=1}^{k} a_i y_i \ .$$

We have to show that for each $n \in \mathbb{N}$

$$p_n(1-s_k) \longrightarrow 0 \text{ as } k \longrightarrow \infty.$$

We first look at

$$p_n(1-s_k) \text{ for } i_n < k \le i_{n+1}.$$

In this case

$$\text{vii)}\ p_n(1-s_k) = p'_n(\pi_n(1-s_k))$$

$$= p'_n(\sum_{j=1}^{i_n} \pi_n(a_j)x_j^{(n)} - \sum_{j=1}^{k} \pi_n(a_j)\pi_n(y_j))$$

$$\le (\sum_{j=1}^{i_n} p'_n(\pi_n(a_j)) \cdot \sum_{i=n+1}^{\infty} \delta_i) + \sum_{j=i_n+1}^{k} p'_n(\pi_n(a_j)) \cdot \sum_{i=n+1}^{\infty} \delta_i$$

$$= (\sum_{i=n+1}^{\infty} \delta_i)(\sum_{j=1}^{k} p_n(a_j)),$$

by v) and vi).

Till now we have not specified the δ_j's. We now choose

$$0 < \delta_n < 1/(2^n \max\{1, \sum_{i=1}^{i_{n+1}} p_n(a_i)\}.$$

Then

$$\text{viii)}\qquad p_n(1-s_k) \le \sum_{j=n+1}^{\infty} 2^{-j} \quad \text{for } i_n < k \le i_{n+1}.$$

Let $k > i_{n+1}$. Then there is $l \ge n$, so that $i_l < k \le i_{l+1}$. Then

$$p_n(1-s_k) \le p_l(1-s_k) \le \sum_{j=l+1}^{\infty} 2^{-j}, \text{ by viii)}.$$

Hence $p_n(1-s_k) \longrightarrow 0$ as $k \longrightarrow \infty$, and we are done. □

<u>(6.1.10) Corollary [ARE 1].</u> Let A be an F-algebra, and let $f_1,\dots,f_r \in A$. Then there are $g_1,\dots,g_r \in A$ such that

$$\sum_{i=1}^{r} f_i g_i = 1$$

iff

$$\{\varphi \in M(A): \varphi(f_1) = \dots = \varphi(f_r) = 0\} = \emptyset.$$

(6.2) The theorem of Arens-Royden

It is an open question, whether the theorem of Arens-Royden for B-algebras (1.4.8) can be extended to F-algebras, i.e. whether for each $f \in C(M(A))^{-1}$ there exists $g \in A^{-1}$ such that $f/\hat{g}$ has a continuous logarithm on M(A). We shall obtain this result for a special class of F-algebras. In general we can only prove a weaker result.

(6.2.1) Let (X,μ) be a hemicompact space. We denote by τ the topology which is generated by the μ-compact subsets of X, i.e. $W \subset X$ is τ-closed iff $W \cap K$ is μ-closed for every μ-compact set $K \subset X$. We denote by kX the set X endowed with the topology τ. Then kX is a hemicompact k-space, cf. (4.1.7)i).

Theorem. Let A be an F-algebra, and let $f \in C(kM(A))^{-1}$. Then there exist $g \in A^{-1}$ and $h \in C(kM(A))$ so that $f = \hat{g}\exp(h)$.

(6.2.2) We first prove a proposition.
Let A, B be B-algebras, and let $\pi: A \longrightarrow B$ be a continuous algebra homomorphism which has dense range. By (3.2.5) the adjoint spectral map $\pi^*: M(B) \longrightarrow M(A)$ is a continuous and injective map. Let $f \in C(M(A))^{-1}$, then $f \circ \pi^* \in C(M(B))^{-1}$ and by (1.4.8) there are $b \in B^{-1}$, $g \in C(M(B))$ so that $f \circ \pi^* = \hat{b}\exp(g)$.

Proposition. Let A, B, π, f, b and g be as above. Then there exist $a \in A^{-1}$ and $h \in C(M(A))$ such that

i) $f = \hat{a}\exp(h)$,

ii) $\|\pi(a)-b\|_B < \varepsilon$, and

iii) $\|h\circ\pi^*-g\|_{M(B)} < \varepsilon$.

<u>Proof.</u> Fix $a_o \in A^{-1}$, $h' \in C(M(A))$ so that $f = \hat{a}_o\exp(h')$ (cf. (1.4.8). Then

$$\hat{b}\exp(g) = f\circ\pi^* = \hat{\pi}(a_o)\exp(h'\circ\pi^*),$$

where $\hat{\pi}(a_o)$ denotes the Gelfand transform of $\pi(a_o)$. So

$$\hat{b}/\hat{\pi}(a_o) = \exp(h'\circ\pi^*-g).$$

(Note that $\pi(a_o) \in B^{-1}$, hence $\hat{\pi}(a_o)$ has no zero on $M(B)$ by (1.2.9).) By (1.4.7) there exists $c \in B$ so that

$$b/\pi(a_o) = \exp(c) \text{ and } \hat{c} = h'\circ\pi^*-g.$$

Choose $x \in A$ so that

$$\|\pi(x)-c\|_B < \varepsilon \text{ and}$$

(*) $$\|\exp(\pi(x))-\exp(c)\|_B < \varepsilon/\|\pi(a_o)\|_B.$$

This is possible by remark (1.4.4)ii) and since π has dense range. Set

$$a = a_o\exp(x) \text{ and } h = h'-\hat{x}.$$

Then

i) $f = \hat{a}\exp(h)$;

ii) $\|\pi(a)-b\|_B = \|\pi(a_o)\pi(\exp(x))-\pi(a_o)\exp(c)\|_B \leq$

$\leq \|\pi(a_o)\|_B\cdot\|\exp(\pi(x))-\exp(c)\|_B < \varepsilon$ by (*);

iii) $\|h\circ\pi^*-g\|_{M(B)} = \|h'\circ\pi^*-\hat{\pi}(x)-g\|_{M(B)} \leq$

$\leq \|h'\circ\pi^*-g-\hat{c}\|_{M(B)}+\|\hat{c}-\hat{\pi}(x)\|_{M(B)} = \|\hat{c}-\hat{\pi}(x)\|_{M(B)} \leq$

$\leq \|c-\pi(x)\|_B < \varepsilon$.

The next to last inequality follows from (1.2.3). □

<u>Proof of the theorem.</u> Let $(p_n)_n$ be a defining sequence of semi-norms. Let A_n, p'_n and $\pi_n: A \longrightarrow A_n$ be as in (3.2.6), and let

$\pi_{k,n}: A_k \longrightarrow A_n$ $(k \geq n)$ be the canonical maps (cf. (3.3.6)). As in (3.2.7) we identify $M(A_n)$ and $\pi_n^*(M(A_n))$.
For $f \in A_n$ set

$$d_n(f) = \inf\{p_n'(f-g) : g \in A_n \setminus (A_n)^{-1}\}.$$

Note that $d_n(f) > 0$ iff $f \in (A_n)^{-1}$ by (1.1.5).

We first choose $b_1 \in (A_1)^{-1}$, $g_1 \in C(M(A_1))$ so that

$$f|_{M(A_1)} = \hat{b}_1 \exp(g_1) \quad \text{(cf. (1.4.8)).}$$

(Recall that $f \in C(kM(A))$ iff $f|_{M(A_n)} \in C(M(A_n))$ for all $n \in \mathbb{N}$.)

We now use the proposition to choose inductively sequences $(b_n)_n$ and $(g_n)_n$ such that for all $n \in \mathbb{N}$

i) $b_n \in (A_n)^{-1}$, $g_n \in C(M(A_n))$;

ii) $f|_{M(A_n)} = \hat{b}_n \exp(g_n)$;

iii) $p_n'(\pi_{n+1,n}(b_{n+1}) - b_n) < 2^{-(n+1)} \cdot \min\{1, d_1(b_1), \ldots, d_n(b_n)\}$;

iv) $\|g_{n+1}|_{M(A_n)} - g_n\|_{M(A_n)} < 2^{-(n+1)}$.

Now fix $n \in \mathbb{N}$, and let $k > l \geq n$. Then

$$(*) \qquad p_n'(\pi_{k,n}(b_k) - \pi_{l,n}(b_l))$$

$$= p_n'(\pi_{k,n}(b_k) - \pi_{k-1,n}(b_{k-1}) + \pi_{k-1,n}(b_{k-1}) - \ldots + \pi_{l+1,n}(b_{l+1}) - \pi_{l,n}(b_l))$$

$$= p_n'(\pi_{k-1,n}(\pi_{k,k-1}(b_k) - b_{k-1}) + \ldots + \pi_{l,n}(\pi_{l+1,l}(b_{l+1}) - b_l))$$

$$\leq p_{k-1}'(\pi_{k,k-1}(b_k) - b_{k-1}) + \ldots + p_l(\pi_{l+1,l}(b_{l+1}) - b_l)$$

$$\leq \sum_{i=l+1}^{k} 2^{-i} \cdot \min\{1, \ldots, d_{i-1}(b_{i-1})\}.$$

It follows that $(\pi_{k,n}(b_k))_{k\geq n}$ is a Cauchy sequence in A_n. Denote the limit by a_n. In particular we have by (*)

$$p'_n(\pi_{k,n}(b_k)-b_n) \leq (\sum_{j=n+1}^{k} 2^{-j})d_n(b_n) < (1/2)d_n(b_n)$$

for all $k \in \mathbb{N}$. So

$$p'_n(a_n-b_n) \leq (1/2)d_n(b_n),$$

i.e. $a_n \in (A_n)^{-1}$.

It follows from iv) that $(g_k|_{M(A_n)})_{k\geq n}$ is a Cauchy sequence in $C(M(A_n))$, hence converges to an element $h_n \in C(M(A_n))$. We have

$$f|_{M(A_n)} = \hat{\pi}_{k,n}(b_k)\exp(g_k|_{M(A_n)}) \quad \text{for all } k\geq n \text{ by ii)},$$

where $\hat{\pi}_{k,n}(b_k)$ denotes the Gelfand transform of $\pi_{k,n}(b_k)$. Hence for each $\varphi \in M(A_n)$

$$f(\varphi) = \lim_{k\to\infty} \hat{\pi}_{k,n}(b_k)(\varphi)\exp(g_k(\varphi)) = \hat{a}_n(\varphi)\exp(h_n(\varphi)),$$

i.e. $f|_{M(A_n)} = \hat{a}_n\exp(h_n)$. It is easily seen that $\pi_{n+1,n}(a_{n+1}) = a_n$ and $h_{n+1}|_{M(A_n)} = h_n$ for all $n \in \mathbb{N}$. So $(a_n)_n$ and $(h_n)_n$ define elements $a \in A$ and $h \in C(kM(A))$ so that $f = \hat{a}\exp(h)$. Since $\hat{a}|_{M(A_n)} = \hat{a}_n$, $\hat{a}$ never vanishes on $M(A)$, hence $a \in A^{-1}$ by (6.1.4).□

Corollary. Let A be an F-algebra. Suppose that M(A) is a k-space, e.g. a locally compact space or a space which satisfies the first axiom of countability. Then for each $f \in C(M(A))^{-1}$ there exist $g \in A^{-1}$ and $h \in C(M(A))$ so that $f = \hat{g}\exp(h)$.

Remarks. i) The theorem was stated first - without proof - by Arens [ARE 3]. The proof above is due to Brooks [BRO 1].

ii) One can use the Arens-Royden theorem to describe $H^1(kM(A),\mathbb{Z})$, the first Cech cohomology group of kM(A) with coefficients in $\mathbb{Z}$. $H^1(kM(A))$ is isomorphic to $A^{-1}/\exp(A)$ for each F-algebra A [BRO 1].

(Note that exp(A) is a subgroup of A^{-1}.) Hence, if M(A) is a k-space, then $H^1(M(A))$ is isomorphic to $A^{-1}/\exp(A)$.
iii) In the next chapter we shall construct an F-algebra A such that M(A) is not a k-space. It is unproved whether $H^1(M(A)) = H^1(kM(A))$ in this example.

CHAPTER 7

AN F-ALGEBRA WHOSE SPECTRUM IS NOT A K-SPACE

The results of (6.2) provide interest to the question whether the spectrum of an F-algebra is always a k-space. The following example due to Dors [DOR] shows that it is not always the case. Furthermore we consider this example also in connection with the following problems:

i) Given an F-algebra A and a subalgebra B. When can all elements of M(B) be extended to elements of M(A)?

ii) Give conditions such that the weak Nullstellensatz is valid for a uF-algebra (A,X).

(7.1) The example of Dors

(7.1.1) We first introduce the concept of dimension for complex spaces.

Let X, Y be topological spaces. We say that a map $\pi: X \longrightarrow Y$ is discrete if $\pi^{-1}(y)$ is a discrete subset of X for each $y \in Y$.

Definition. Let A be an analytic subset of an open subset $U \subset \mathbb{C}^n$, and let $x \in A$. Then

$$\dim_x A = \min\{k \in \mathbb{N} : \text{There exists an open neighbourhood } V \text{ of } x \text{ in } A \text{ and a discrete holomorphic map } \pi: V \longrightarrow \mathbb{C}^k\}.$$

is called the dimension of A at x.

If $\dim_x A = k$ for all $x \in A$ we say that A is of pure dimension k.

Remarks. i) Let $\dim_x A = k$, then there is a neighbourhood U of x in A such that $\dim_y A \le k$ for all $y \in U$.

ii) Note that $\dim_x A \le n$ if A is an analytic subset of a domain in $\mathbb{C}^n$.

iii) If x is a regular point, i.e. if there exists a neighbourhood U of x in A such that U is biholomorphically equivalent to some domain in $\mathbb{C}^k$, then $\dim_x A=k$. One can show that

$$\dim_x A = \sup\{k \in \mathbb{N}: \text{In every open neighbourhood of } x \text{ in } A \text{ there is a regular point } y \text{ such that } \dim_y A=k\}.$$

(Recall from remark (2.3.1)ii) that the set of all regular points lies dense in A.)

iv) Let X be a reduced complex space, and let $x \in X$. Let (U_i,φ_i,X_i) be a chart around x (cf. definition (2.3.1)c)), then set

$$\dim_x X = \dim_{\varphi_i(x)} X_i.$$

It can be shown that the definition is independent of the chosen charts.

v) As a consequence of the maximum principle we see that a compact analytic subset $A \subset U \subset \mathbb{C}^n$ consists of finitely many points. Hence $\dim_x A=0$ for all $x \in A$.

vi) Let $A \subset G \subset \mathbb{C}^n$ $(n\ge2)$ be an analytic subset such that $\dim_x A \le n-2$ for all $x \in A$. Then every $f \in Hol(G\setminus A)$ can be extended to an element of Hol(G), i.e. there exists $\tilde{f} \in Hol(G)$ such that $\tilde{f}|_{G\setminus A}=f$ (Riemann extension theorem, second form).

vii) Let A be a proper analytic subset of G. Then $f \in Hol(G\setminus A)$ is called locally bounded if each $a \in A$ has a neighbourhood U in G so that

$$\sup\{|f(x)|: x \in U\setminus A\} < \infty.$$

If $f \in Hol(G\setminus A)$ is locally bounded, then f can be extended to a holomorphic function $\tilde{f}$ on G (Riemann extension theorem, first form).

viii) Let $U \subset \mathbb{C}^n$ $(n\ge2)$ be an open set, and let $f \in Hol(U)$. If

$$V(f)=\{x \in U: f(x) = 0\}$$

is not empty, then V(f) is an analytic subset of U such that $\dim_x V(f) \ge n-1$ for all $x \in V(f)$.

In fact the following is true: Let $A \subset U$ be an analytic subset such that $\dim_x A \geq k$ for a point $x \in A$. Let $W \subset U$ be an open neighbourhood of x, and let $f \in Hol(W)$ such that $f(x)=0$. Then
$A' = A \cap \{y \in W: f(y) = 0\}$ is an analytic subset of a suitable open set and $\dim_x A' \geq k-1$, cf. [G/R], p. 115.

ix) Now consider the following situation: Let $f_1, \ldots, f_r \in Hol(\mathbb{C}^k)$, and let

$$A = \{x \in \mathbb{C}^k: f_1(x) = \ldots = f_r(x) = 0\}$$

be nonempty. For $x \in A$ we have $\dim_x V(f_1) \geq n-1$ and $\dim_x V(f_1) \cap V(f_2) \geq n-2$ by viii). Continuing in this manner we see that

$$\dim_x A \geq \max\{n-r, 0\}.$$

Hence, if $n > r$, A cannot be compact.

<u>(7.1.2) Example.</u> Set $\mathbb{C}^{\mathbb{N}} = \prod_{n \in \mathbb{N}} \mathbb{C}$ (countable many copies with the product topology),

$$W_k = \{(x_n)_n \in \mathbb{C}^{\mathbb{N}}: x_n = 0 \text{ for } j>k\}$$

and

$$K_k = \{(x_n)_n \in W_k: |x_i| \leq k \text{ for } 1 \leq i \leq k\}, \quad k \in \mathbb{N}.$$

Denote by z_k the projection

$$\mathbb{C}^{\mathbb{N}} \longrightarrow \mathbb{C}, \ (x_n)_n \longrightarrow x_k.$$

Define

$$P(K_k) = \{f: K_k \longrightarrow \mathbb{C}, \ f \text{ can be approximated uniformly on } K_k \text{ by polynomials in } z_1, \ldots, z_k\}.$$

Then $P(K_k)$ is a uB-algebra and

(*) $\qquad M(P(K_k)) = K_k,$

since K_k (interpreted as a subset of $\mathbb{C}^k$) is polynomially convex (cf. (1.2.4)ii)). Set $A = \varprojlim P(K_k)$. Then A is a uF-algebra,

$$M(A) = \cup K_k \quad \text{(as sets)}$$

and $(K_k)_k$ is an admissible exhaustion of M(A) with respect to the Gelfand topology, cf. (3.3.4).

Set

$$X = \{(x_n)_n \in \bigcup K_k : |x_i|<1 \text{ for all } i \in \mathbb{N}\}.$$

Clearly $X \cap K_k$ is an open set (with respect to the induced Gelfand topology on K_k, which equals the original topology on K_k by (*)) for each $k \in \mathbb{N}$.

Claim. X is not Gelfand open, i.e. M(A) is not a k-space.

Suppose X is Gelfand open. We have $0 = (0,0,\ldots.) \in X$, hence there are $f_1,\ldots,f_r \in A$ so that

(**) $\{y \in M(A): |\hat{f}_i(y)-\hat{f}_i(0)| < 1, i=1,\ldots,r\} \subset X$

by (3.2.2)i). Each element $\hat{f} \in \hat{A}$ can be interpreted as a complex-valued function on $\bigcup K_n$, which can be approximated uniformly on each K_n by polynomials. In particular we can interpret $\hat{f}|_{W_k}$ as holomorphic function on $\mathbb{C}^k$ for each $k \in \mathbb{N}$.

By (7.1.1)ix) the common zero set of holomorphic functions $g_1,\ldots,g_r \in Hol(\mathbb{C}^k)$ is unbounded if $k>r$. Hence

$$\{y \in W_{r+1}: \hat{f}_i|_{W_{r+1}}(y) = \hat{f}_i(0), i=1,\ldots r\}$$

is unbounded, a contradiction to (**).

Remarks.i) Another example is due to Hayes and Vigue [H/V 2]. They constructed even a reduced complex space X with countable topology, such that M(Hol(X)) is not a k-space.

ii) It was proved by Warner [WAR] that the spectrum M(A) of an F-algebra is a k-space iff C(M(A)) is an F-algebra.

(7.2) Surjectivity of the transpose map

Let A,B be F-algebras, and let $T: A \longrightarrow B$ be a continuous algebra homomorphism. We showed in (3.2.5) that the transpose map

$$T^*: M(B) \longrightarrow M(A),\ \varphi \longrightarrow \varphi\circ T,$$

is injective if T has dense range. In this section we deal with the surjectivity of T^*.

(7.2.1) For $n \in \mathbb{N}$ set

$$U_n(A) = \{(f_1,\dots,f_n) \in A^n: \text{There is } (g_1,\dots,g_n) \in A^n \text{ so that } \sum_{i=1}^n f_i g_i = 1\}.$$

For every $n \in \mathbb{N}$ we have a natural mapping

$$T_n: A^n \longrightarrow B^n,\ (f_1,\dots,f_n) \longrightarrow (T(f_1),\dots,T(f_n)).$$

Note that $T_n(U_n(A)) \subset U_n(B)$, since we always assume that $T(1) = 1$.

Further denote by $U_\infty(A)$ the set of all sequences $(f_n)_n$ in A so that there is another sequence $(g_n)_n$ in A such that

$$\sum_{i\in\mathbb{N}} g_i f_i = 1.$$

Set

$$T_\infty: \prod_{n\in\mathbb{N}} A \longrightarrow \prod_{n\in\mathbb{N}} B,\ (f_n)_n \longrightarrow (T(f_n))_n.$$

Theorem. Let A, B be F-algebras, and let $T: A \longrightarrow B$ be a continuous algebra homomorphism with $T(1)=1$. Then $T^*(M(B)) = M(A)$ iff

$$(T_\infty)^{-1}(U_\infty(B)) = U_\infty(A).$$

Proof. First note that $(f_n)_n \in U_\infty(A)$ (resp. $(f_1,\dots,f_n) \in U_n(A)$) iff for each $\varphi \in M(A)$ there exists $i \in \mathbb{N}$ (resp. $i \in \{1,\dots,n\}$) so that $\varphi(f_i) \neq 0$, cf. (6.1.9).

We always have $U_\infty(A) \subset (T_\infty)^{-1}(U_\infty(B))$.

i) Suppose that $T^*(M(B)) = M(A)$.

Let $(f_n)_n \in (\prod_{n\in\mathbb{N}} A)\backslash U_\infty(A)$ be an arbitrary element. Then there exists $\varphi \in M(A)$ so that $\varphi(f_n) = 0$ for all $n \in \mathbb{N}$. Choose $\psi \in M(B)$ such that

$T^*(\psi) = \varphi$, i.e. $\psi \circ T = \varphi$, then $\psi(T(f_n)) = 0$ for all $n \in \mathbb{N}$, i.e.

$$(T(f_n))_n \in (\prod_{n\in\mathbb{N}} B)\setminus U_\infty(B).$$

ii) Suppose there is $\varphi \in M(A)\setminus T^*(M(B))$.

Then we find for each $\psi \in M(B)$ an element $f_\psi \in A$ such that

$$\psi(T(f_\psi)) \neq 0 \text{ and } \varphi(f_\psi) = 0.$$

By continuity there exists a neighbourhood U_ψ of ψ in M(B) such that

$$\tilde{\psi}(T(f_\psi)) \neq 0 \quad \text{for all } \tilde{\psi} \in U_\psi.$$

Since M(B) is hemicompact there exists a sequence $(\psi_i)_i$ in M(B) so that

$$M(B) = \bigcup_{i\in\mathbb{N}} U_{\psi_i}.$$

Hence we have

α) $\varphi(f_{\psi_i}) = 0$ for all $i \in \mathbb{N}$ and

β) for each $\tilde{\psi} \in M(B)$ there is $n \in \mathbb{N}$ such that $\tilde{\psi}(T(f_{\psi_n})) \neq 0$.

It follows that $(T(f_{\psi_i}))_i \in U_\infty(B)$, while $(f_{\psi_i})_i \in \prod_{i\in\mathbb{N}} A \setminus U_\infty(A)$.

□

(7.2.2) We remark that $(T_\infty)^{-1}(U_\infty(B)) = U_\infty(A)$ implies that

(*) $\qquad (T_n)^{-1}(U_n(B)) = U_n(A)$ for all $n \in \mathbb{N}$.

This leads to the question whether the weaker condition (*) already guarantees the surjectivity of T^*.

We shall show below that in general this is not the case. However we have the following result.

Corollary. Let A, B be F-algebras, and let $T: A \longrightarrow B$ be a continuous algebra homomorphism such that T(1)=1. Suppose that (*) is satisfied. Then T^* is surjective if M(B) is compact, e.g. B is a B-algebra, or if there is $(f_1,\dots,f_n) \in A^n$ such that the map

$$\hat{F}: M(A) \longrightarrow \mathbb{C}^n,\ \varphi \longrightarrow (\hat{f}_1(\varphi),\dots,\hat{f}_n(\varphi)),$$

is injective, e.g. A is rationally finitely generated.

Proof. i) Let M(B) be compact.
Suppose that T^* is not surjective. We use the same notations as in the proof of ii) of theorem (7.2.1). Repeating this proof we see that there are finitely many points $\psi_1,\dots,\psi_r$ so that

$$M(B) = \bigcup_{i=1}^{r} U_{\psi_i}.$$

It follows that $(T(f_{\psi_1}),\dots,T(f_{\psi_r})) \in U_r(B)$, while

$(f_{\psi_1},\dots,f_{\psi_r}) \in A^r \setminus U_r(A)$, a contradiction to (*).

ii) Let $\hat{F}$ be injective.
Suppose there is $\varphi \in M(A)\setminus T^*(M(B))$. Then

$$\{\varphi\} = \{\psi \in M(A): \varphi(f_i) = \psi(f_i),\ i=1,\dots,n\}.$$

Hence

$$(f_1-\varphi(f_1),\dots,f_n-\varphi(f_n)) \in A^n\setminus U_n(A),$$

but

$$(T(f_1-\varphi(f_1)),\dots,T(f_n-\varphi(f_n))) \in U_n(B),$$

a contradiction to (*).
(Actually we have proved a stronger result. Namely if $\hat{F}$ is injective, then T^* is surjective if $(T_n)^{-1}(B_n) = A_n$.) □

(7.2.3) Finally we use the example of Dors (7.1.2) to show that condition (*) of (7.2.2) in general does not imply the surjectivity of T^*.

We use the same notations as in (7.1.2). Set

$$D_k = \{(x_n)_n \in K_k: |x_i|<1/2,\ i=1,\dots,k\}$$

and

$$E_k = K_k\setminus D_k.$$

E_k is a compact set for each $k \in \mathbb{N}$ and

$$\ldots \longrightarrow C(E_{k+1}) \xrightarrow{r_k} C(E_k) \longrightarrow \ldots$$

constitutes a dense projective system of uB-algebras with respect to the natural restriction mappings

$$r_k: C(E_{k+1}) \longrightarrow C(E_k),\ f \longrightarrow f|_{E_k}.$$

Set $B = \varprojlim C(E_n)$. Then B is a uF-algebra by (4.1.3) and

$$M(B) = \bigcup_{n\in\mathbb{N}} M(C(E_n)) = \bigcup_{n\in\mathbb{N}} E_n$$

as sets by (1.2.4)i) and (3.3.4). $\hat{B}$ is the algebra of all complex-valued functions f on $\bigcup E_n$ such that $f|_{E_k}$ is continuous for all $k \in \mathbb{N}$.

Let $A = \varprojlim P(K_n)$ be the algebra of Dors example. Interpret each $\hat{f} \in \hat{A}$ as complex-valued function on $\bigcup K_n$, which can be approximated uniformly on each K_n by polynomials. Then

$$T: \hat{A} \longrightarrow \hat{B}, \hat{f} \longrightarrow \hat{f}|_{\bigcup E_n},$$

defines a continuous homomorphism and

$$T^*(M(\hat{B})) = \bigcup E_n \neq \bigcup K_n = M(\hat{A}),$$

i.e. T^* is not surjective. (Note that we can naturally identify M(A) and $M(\hat{A})$.)

Now fix $k \in \mathbb{N}$, and let $(\hat{g}_1,\ldots,\hat{g}_k) \in \hat{A}^k \backslash U_k(\hat{A})$. Hence there exists $x \in \bigcup K_n$ such that $\hat{g}_1(x) = \ldots = \hat{g}_k(x) = 0$, w.l.o.g. $x \in K_1$. The set

$$V = \{y \in W_{k+1}: \hat{g}_1(y) = \ldots = \hat{g}_k(y) = 0\}$$

is not empty and we conclude as in (7.1.2) that V (interpreted as a subset of $\mathbb{C}^{k+1}$) is unbounded, i.e. there exists

$$y \in (\bigcup E_n) \cap W_{k+1}$$

so that $\hat{g}_1(y) = \ldots = \hat{g}_k(y) = 0$. It follows that

$$(T(\hat{g}_1),\dots,T(\hat{g}_k)) \in \hat{B}^k \setminus U_k(\hat{B})$$

Thus

$$U_k(\hat{A}) = (T_k)^{-1}(U_k(\hat{B})) \text{ for all } k \in \mathbb{N}.$$

Remark. Corach and Suarez [C/S] proved that the condition (*) is equivalent to the surjectivity of T^* in the case that A and B are B-algebras. The proof of theorem (7.2.1) is a suitable modification of their proof.

(7.3) The weak Nullstellensatz

(7.3.1) Definition. The weak Nullstellensatz is valid for a uF-algebra (A,X) if for every finite set of functions $f_1,\dots,f_m \in A$ with no common zero in X there exist functions $g_1,\dots,g_m \in A$ such that

$$\sum_{i=1}^{m} f_i g_i = 1.$$

(7.3.2) Remarks. i) We use the terminology of Hayes [HAY 2]. This terminology is motivated by a theorem in algebraic geometry referred to as the weak Nullstellensatz. It states that every proper ideal in in a polynomial ring $K[T_1,\dots,T_n]$ in n variables with coefficients in an algebraically closed field K has at least one zero in K^n. Since $K[T_1,\dots,T_n]$ is Noetherian, this is equivalent to saying that finitely many elements in $K[T_1,\dots,T_n]$ without common zero never generate a proper ideal.

ii) Recall from (2.2.7) that for domains $G \subset \mathbb{C}^n$ the following is true: The weak Nullstellensatz is valid for (Hol(G),G) if G is a domain of holomorphy resp. iff M(Hol(G))=G. This result does not hold in our general setting as we shall see below.

iii) It follows from (6.1.10) that the weak Nullstellensatz is valid for $(\hat{A},M(A))$ if A is a uF-algebra.

(7.3.3) Theorem. The weak Nullstellensatz is valid for a uF-algebra iff

$$F(X) = \hat{F}(M(A))$$

for all $F \in A^m$ and all $m \in \mathbb{N}$.
(For $(f_1,\dots,f_m) \in A^m$ we set $F(X) = \{(f_1(x),\dots,f_m(x)) \in \mathbb{C}^m: x \in X\}$ and

$$\hat{F}(M(A)) = \{(\hat{f}_1(x),\dots,\hat{f}_m(x)) \in \mathbb{C}^m: x \in M(A)\}.)$$

Proof. The inclusion "$\subset$" is always true.
i) Suppose that the weak Nullstellensatz is valid for (A,X).
Assume there is $m \in \mathbb{N}$ and $f \in A^m$ such that

$$\hat{F}(M(A))\backslash F(X) \neq \emptyset.$$

We choose a point $(\lambda_1,\dots,\lambda_m)$ in this set. The functions $f_1-\lambda_1,\dots,$ $f_m-\lambda_m$ have no common zero in X, hence there are $g_1,\dots,g_m \in A$ so that

$$\sum_{i=1}^{m} (f_i-\lambda_i)g_i = 1.$$

This implies

$$\sum_{i=1}^{m} (\hat{f}_i-\lambda_i)\hat{g}_i = 1,$$

a contradiction to the fact that $\hat{f}_1-\lambda_1,\dots,\hat{f}_m-\lambda_m$ have a common zero in M(A).
ii) Now let $F(X) = \hat{F}(M(A))$ for all $f \in A^m$.
Let $g_1,\dots,g_m \in A$ be without common zero in X. Then $\hat{g}_1,\dots,\hat{g}_m$ have no common zero in M(A), hence by (6.1.10) there are $h_1,\dots,h_m$ so that

$$\sum_{i=1}^{m} h_i g_i = 1.$$

□

(7.3.4) Corollary. i) The weak Nullstellensatz is valid for a uF-algebra (A,X) if the structure map $j: X \longrightarrow M(A)$, $x \longrightarrow \varphi_x$, (cf. (4.2.1)) is surjective.
ii) If (A,X) is a uF-algebra and there is $F \in A^m$ such that

$\hat{F}: M(A) \longrightarrow \mathbb{C}^m$ is injective, e.g. if A is rationally finitely generated, then the weak Nullstellensatz is valid iff j is surjective.

Proof. i) This is a direct consequence of theorem (7.3.3).
ii) If j is not surjective we find a point $x \in M(A)\backslash j(X)$. Since $\hat{F}$ is injective we have

$$\hat{F}(x) \in \hat{F}(M(A))\backslash F(X),$$

and our assertion follows from (7.3.3). □

(7.3.5) We use the example of Dors (7.1.2) to show that in general the converse of statement i) is false.
We use the notations of (7.1.2) and (7.2.3):

$$W_k = \{(x_n)_n \in \mathbb{C}^{\mathbb{N}}: x_j = 0, j>k\},$$

$$K_k = \{(x_n)_n \in W_k: |x_i| \leq k, i=1,\ldots,k\},$$

$$D_k = \{(x_n)_n \in K_k: |x_i| < 1/2, i=1,\ldots,k\}$$

and

$$E_k = K_k\backslash D_k.$$

Let $A = \varprojlim P(K_n)$ be the algebra in the example of Dors. Recall that

$$M(A) = \cup K_k \quad \text{as sets},$$

that $(K_k)_k$ is an admissible exhaustion of M(A) with respect to the Gelfand topology, and that $\hat{A}$ is the algebra of all complex-valued functions on $\cup K_k$ which can be approximated uniformly on each K_n by polynomials. Set $X = \cup E_k$, and endow X with the relative Gelfand topology. Note that $(E_k)_k$ is an admissible exhaustion of X, i.e. X is a hemicompact space. Consider the restriction mapping

$$r: \hat{A} \longrightarrow \hat{A}|_X, \hat{f} \longrightarrow \hat{f}|_X.$$

Clearly r is a surjective algebra homomorphism. Let $r(\hat{f}) = 0$. Then

$$\hat{f}|_{W_k\backslash E_k} \equiv 0 \quad \text{for every } k \in \mathbb{N}.$$

Since we can interpret $\hat{f}|_{W_k}$ as a holomorphic function on $\mathbb{C}^k$, we have $\hat{f}|_{W_k} \equiv 0$ by the identity theorem, i.e. $\hat{f} \equiv 0$, and r is injective. Moreover it follows from the maximum modulus principle for holomorphic functions that

$$\|\hat{f}\|_{K_k} = \|\hat{f}\|_{E_k} \quad \text{for all } k \in \mathbb{N},\ \hat{f} \in \hat{A},$$

i.e. r is a topological homomorphism. So $(\hat{A}|_X, X)$ is a uF-algebra, which is isomorphic to $(\hat{A}, M(A))$ and we see that

$$j: X \longrightarrow M(\hat{A}|_X) = M(A)$$

is not surjective.

Now let $\hat{g}_1|_X, \ldots, \hat{g}_k|_X \in \hat{A}|_X$ be arbitrary elements without common zero on X.Then we see as in example (7.2.3) that $\hat{g}_1, \ldots, \hat{g}_k$ don't have a common zero on $M(A)=M(\hat{A})$, hence there are $\hat{h}_1, \ldots, \hat{h}_k \in \hat{A}$ so that

$$\sum_{i=1}^{k} \hat{h}_i \hat{g}_i = 1,$$

by (6.1.10), i.e. the weak Nullstellensatz is valid for $(\hat{A}|_X, X)$.

<u>(7.3.6) Remarks.</u> i) Nevertheless the validity of the weak Nullstellensatz implies that the structure map j has dense range in M(A). Otherwise there would be a point $x \in M(A)$ and a Gelfand open neighbourhood U of x,

$$U = \{y \in M(A) : |\hat{f}_i(x) - \hat{f}_i(y)| < 1,\ i=1,\ldots,r\}, \quad f_1 \ldots, f_r \in A,$$

(cf. (3.2.2)i)) such that

$$U \subset M(A) \setminus j(X).$$

But this would imply that

$$\hat{F}(x) = (\hat{f}_1(x), \ldots, \hat{f}_n(x)) \in \hat{F}(M(A)) \setminus F(X),$$

and we get a contradiction to theorem (7.3.3).

ii) The converse of remark i) is false.
For instance let $\overline{D} = \{|x| \le 1\}$, $X = \overline{D} \setminus \{0\}$, and let A be the algebra of all functions on X which can be approximated uniformly by polynomials on each compact subset of X. Then (A,X) is a uF-algebra which is algebraically and topologically isomorphic to the algebra $P(\overline{D})$. We have $M(P(\overline{D})) = \overline{D}$ by (1.2.4)ii), hence

$$j: X \longrightarrow M(A) = \overline{D},$$

has dense range, but

$$z(X) \neq \hat{z}(M(A))$$

for the map

$$z: \overline{D} \longrightarrow \mathbb{C},\ x \longrightarrow x.$$

So the weak Nullstellensatz is not valid for (A,X) by (7.3.3).
iii) In [HAY 2] Hayes constructed a hemicompact reduced complex space X so that the weak Nullstellensatz is valid for (Hol(X),X) but j is not surjective.

CHAPTER 8

SEMISIMPLE F-ALGEBRAS

An F-algebra A is called semisimple if the Gelfand map $\Gamma: A \longrightarrow \hat{A}$, $f \longrightarrow \hat{f}$, is injective. In particular each uF-algebra is semisimple by the proof of (4.1.3). In this chapter we prove two results for semisimple F-algebras. Firstly we show that each semisimple F-algebra has a unique topology as an F-algebra. This extends the analogous result for semisimple B-algebras (1.2.12). Secondly we prove that each derivation D on A is automatically continuous. (By a derivation we mean a linear mapping $D: A \longrightarrow A$ so that $D(fg) = f\,D(g)+gD(f)$ for all $f,g \in A$.) We remark that, opposite to the situation for semisimple B-algebras, there exist nontrivial derivations on semisimple F-algebras. Both results are due to Carpenter [CAR 1] and [CAR 5].

(8.1) Uniqueness of topology

<u>(8.1.1) Definition.</u> The radical of an algebra A is the intersection of all maximal ideals in A. It is denoted by rad A. A is called semisimple if rad A = {0}.

<u>(8.1.2) Proposition.</u> An F-algebra A is semisimple iff the Gelfand map

$$\Gamma: A \longrightarrow \hat{A},\ f \longrightarrow \hat{f},$$

is injective.

<u>Proof.</u> It suffices to show that

$$\text{rad } A = \bigcap_{\varphi \in M(A)} \ker \varphi .$$

Certainly rad A $\subset \bigcap$ ker φ.

Now let $x \in \ker \varphi$ for all $\varphi \in M(A)$. Then

(*) $\qquad \varphi(1-xy) = 1$ for each $\varphi \in M(A)$, and all $y \in A$,

hence $1-xy$ is invertible by (6.1.4) for each $y \in A$.

Suppose there is a maximal ideal M in A so that $x \in A\backslash M$.

The smallest ideal which contains M and x is A. Hence there is $y \in A$ and $m \in M$, so that $1 = xy + m$, i.e. $1 - xy \in M$, and we get a contradiction to (*). □

(8.1.3) Remarks . i) Each uF-algebra is semisimple by the proof of (4.1.3).

ii) Let $A = C^{\infty}([0,1])$ (cf. (3.3.5)). Then $M(A) = [0,1]$ by (3.3.5) and A is not a uF-algebra by (4.1.4)iv). Let $f \in$ rad A, i.e. $\hat{f}(x)=0$ for all $x \in M(A)$, hence $f(x)=0$ for all $x \in [0,1]$. So $f\equiv 0$ and A is semisimple by (8.1.2).

(8.1.4) Proposition. Let A be an F-algebra. Let $(\varphi_i)_i$ be a sequence of distinct points in M(A). Then there exists $h \in A$ so that $(\varphi_i(h))_i$ is a sequence of distinct points in the plane.

Proof. For $i \neq j$ set

$$A_{ij} = \{f \in A: \varphi_i(f) = \varphi_j(f)\}.$$

Obviously A_{ij} is a closed subset of A. Let $(p_n)_n$ be a defining sequence of seminorms for A, cf. (3.1.7). Suppose that A_{ij} has non-empty interior. Then there are $f \in A_{ij}$, $n \in \mathbb{N}$ such that

$$\{g \in A: p_n(f-g) < 1/n\} \subset A_{ij}.$$

Choose $u \in A$ such that $\varphi_i(u) = 0$ and $\varphi_j(u) = 1$ by (3.2.3). If $p_n(u)=0$ then $p_n(f-(f+u)) < 1/n$, but $f+u \in A\backslash A_{ij}$, and we get a contradiction. If $p_n(u) \neq 0$, then

$$p_n(f-f+(u/2np_n(u))) < 1/n,$$

but $f-(u/2np_n(u)) \in A \setminus A_{ij}$, a contradiction. Hence A_{ij} is nowhere dense. By Baire's theorem there exists

$$h \in A \setminus \bigcup_{i=1}^{\infty} \bigcup_{j>i} A_{ij}. \qquad \square$$

Corollary. Let A be an F-algebra, and let $(\varphi_n)_n$ be a sequence of distinct points of M(A). Then, for each $k \in \mathbb{N}$, there exists $f_k \in A$ such that

$$\varphi_j(f_k) = 0 \text{ for } j<k \text{ and } \varphi_j(f_k) \neq 0 \text{ for } j \geq k.$$

Proof. Choose f according to the proposition. For $j \in \mathbb{N}$ set

$$g_j = f - \varphi_j(f).$$

Then $f_k = g_1 \cdot \ldots \cdot g_{k-1}$. $\square$

Remark. The lemma is due to Carpenter [CAR 1]. The proof above is different from his proof.

(8.1.5) Theorem [CAR 1]. Let A be a semisimple F-algebra. Then A has a unique topology as an F-algebra.

Proof. Denote by τ the given topology on A. Assume that A is an F-algebra with respect to a second topology τ_1, determined by a sequence of seminorms $(p_n)_n$, with $p_n \leq p_{n+1}$ for all $n \in \mathbb{N}$. We denote by $M(A,\tau_1)$ resp. $M(A,\tau)$ the spectrum of A with respect to τ_1 resp. τ.
Claim i). Let $K \subset M(A,\tau)$ be a compact set. Then all but finitely many elements of K are continuous with respect to τ_1.
Assume the contrary. Then there is a sequence $(\varphi_n)_n$ of distinct points in K, every φ_n discontinuous on (A,τ_1). We use (8.1.4) to obtain a sequence $(f_n)_n$ of elements of A such that

$$\varphi_k(f_j) = 0 \ (k<j),$$

and

$$\varphi_k(f_j) \neq 0 \ (k \geq j),$$

and set

$$a_k = \varphi_k(f_1 \cdot \ldots \cdot f_k),$$

hence $a_k \neq 0$. Since φ_k is discontinuous, we can successively choose $g_k \in A$ so that

a) $\max\{p_k(f_j \cdot \ldots \cdot f_k \cdot g_k),\ 1 \leq j \leq k\} < 2^{-k}$

and

b) $|a_k \varphi_k(g_k)| > k + |\varphi_k(\sum_{j=1}^{k-1} f_1 \cdot \ldots \cdot f_j \cdot g_j)|$.

Set

$$h_j = \sum_{i=j}^{\infty} f_j \cdot \ldots \cdot f_i \cdot g_i\ ,$$

then $h_j \in A$ for all $j \in \mathbb{N}$ by a) and by the assumption "$p_n \leq p_{n+1}$".
We have

$$h_1 = \sum_{j=1}^{k} f_1 \cdot \ldots \cdot f_j \cdot g_j + f_1 \cdot \ldots \cdot f_{k+1}(g_{k+1} + h_{k+2})$$

for all $k \in \mathbb{N}$. Since

$$\varphi_k(f_{k+1}) = 0,$$

we get

$$\varphi_k(h_1) = \varphi_k(\sum_{j=1}^{k-1} f_1 \cdot \ldots \cdot f_j \cdot g_j) + a_k \varphi_k(g_k),$$

and so

$$|\varphi_k(h_1)| \geq k \text{ for each } k \in \mathbb{N} \text{ by b)}.$$

This is a contradiction, since $\hat{h}_1$ is bounded on the compact set K.
Claim ii). The set $S = \{\varphi \in M(A,\tau) : \varphi \in M(A,\tau_1)\}$ is dense in $M(A,\tau)$.

Choose an arbitrary point φ of $M(A,\tau)$. Assume first that φ is not an isolated point of $M(A,\tau)$. By (6.1.6) there is a compact set $K \subset M(A,\tau)$ such that φ is not an isolated point of K. Hence φ lies in the closure of S by i).
If φ is isolated in $M(A,\tau)$, then $\varphi \in M(A,\tau_1)$ by (10.1.1).
Claim iii). The identity map id: $M(A,\tau) \longrightarrow M(A,\tau_1)$, $f \longrightarrow f$, is continuous.

Let $(f_n)_n$ be a sequence in A such that $(f_n)_n$ converges to f with respect to τ and to g with respect to τ_1. Then

$$\varphi(f) = \varphi(g) \text{ for all } \varphi \in S.$$

It follows by ii) and since A is semisimple that f=g, i.e. the graph of id is closed. This implies the continuity of id by the closed graph theorem, cf. B.2.

iv) It follows now from the open mapping theorem for Fréchet spaces that id is a homeomorphism.

□

(8.1.6) Remark. If we knew that each homomorphism $\varphi: A \longrightarrow \mathbb{C}$ were continuous (cf. chapter 10), the theorem would follow immediately from part iii) of the proof above.

(8.1.7) Corollary [CAR 1]. Let A be a semisimple F-algebra, and let $\varphi: A \longrightarrow \mathbb{C}$ be a homomorphism. If there is an F-algebra topology for A with respect to which φ is continuous, then φ is already continuous on A.

(8.2) Continuity of derivations

(8.2.1) Definition. Let A be an F-algebra. A linear mapping $D: A \longrightarrow A$ is called derivation if

$$D(ab) = D(a)b+aD(b) \text{ for all } a, b \in A.$$

If A is a semisimple B-algebra, then due to results of Singer, Wermer, Curtis and Johnson (cf. [JOH]) there exists only the trivial derivation

$$D: A \longrightarrow A, \; a \longrightarrow 0.$$

Opposite to this result there are semisimple F-algebras which admit nontrivial derivations. For instance let $G \subset \mathbb{C}$ be a domain, then

$$D: \mathrm{Hol}(G) \longrightarrow \mathrm{Hol}(G), \; f \longrightarrow f',$$

where f' denotes the derivative of f, is a derivation. We shall prove in this section that each derivation on a semisimple F-algebra is automatically continuous. The proof of this result runs along the lines of the proof of (8.1.5), and is due to Carpenter [CAR 5] and Johnson [JOH].

(8.2.2) Proposition [JOH]. Let A be an F-algebra, and let $D: A \longrightarrow A$ be a derivation. Let $(\varphi_i)_i$ be a sequence of distinct points in M(A) so that the functionals $\varphi_i \circ D: A \longrightarrow \mathbb{C}$ are discontinuous for all $i \in \mathbb{N}$. Then there exists $c \in A$ such that $\widehat{D}(c)$, the Gelfand transform of D(c), is unbounded on $(\varphi_i)_i$.

Proof. First note that $D(c\cdot 1) = 0$ for all $c \in \mathbb{C}$.
Let $(p_i)_i$ be a defining sequence of seminorms for A (cf. (3.1.7) and recall in particular that $p_n \le p_{n+1}$ and $p_n(1)=1$ for all $n \in \mathbb{N}$). W.l.o.g. we can assume that each φ_i is continuous with respect to p_i, i.e.

(*) $\qquad |\varphi_i(f)| \le p_i(f)$ for all $f \in A$, $i \in \mathbb{N}$.

According to (8.1.4) we can choose a sequence $(f_k)_k$ in A so that

$$\varphi_j(f_k) = 0 \text{ for } j<k,$$

and

$$\varphi_j(f_k) \neq 0 \text{ for } j\ge k.$$

Set

$$a_k = \varphi_k(f_1^2\cdot\ldots\cdot f_k^2),$$

then $a_k \neq 0$ for all $k \in \mathbb{N}$.
Next we construct by induction a sequence $(g_k)_k$ in A so that

i) $\max\{p_k(f_j^2\cdot\ldots\cdot f_k^2\cdot g_k),\ 1\le j\le k\} < 2^{-k}$,

ii) $\varphi_k(g_k) = 0$,

iii) $|a_k\varphi_k\circ D(g_k)| > k + |\varphi_k\circ D(\sum_{j=1}^{k-1} f_1^2\cdot\ldots\cdot f_j^2\cdot g_j)|$.

Since in the first step we use the same arguments as in the induction step, we suppose that $g_1,\dots,g_{k-1}$ have already been chosen. Since $\varphi_k\circ D$ is discontinuous there exists $h_k \in A$ so that

$$\text{i)'}\ \max\{p_k(f_j^2\cdot\ldots\cdot f_k^2)p_k(h_k),\ 1\le j\le k\} < 2^{-(k+1)}$$

$$\text{ii)'}\ |a_k\varphi_k\circ D(h_k)| > k + |\varphi_k\circ D(\sum_{j=1}^{k-1} f_1^2\cdot\ldots\cdot f_j^2\cdot g_j)|.$$

Set $g_k = h_k-\varphi_k(h_k)\cdot 1$. Then g_k satisfies ii) and iii), since $D(g_k) = D(h_k)$. Moreover

$$p_k(f_j^2\cdot\ldots\cdot f_k^2(h_k-\varphi_k(h_k)\cdot 1))$$

$$\le p_k(f_j^2\cdot\ldots\cdot f_k^2)p_k(h_k-\varphi_k(h_k)\cdot 1)$$

$$\le p_k(f_j^2\cdot\ldots\cdot f_k^2)\cdot 2p_k(h_k) < 2^{-k}.$$

The next to last inequality follows from the triangle inequality and from (*).

Set

$$c_j = \sum_{i=j}^{\infty} f_j^2\cdot\ldots\cdot f_i^2\cdot g_i\ ,$$

then $c_j \in A$ for all $j \in \mathbb{N}$ by i) and by the assumption "$p_n\le p_{n+1}$". We have

$$c_1 = \sum_{j=1}^{k} f_1^2\cdot\ldots\cdot f_j^2\cdot g_j + f_1^2\cdot\ldots\cdot f_{k+1}^2(g_{k+1}+c_{k+2}) \quad \text{for all } k \in \mathbb{N},$$

and

$$\varphi_k\circ D(c_1) = \varphi_k\circ D(\sum_{j=1}^{k} f_1^2\cdot\ldots\cdot f_j^2\cdot g_j) +$$

$$+ \varphi_k(f_{k+1})\varphi_k\circ D(f_1^2\cdot\ldots\cdot f_k^2\cdot f_{k+1}(g_{k+1}+c_{k+2})) +$$

$$+ \varphi_k\circ D(f_{k+1})\varphi_k(f_1^2\cdot\ldots\cdot f_k^2\cdot f_{k+1}(g_{k+1}+c_{k+2}))$$

$$= \varphi_k \circ D\left(\sum_{j=1}^{k} f_1^2 \cdot \ldots \cdot f_j^2 \cdot g_j \right),$$

since $\varphi_k(f_{k+1}) = 0$. So

$$\varphi_k \circ D(c_1) = \varphi_k \circ D\left(\sum_{j=1}^{k-1} f_1^2 \cdot \ldots \cdot f_j^2 \cdot g_j \right) + \varphi_k \circ D(f_1^2 \cdot \ldots \cdot f_k^2 \cdot g_k)$$

$$= \varphi_k \circ D\left(\sum_{j=1}^{k-1} f_1^2 \cdot \ldots \cdot f_j^2 \cdot g_j \right) + a_k \varphi_k \circ D(g_k)$$

by ii). It follows by iii) that

$$|\varphi_k \circ D(c_1)| > k \text{ for all } k \in \mathbb{N}.$$

□

(8.2.3) Theorem [CAR 5]. Let A be a semisimple F-algebra, and let $D: A \longrightarrow A$ be a derivation. Then D is continuous.

Proof. By the closed graph theorem for Fréchet spaces it suffices to show that the graph of D is closed.

Let $(a_n)_n$ be a sequence in A such that $a_n \longrightarrow a$ and $D(a_n) \longrightarrow b$ as $n \longrightarrow \infty$. Then we have to show that $D(a) = b$, or equivalently - since A is semisimple - that $\widehat{D}(a) = \hat{b}$, where $\widehat{D}(a)$ denotes the Gelfand transform of D(a). Set

$$S = \{\varphi \in M(A) : \varphi \circ D : A \longrightarrow \mathbb{C} \text{ is continuous}\}.$$

Then our assertion follows if S is a dense subset of M(A).

Let $\varphi \in M(A)$ be an arbitrary point.

Case 1: φ is an isolated point of M(A).

By Shilov's idempotent theorem (6.1.5) there is $e \in A$ so that

$$\hat{e}(\varphi) = 0 \text{ and } \hat{e}\,|_{M(A)\setminus\{\varphi\}} \equiv 1.$$

Let $a \in A$ be an arbitrary element. Then $a - \varphi(a) \cdot 1 \in \ker \varphi$, and

$$\hat{e}\,(a - \varphi(a) \cdot 1)^\wedge = (a - \varphi(a) \cdot 1)^\wedge,$$

where $(a-\varphi(a)\cdot 1)^\wedge$ denotes the Gelfand transform of $(a-\varphi(a)\cdot 1)$. So

$$e(a-\varphi(a)\cdot 1) = a-\varphi(a)\cdot 1,$$

since A is semisimple. Thus

$$\varphi\circ D(a) = \varphi\circ D(a-\varphi(a)\cdot 1) = \varphi\circ D((a-\varphi(a)\cdot 1)e) = 0,$$

since $a-\varphi(a)\cdot 1 \in \ker \varphi$ and $e \in \ker \varphi$. It follows that $\varphi\circ D(a) = 0$ for all $a \in A$, so $\varphi\circ D \in S$.

Case 2: φ is not an isolated point of M(A).

Recall the definition of the algebras A_n from (3.2.6). Suppose there exists an open neighbourhood U of φ so that $U \cap S = \emptyset$, i.e. S is not dense.

By (3.3.6) and (3.3.4) $M(A_n)$ is an admissible exhaustion of M(A), and by (6.1.6) there exists $k \in \mathbb{N}$ such that φ is not an isolated point of $M(A_k)$. Choose a sequence $(\varphi_i)_i$ of distinct points in $M(A_k) \cap U$. Then $\varphi_i\circ D$ is not continuous for all $i \in \mathbb{N}$. Thus we find $c \in A$ by (8.2.2) such that $\widehat{D}(c)$ is unbounded on $(\varphi_i)_i$. This is a contradiction, since $M(A_k)$ is compact. □

CHAPTER 9

SHILOV BOUNDARY AND PEAK POINTS FOR F-ALGEBRAS

In this chapter we define the Shilov boundary for F-algebras. There are significant differences to the results in the B-algebra case. For instance an element $\hat{f} \in \hat{A}$ which vanishes on the Shilov boundary does not vanish identically in general. In the second part we introduce the notion of peak points for F-algebras. We show that an analogue to Rossi's maximum principle does not hold for F-algebras.

(9.1) The Shilov boundary of an F-algebra

(9.1.1) Let A be an F-algebra with defining sequence of seminorms $(p_n)_n$. Denote by P'(A) the set of all multiplicative seminorms p on A such that p(1)=1, and set

$$P(A) = \{p \in P'(A): \text{There exists } k \in \mathbb{N} \text{ and } c_k>0 \text{ such that } p(f) \le c_k p_k(f) \text{ for all } f \in A\}.$$

Let $(q_n)_n$ be another defining sequence of seminorms. Then for each $k \in \mathbb{N}$ there exists $l(k) \in \mathbb{N}$ so that

$$\{f \in A: q_{l(k)}(f) < 1/l(k)\} \subset \{f \in A: p_k(f) < 1/k\}.$$

(cf. (3.1.6)). Let $g \in A$ be an element such that $q_{l(k)}(g) > 0$. Then

$$p_k(g/(l(k)q_{l(k)}(g))) < 1/k,$$

hence

$$p_k(g) \le (l(k)/k)q_{l(k)}(g).$$

This inequality remains valid if $q_{l(k)}(g) = 0$. It follows from this that the definition of $P(A)$ does not depend on the choice of the defining seminorms $(p_n)_n$.

Let $p \in P(A)$ be an arbitrary element. Set $I_p = \{f \in A: p(f) = 0\}$ and denote by A_p the completion of the algebra A/I_p with respect to the norm

$$p'(a+I_p) = p(a).$$

Then A_p is a B-algebra and - in the same way as in (3.2.6)(*) - we can identify $M(A_p)$ with the compact set

$$\{\varphi \in M(A): |\varphi(f)| \le p(f) \text{ for all } f \in A\}.$$

<u>(9.1.2)</u> Recall that γA_p denotes the Shilov boundary of A_p (cf. (1.3.5)). If $\varphi \in M(A)$ is an arbitrary point we set

$$P_\varphi = \{p \in P(A): |\varphi(f)| \le p(f) \text{ for all } f \in A\}$$

Note that $\varphi \in M(A_p)$ for all $p \in P_\varphi$.

<u>Definition.</u> Let A be an F-algebra. A point $\varphi \in M(A)$ is called independent , provided that

$$\varphi \in \bigcap_{p \in P_\varphi} \gamma A_p .$$

The closure of the set of all independent points will be denoted by ΓA and is called the Shilov boundary of A

<u>(9.1.3) Proposition.</u> Let A be an F-algebra, and let $(p_n)_n$ be a defining sequence of seminorms. Suppose that $M(A)$ is locally compact. Then

$$\Gamma A = \text{closure}\,\Big(\bigcup_{k=1}^{\infty} \bigcap_{n \ge k} \gamma A_n \Big).$$

Here, as usual, we set $A_n = A_{p_n}$.

<u>Proof.</u> i) Let $\varphi \in M(A)$ be an independent point. Since $(M(A_n))_n$ is an admissible exhaustion of $M(A)$ (cf. (3.3.4) and (3.3.6)) there exists

$k \in \mathbb{N}$ such that $\varphi \in M(A_n)$ for all $n \geq k$. Hence

$$\varphi \in \bigcap_{n \geq k} \gamma A_n$$

by the definition of an independent point and by (3.2.6)(*).

ii) Let $k \in \mathbb{N}$, and let $\varphi \in \bigcap_{n \geq k} \gamma A_n$ be an arbitrary element. Our assertion is proved if we can show that $\varphi \in \Gamma A$.

Since $(M(A_n))_n$ is an admissible exhaustion of $M(A)$, we can assume that $M(A_k)$ is a compact neighbourhood of φ. Let $W \subset M(A)$ be an open neighbourhood of φ such that $W \subset M(A_k)$. Denote by $\overline{A}_k$ the closure of the algebra $\hat{A}_k$ in $C(M(A_k))$ (with respect to the sup norm $\|\cdot\|_{M(A_k)}$). Then $\overline{A}_k$ is a uB-algebra. The map $\Gamma: A_k \longrightarrow \overline{A}_k$, $f \longrightarrow \hat{f}$, defines a continuous algebra homomorphism by (1.2.3) which has dense range. Hence the adjoint spectral map

$$\Gamma^*: M(\overline{A}_k) \longrightarrow M(A_k),\ \varphi \longrightarrow \varphi \circ \Gamma,$$

is injective and continuous by (3.2.5). The same is true for the structure map

$$j: M(A_k) \longrightarrow M(\overline{A}_k),\ x \longrightarrow \varphi_x,$$

by (1.2.4). Since $j \circ \Gamma^* = id_{M(\overline{A}_k)}$, and $\Gamma^* \circ j = id_{M(A_k)}$ we can identify $M(A_k)$ and $M(\overline{A}_k)$ as topological spaces. Using (1.3.6), we easily show that $\gamma A_k = \gamma \overline{A}_k$.

By (1.3.12) W contains a strong boundary point ψ of $\overline{A}_k$, so for each neighbourhood U of ψ in $M(A_k)$ there exists $g_U \in \overline{A}_k$ such that

$$(*) \qquad |\hat{g}_U(\psi)| > \|\hat{g}_U\|_{M(A_k)\setminus U}.$$

Let $p \in P_\psi$ be an arbitrary element, and let V be an arbitrary open neighbourhood of ψ in $M(A_p)$. Choose an open neighbourhood U of ψ in $M(A)$, $U \subset M(A_k)$, such that $U \cap M(A_p) \subset V$, and let g_U be as in (*), w.l.o.g. $g_U \in A_k$. The natural map

$$A \longrightarrow A_k,\ h \longrightarrow h + \{p_k(h) = 0\},$$

has dense range and the map

$$A_k \longrightarrow \widehat{A}_k, \; h \longrightarrow \hat{h},$$

is continuous by (1.2.4). Hence we find $g \in A$ such that

$$|\hat{g}(\psi)| > \|\hat{g}\|_{M(A_k)\setminus U}.$$

It follows that $U \cap M(A_p)$ contains a local peak set for A_p. By Rossi's maximum principle (1.4.9) V contains a peak set for A_p, so $V \cap \gamma A_p \neq \emptyset$. Since V was taken to be arbitrary and γA_p is closed, we see that $\psi \in \gamma A_p$, hence

$$\psi \in \bigcap_{p \in P_\psi} \gamma A_p \subset \Gamma A.$$

Repeating the last argument for W and ΓA we conclude that $\varphi \in \Gamma A$ as desired.

□

<u>(9.1.4) Remark and notation.</u> Let A be a B-algebra. It follows from (9.1.3) that $\Gamma A = \gamma A$ in this case. So we shall denote henceforth the Shilov boundary of an F-algebra A by γA, too.

<u>(9.1.5) Example.</u> Proposition (9.1.3) fails to be true if $M(A)$ is not locally compact.

For $n \in \mathbb{N}$ define the following sets in $\mathbb{C}^2$:

$$I_n = \{(t,t/n) : t \in [0,1]\},$$

$$K_n = \{(z,w) : \prod_{i=1}^{n} (w-(z/i)) = 0, \; |z| \le 1, \; |w| \le 1\}$$

and

$$K_n' = K_n \cup I_{n+1}.$$

i) We claim that K_n and K_n' are polynomially convex compact sets.

This is clearly true for K_n. Let $(x,y) \in \mathbb{C}^2 \setminus K_n'$.

Case 1: $(x,y) \in \mathbb{C}^2 \setminus K_{n+1}$.

Then, since K_{n+1} is polynomially convex, we find a polynomial q

such that $|q(x,y)| > \|q\|_{K'_n}$.

Case 2: $(x,y) \in K_{n+1}$.

Then $x \in \mathbb{C}\setminus[0,1]$ and we find a polynomial q in one variable so that $|q(x)| = 1 > \|q\|_{[0,1]}$, since [0,1] is polynomially convex, cf. (1.2.4).

Set $r(z,w) = q(z)$, then

$$|r(x,y)| > \|r\|_{I_{n+1}}.$$

Furthermore consider

$$s(z,w) = \prod_{i=1}^{n} (w-(z/i)).$$

Note that $r(x,y) \neq 0$. Thus $|p(x,y)| > \|p\|_{K'_n}$ for

$$p(z,w) = s(z,w)(r(z,w))^m,$$

m sufficiently large.

ii) We note that

$$(*) \qquad \ldots \subset K_n \subset K'_n \subset K_{n+1} \subset K'_{n+1} \subset \ldots \quad .$$

Set $A = \varprojlim P(K'_n)$. Then A is a doubly generated uF-algebra. Note that a defining sequence of seminorms is given by $(\|\cdot\|_{K'_n})_n$ and that

$$A_{\|\cdot\|_{K'_n}} = P(K'_n).$$

We have $M(A) = \bigcup K'_n$ (as sets) and $(K'_n)_n$ is an admissible exhaustion of M(A) by (5.1.6).

iii) Claim: $(0,0) \in \gamma P(K'_n)$ for all $n \in \mathbb{N}$.

Let $(x,y) \in I_{n+1}\setminus\{(0,0)\}$ be an arbitrary point, and let U be an open neighbourhood of (x,y) in K'_n. Set

$$p(z,w) = \prod_{i=1}^{n} (w-(z/i)),$$

and

$$q(z,w) = 1-(z-x)^2.$$

Note that $p(x,y) \neq 0$, and that

$$|q(x,y)| = 1 > |q(x',y')| \quad \text{for all } (x',y') \in I_{n+1}\setminus\{(x,y)\}.$$

Hence - for sufficiently large m -

$$|(q(x,y)^m \cdot p(x,y)| > \|q^m \cdot p\|_{K'_n \setminus U}.$$

It follows that $\gamma P(K'_n) \cap U \neq \emptyset$. Since U was chosen to be arbitrary, and since $\gamma P(K'_n)$ is closed, we see that $(x,y) \in \gamma P(K'_n)$, hence $I_{n+1} \setminus \{(0,0)\} \subset \gamma P(K'_n)$ and finally $(0,0) \in \gamma P(K'_n)$.

iv) Define

$$U = \{(z,w) \in M(A) = \cup K'_n : |z|<1/2,\ |w|<1/2\}.$$

Then U is a Gelfand open set. We claim that $U \cap \gamma A = \emptyset$. Suppose the contrary. Then there exists $(x,y) \in U$ such that

$$(x,y) \in \bigcap_{p \in P_{(x,y)}} \gamma A_p.$$

By (*) there is $n \in \mathbb{N}$, so that $(x,y) \in K_n$. Clearly $\|\cdot\|_{K_n} \in P_{(x,y)}$ and $A_{\|\cdot\|_{K_n}} = P(K_n)$. Let $f \in P(K_n)$ be an arbitrary element. Note that

$$X_n = \{(z,w) : \prod_{i=1}^{n} (w-(z/i)) = 0,\ |z|<1,\ |w|<1\}$$

is an analytic subset of $\{|z|<1, |w|<1\}$, and that $f|_{X_n} \in Hol(X_n)$ by (2.1.7),(2.3.1). Hence - by the maximum principle for holomorphic functions and by continuity of f -

$$|f(x,y)| \leq \|f\|_{K_n \cap \{|z|=1 \text{ or } |w|=1\}}.$$

So (x,y) is not contained in the Shilov boundary for $P(K_n)$, and we get a contradiction. □

<u>(9.1.6) Proposition.</u> Let A be an F-algebra, and let $K \subset M(A)$ be an open and compact subset. Then $\gamma A \neq \emptyset$.

<u>Proof.</u> By (6.1.5) there exists $e \in A$ such that $\hat{e}|_K \equiv 1$ and $\hat{e}|_{M(A)\setminus K} \equiv 0$.

Denote by B the completion of the algebra $\hat{A}|_K$ with respect to the sup norm $\|\cdot\|_K$. Then B is a uB-algebra and K contains a strong boundary point x for B by (1.3.2)ii) and (1.3.12). Let U be an arbitrary open neighbourhood of x in K. Since $\hat{A}|_K$ is a dense subalgebra of B we find $f_U \in A$ such that

$$|\hat{f}_U(x)| > |\hat{f}_U(y)| \text{ for all } y \in K\backslash U.$$

Thus

$$|(ef_U)^\wedge(x)| > |(ef_U)^\wedge(y)| \text{ for all } y \in M(A)\backslash U,$$

where $(ef_U)^\wedge$ denotes the Gelfand transform of ef_U. It follows from this observation that $x \in \gamma A$. □

<u>(9.1.7) Proposition.</u> Let A be an F-algebra such that $\gamma A = \emptyset$. Then each element of $\hat{A}$ satisfies the maximum principle on M(A), i.e.

$$(*) \qquad \|\hat{f}\|_K = \|\hat{f}\|_{\partial K}$$

for all $\hat{f} \in \hat{A}$ and all compact subsets $K \subset M(A)$. (Here ∂K denotes the boundary of K.)
Vice versa, if M(A) is locally compact and (*) is satisfied, then $\gamma A = \emptyset$.

<u>Proof.</u> i) Let $\gamma A = \emptyset$. By (9.1.6) $\partial K \neq \emptyset$ for all compact subsets $K \subset M(A)$.
Suppose there exists a compact set $K \subset M(A)$ and $\hat{f} \in \hat{A}$ so that

$$\|\hat{f}\|_K > \|\hat{f}\|_{\partial K}.$$

Choose a seminorm $p \in P(A)$ such that $K \subset M(A_p)$. Then by Rossi's maximum principle (1.4.9) - there exists $\varphi \in \text{int } M(A_p) \cap \gamma A_p$, where int $M(A_p)$ denotes the interior of $M(A_p)$ with respect to the Gelfand topology on M(A). Repeating the proof of proposition (9.1.3)ii) (replace A_k by A_p) one sees that $\varphi \in \gamma A$. This is impossible.
ii) Let $(p_n)_n$ be a defining sequence of seminorms. Assume (*), then

$$\gamma A_{p_n} \subset \partial M(A_{p_n}) \text{ for all } n \in \mathbb{N}.$$

Since M(A) is locally compact we have for each $k \in \mathbb{N}$

$$\bigcap_{n\geq k} \partial M(A_{p_n}) = \emptyset,$$

so

$$\bigcap_{n\geq k} \gamma A_{p_n} = \emptyset,$$

hence $\gamma A = \emptyset$ by (9.1.3). □

<u>(9.1.8) Example.</u> We construct a uF-algebra A such that (*) of (9.1.7) is satisfied but $\gamma A \neq \emptyset$ (hence M(A) is not locally compact).

For $n \in \mathbb{N}$ set

$$K_n = \{z \in \mathbb{C}: |\mathrm{Im}(z)| \leq 1-(1/n),\ |\mathrm{Re}(z)| \leq 1-(1/n)\} \cup [0,1].$$

Every K_n is polynomially convex (cf. (1.2.4)ii)). Set $A = \varprojlim P(K_n)$. A is a singly generated uF-algebra,

$$M(A) = \cup K_n = D \cup \{1\} \quad \text{as sets},$$

where D denotes an open square, and $(K_n)_n$ is an admissible exhaustion of M(A) with respect to the Gelfand topology by (5.1.6).
<u>Claim</u>. The point $1 \in M(A)$ is a boundary point of each K_n with respect to the Gelfand topology, hence 1 is a boundary point of each compact subset which contains it.
Suppose the contrary. Then by (3.2.2)i) there are $f_1,\ldots,f_r \in A$ and $k \in \mathbb{N}$ such that

$$U = \{y \in D \cup \{1\}: |\hat{f}_i(1)-\hat{f}_i(y)| < 1,\ i=1,\ldots,r\} \subset K_k.$$

Since for each $m \in \mathbb{N}$ $\hat{f}_i|_{K_m}$ is continuous with respect to the euclidean topology (5.1.7)ii), we find a point $y_1 \in K_k\setminus\{1\}$, $\mathrm{Re}(y_1) > 1-(1/k)$, so that

$$|\hat{f}_i(y_1)-\hat{f}_i(1)| < 1/4 \ (i=1,\ldots r).$$

Choose $l \in \mathbb{N}$ such that $\mathrm{Re}(y_1)<1-(1/l)$, i.e. $y_1 \in \mathrm{int}\, K_l$ (euclidean topology). Hence - by continuity of $\hat{f}_i|_{K_l}$ - there exists $y_2 \in K_l \backslash K_k$ such that

$$|\hat{f}_i(y_2)-\hat{f}_i(y_1)| < 1/4 \ (i=1,\ldots,r).$$

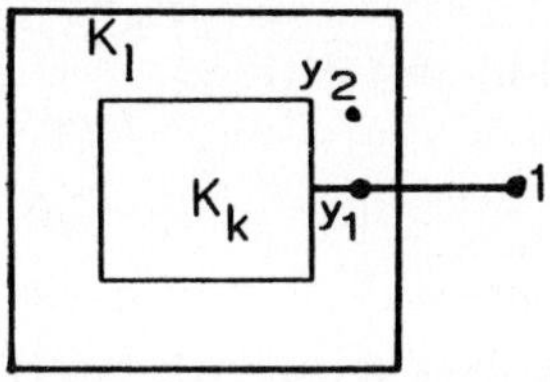

It follows that $y_2 \in U$, but $y_2 \in M(A)\backslash K_k$. This contradicts our assumption.

We note that on D euclidean topology and Gelfand topology coincide by (5.1.7)ii). Every $\hat{f} \in \hat{A}$ is a complex-valued function on $D \cup \{1\}$ which can be approximated uniformly on each K_n by polynomials (cf. (5.1.7)i)). So $\hat{f}|_D \in \mathrm{Hol}(D)$, for every $\hat{f} \in \hat{A}$. It follows from this and from the considerations above that

$$\|\hat{f}\|_K = \|\hat{f}\|_{\partial K}$$

for each $\hat{f} \in \hat{A}$ and each compact subset $K \subset M(A)$. The point $1 \in M(A)$ is a peak point for A (cf. (9.2)) with peaking function

$$\hat{h}: M(A) \longrightarrow \mathbb{C},\ x \longrightarrow \exp(x),$$

note that $h \in A$ by (1.4.5), and we conclude that $\gamma A = \{1\}$.

<u>(9.1.9) Remarks.</u> i) In the case of B-algebras the Shilov boundary is a determining set in the following sense: If $\hat{f} \in \hat{A}$ vanishes on γA then $\hat{f} \equiv 0$.

This turns out to be false for F-algebras. Let A and $\hat{h}$ be as in (9.1.8). Then $\hat{f} = \hat{h}-\exp(1) \in \hat{A}$ and $\hat{f}|_{\gamma A} = 0$, while $\hat{f}(y) \neq 0$ for every $y \in M(A)\backslash\gamma A$.

ii) The notion of independent points was introduced by Rickart [RIC 1]. He considered more general algebras ("natural systems"). Our definition (in the case of F-algebras) is equivalent to his definition. This was proved by Brooks [BRO 4].
Various other authors have given generalizations of the Shilov boundary. We mention Brooks [BRO 4], [BRO 8], [BRO 9], who also proved relations between the different definitions.

(9.2) Peak points for F-algebras

(9.2.1) Definition. Let A be an F-algebra. A point $x \in M(A)$ is said to be a local peak point for A if there exists a neighbourhood U of x in M(A) and an element $f \in A$ such that

$$\hat{f}(x) = 1 \text{ and } |\hat{f}(y)|<1 \text{ if } y \in U\setminus\{x\}.$$

We call x a (global) peak point for A if U can be taken to be M(A).

(9.2.2) For Banach algebras Rossi's maximum principle (1.4.9) shows that every local peak point is a global peak point. This result is not valid for F-algebras. We construct a doubly generated uF-algebra with locally compact spectrum which has local but no global peak points.

Example. We define the following sets in $\mathbb{C}^2$.

$$X_o = \{(x,y): |x|\leq 1, y=1\},$$
$$X_1 = \mathbb{C} \times \{0\},$$
$$X_2 = \{0\} \times \mathbb{C},$$
$$X_n = \{1/n\} \times \mathbb{C} \quad \text{for } n>2,$$
$$X = \cup X_n, \text{ and}$$
$$K_n = X \cap \{(x,y): |x|\leq n, |y|\leq n\}.$$

Then X is locally compact and $\ldots \subset K_n \subset K_{n+1} \subset \ldots$ is an admissible exhaustion of X with respect to the euclidean topology. We may illustrate X in the following way.

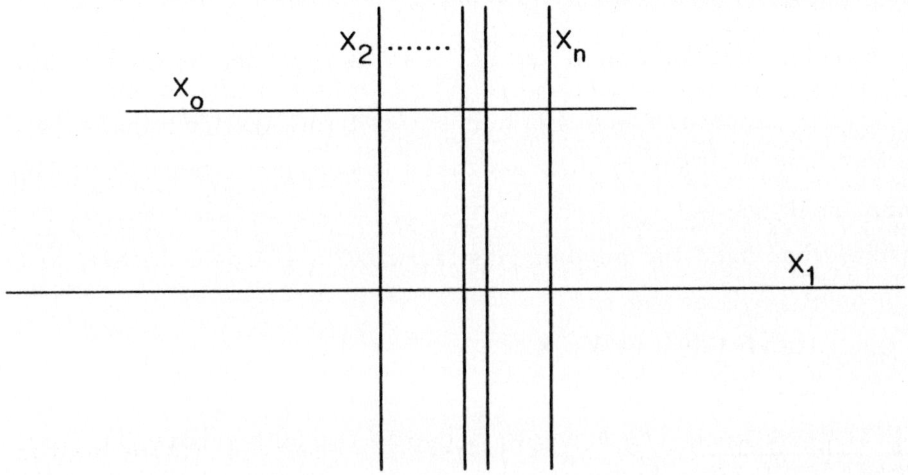

<u>Claim</u>. Each K_n is polynomially convex.

To prove this let $x=(x_1,x_2) \in \mathbb{C}^2\setminus K_n$.

Case 1: $|x_1|>n$ or $|x_2|>n$. Then

$$|z_1(x)|>\|z_1\|_{K_n} \text{ or } |z_2(x)|>\|z_2\|_{K_n},$$

where z_1, z_2 denote the coordinate functions.

Case 2: $|x_1|\le n$ and $|x_2|\le n$, $x_2\ne 1$. Since $x \in \mathbb{C}^2\setminus K_n$, there is $m \in \mathbb{N}$ such that $|x_1|>1/m$. For sufficiently large $l \in \mathbb{N}$ we have $|p(x)|>\|p\|_{K_n}$ for the polynomial

$$p(z_1,z_2) = \prod_{i=3}^{m} (z_1-(1/i))z_2(z_2-1)(z_1/|x_1|)^l.$$

The case $|x_1|\le n$, $|x_2|\le n$ and $x_2=1$ can be treated in a similar way.

Since K_n is polynomially convex we can identify K_n with the spectrum of $P(K_n)$ by (1.2.4)ii). We consider the algebra $\varprojlim P(K_n) = A$. The spectrum of A can be identified as a set with $\cup M(P(K_n)) = \cup K_n = X$ by (5.1.6). Since K_n is an admissible exhaustion of X with respect to the euclidean topology, the Gelfand topology and the induced euclidean topology are equivalent on X by (5.1.7)ii).

Now let $\hat{f} \in \hat{A}$ be a function such that $|\hat{f}(x)| \le 1$ for all $x \in X$. Since $\hat{f}$

is the uniform limit of polynomials, $\hat{f}|_{X_i}$ (i≥1) can be regarded as holomorphic functions on $\mathbb{C}$ and $\hat{f}|_{X_o}$ can be regarded as an element of the disc algebra. Since X is connected it follows from Liouville's theorem and the identity theorem that $\hat{f}$ is constant. Hence A has no global peak points.

On the other hand the polynomials $(z_1+x_1)/2x_1 \in \hat{A}$ peak locally at each point $(x_1,1)$ of the set

$$\{(u,v): |u|=1,\ v=1\} \subset X.$$

(9.2.3) Remarks. i) The example is due to the author [GOL 1]. For another example cf. [MEY 1].

ii) In [CAR 4] Carpenter introduced the notion of accessible points in the spectrum of a uF-algebra A. This notion enables us to characterize the peak points of A.

For $0<\vartheta<\pi$ and $r>0$ set

$$S(\vartheta,r) = \{z \in \mathbb{C}: |z|<r,\ -\vartheta/2 \le \arg z \le \vartheta/2\}$$

and

$$S_o(\vartheta,r) = S(\vartheta,r)\setminus\{0\}.$$

Let $A = \varprojlim A_n$ be a projective limit representation of A. Then $x_o \in M(A)$ is called

α) locally accessible if there are a neighbourhood U of x_o, $f \in A$, ϑ and $r>0$ such that

$$\hat{f}^{-1}(0) \cap U = \{x_o\} \text{ and } \hat{f}(U) \cap S_o(\vartheta,r) = \emptyset;$$

β) accessible if there are $f \in A$ and positive constants $\vartheta, r_1, r_2, \ldots$ such that

$$\hat{f}^{-1}(0) = \{x_o\} \text{ and } \hat{f}(M(A_i)) \cap S_o(\vartheta,r_i) = \emptyset \text{ for } i=1,2,\ldots ;$$

γ) completely accessible if there are $f \in A$ and positive constants ϑ, r such that

$$\hat{f}^{-1}(0) = \{x_o\} \text{ and } \hat{f}(M(A)) \cap S_o(\vartheta,r) = \emptyset.$$

We see at once that completely accessible points are accessible, and that accessible points are locally accessible if M(A) is locally compact.

Carpenter proved that x is a peak point for A iff it is completely accessible and that locally accessible points are accessible. Our example above shows that (locally) accessible points in general are not completely accessible.

iii) Let B be a uB-algebra with metrizable spectrum M(B). It follows from (1.3.12) that in this case γB is the closure of the set of all peak points.

In general an analogous statement for uF-algebras is false. Consider for instance the example above. It was shown that the algebra A has no global peak points. The spectrum M(A) = X is metrizable, since Gelfand and euclidean topology coincide on X.

Let x be an arbitrary point of the set $\{(u,v): |u|=1,\ v=1\} \subset X$, and let $K \subset M(A)$ be an arbitrary compact set which contains x. Then $\hat{A}_K$ is topologically and algebraically isomorphic to $P(\hat{K})$, where $\hat{K}$ denotes the polynomially convex hull of K. Recall that x is a local peak point for A, hence x is a local peak point for $P(\hat{K})$, so x is a global peak point for $P(\hat{K})$ by Rossi's maximum principle (1.4.9), i.e. $x \in \gamma P(\hat{K})$. Hence

$$x \in \bigcap_{\substack{x \in K,\\ K \subset M(A) \text{ compact}}} \gamma \hat{A}_K ,$$

and it follows that $x \in \gamma A$. It is easy to show that

$$\gamma A = \{(u,v): |u| = 1,\ v = 1\}.$$

CHAPTER 10

MICHAEL'S PROBLEM

Let A be an F-algebra. Recall from (3.2.1) that we denote by S(A) the set of all nonzero algebra homomorphisms $\varphi: A \longrightarrow \mathbb{C}$, continuous or not. We endow S(A) with the Gelfand topology, that is the coarsest topology such that all Gelfand transforms $\hat{f} \in \hat{A}$ are continuous functions on S(A). A basis for the neighbourhood system of a point $\varphi \in S(A)$ is given by all sets of the form

$$\{\psi \in S(A): |\psi(f_i)-\varphi(f_i)| < 1,\ i=1,\dots,r,\ f_1,\dots,f_r \in A\},$$

cf. (3.2.2)i).

We know by (1.2.1) that S(A) = M(A) for B-algebras. Michael [MIC] posed the question as whether the same is true for F-algebras. This question is still unanswered. Arens [ARE 1] proved that S(A) = M(A) if A is a rationally finitely generated F-algebra.

Besides some other results we sketch in this chapter an approach to this problem due to Dixon and Esterle [D/E], which leads to an interesting problem in the theory of several complex variables.

(10.1) Results on the automatic continuity of characters

(10.1.1)Lemma. M(A) is a dense subset of S(A) for each F-algebra A.

Proof. Suppose the lemma is false. Then there are $\varphi \in S(A)$, $f_1,\dots,f_n \in A$ such that

$$\{\psi \in S(A): |\psi(f_i)-\varphi(f_i)| < 1,\ i=1,\dots,n\} \cap M(A) = \emptyset .$$

Set $\lambda_i = \varphi(f_i)$ and $g_i = f_i - \lambda_i$, $i=1,\dots,n$. Then $\hat{g}_1,\dots,\hat{g}_n$ have no common zero on M(A), hence by (6.1.10) there are $h_1,\dots,h_n \in A$ such that

$$\sum_{i=1}^{n} h_i g_i = 1.$$

But this leads to a contradiction, since $g_1,\dots,g_n \in \ker \varphi$, which is a proper ideal in A.

□

<u>(10.1.2) Remark.</u> One easily sees that if X is a topological space and Y is a k-space, then every continuous map $f: X \longrightarrow Y$ which is proper is closed.

Now consider the map

$$M(A) \longrightarrow S(A),\ \varphi \longrightarrow \varphi.$$

It is continuous and it is also proper. (Otherwise there would be a compact set $K \subset S(A)$ such that $K \cap M(A)$ is not compact. Copying the proof of (4.3.6) - using (3.2.6) - we find $f \in A$ so that $\hat{f}$ is unbounded on $K \cap M(A)$, a contradiction to the compactness of K.) Hence, if S(A) is a k-space, then the map is closed and we have $M(A) = S(A)$ by (10.1.1).

<u>(10.1.3)</u> Let $(f_1,\dots,f_n) \in A^n$ Then set

$$\hat{F}: S(A) \longrightarrow \mathbb{C}^n\ ,\ \varphi \longrightarrow (\hat{f}_1(\varphi),\dots,\hat{f}_n(\varphi)).$$

<u>Proposition.</u> Let A be an F-algebra, and let $(f_1,\dots,f_n) \in A^n$. Then

$$\hat{F}(M(A)) = \hat{F}(S(A)).$$

<u>Proof.</u> Suppose there exists $\lambda \in \hat{F}(S(A)) \setminus \hat{F}(M(A))$. Set $g_i = f_i - \lambda_i \cdot 1$ $(i=1,\dots,n)$ and continue as in the proof of (10.1.1).

□

<u>(10.1.4) Theorem.</u> Let A be an F-algebra. Suppose there are $(f_1,\dots,f_n) \in A^n$ such that for each $\varphi \in M(A)$ the set

$$\hat{F}^{-1}(\hat{F}(\varphi)) \cap M(A)$$

is at most countable, then $S(A) = M(A)$.

Proof. Let $\psi \in S(A)$ be an arbitrary element. Choose $\varphi_o \in M(A)$ by (10.1.3) such that $\hat{F}(\varphi_o) = \hat{F}(\psi)$. The set

$$S = \hat{F}^{-1}(\hat{F}(\varphi_o)) \cap M(A)$$

is at most countable by our hypothesis, hence - by (8.1.4) - we find $f_{n+1} \in A$ so that $\hat{f}_{n+1}$ is injective on S. (If f is finite use (3.2.3).) So, using again (10.1.3), there is exactly one element $\varphi_1 \in M(A)$ such that

(*) $\qquad \varphi_1(f_i) = \psi(f_i), \quad i=1,\dots,n+1.$

Let $g \in A$ be an arbitrary element. Then $\varphi_1(g) = \psi(g)$ by (10.1.3) and (*). But this implies $\varphi_1 = \psi$. □

Remark. The theorem is due to Zelazko [ZEL 2], the proof is as in [GOL 3].

(10.1.5) Theorem [MIC],[ARE 1]. Let A be an F-algebra. Suppose there are $(f_1,\dots,f_k) \in A^k$ such that for each $z \in \mathbb{C}^k$ the set

$$\hat{F}^{-1}(z) \cap M(A)$$

is compact. Then $M(A) = S(A)$.

Proof. Let $\psi \in S(A)$ then, by the hypothesis,

$$K = \{\varphi \in M(A): \hat{F}(\psi) = \hat{F}(\varphi)\}$$

is compact. By (10.1.3) K is not empty.

For each $b \in A$ the set

$$U_b = \{\varphi \in M(A): \varphi(b) \neq \psi(b)\}$$

is open. Suppose that $\bigcup_{b\in A} U_b$ is an open covering of K, then there are $b_1,\dots,b_r \in A$ such that

(*) $\qquad K \subset U_{b_1} \cup \dots \cup U_{b_r}.$

By (10.1.3) there exists $\varphi_o \in M(A)$ so that

i) $\varphi_o(b_i) = \psi(b_i)$ $(i=1,\dots,r)$, and

ii) $\varphi_o(f_i) = \psi(f_i)$ $(i=1,\dots,k)$.

So $\varphi_o \in K$ by ii), which is a contradiction to i) and (*).
Hence there exists

$$\varphi \in K \setminus \bigcup_{b \in A} U_b,$$

i.e. $\varphi = \psi$.

□

(10.1.6) Corollary [ARE 1]. If A is a rationally finitely generated F-algebra then $M(A) = S(A)$.

Proof. Let $f_1,\dots,f_n$ be rational generators for A, then for each $z \in \mathbb{C}^n$ the set $M(A) \cap \hat{F}^{-1}(z)$ is either empty or consists of a single point by (5.2.3). So we get the desired result from the theorem above. □

(10.1.7) Examples. i) Let $A = C^{\infty}([0,1])$ (cf. (3.3.5)). Then $M(A) = [0,1]$ by (3.3.5). Hence $M(A) = S(A)$ by (10.1.4), since

$$\hat{z} : [0,1] \longrightarrow \mathbb{C},\ t \longrightarrow t,$$

defines an element of $\hat{A}$.

ii) Let X be a hemicompact k-space. Then C(X) is a uF-algebra by (4.1.4)ii). We claim that

$$X = M(C(X)) = S(C(X)).$$

The first equality was already established in (4.1.7)i). Let $(K_n)_n$ be an admissible exhaustion of X and suppose that there is

$$\psi \in S(C(X))\setminus X.$$

S(C(X)) is a completely regular space, since it carries the initial topology generated by all Gelfand transforms $\hat{f}$, $f \in C(X)$ (cf. [BNS] (0.2.2),(0.2.3)). Hence we find for each $n \in \mathbb{N}$ an element

$f_n \in C(S(C(X)))$ such that

i) $f_n(\psi) = 0$,

ii) $f_n(S(C(X))) = [0,2^{-n}]$,

iii) $f_n|_{K_n} = 2^{-n}$.

Note that $f_n|_X \in C(X)$. By iii)

$$f = \sum_{n\in\mathbb{N}} f_n|_X$$

defines an element of $C(X)$ and $f(\varphi) \neq 0$ for all $\varphi \in X$ by ii) and iii). Consider the Gelfand transform $\hat{f} \in C(S(C(X)))$. Suppose that $|\hat{f}(\psi)| > \varepsilon > 0$. Choose $n \in \mathbb{N}$ so that $\sum_{i>n} 2^{-i} < \varepsilon/4$. Let

$$U = \{\psi' \in S(C(X)): |\hat{f}(\psi')|>\varepsilon/2 \text{ and } |\sum_{i=1}^{n}\hat{f}_i(\psi')|<\varepsilon/4\}.$$

Then U is an open neighbourhood of ψ and we can choose a point $\varphi \in M(A) \cap U$ by (10.1.1). Hence

$$\hat{f}(\varphi) = f(\varphi) = \sum_{i=1}^{n} f_i(\varphi) + \sum_{i>n} f_i(\varphi) \le \varepsilon/2,$$

a contradiction to $\varphi \in U$. So $\hat{f}(\varphi) \neq 0$, for all $\varphi \in X$, and $\hat{f}(\psi) = 0$, and we get a contradiction to (10.1.3).

iii) A uF-algebra A is called Stein algebra if there exists a reduced Stein space X (cf. (2.3.2)) such that A is topologically and algebraically isomorphic to Hol(X), the algebra of all holomorphic functions on X endowed with the compact open topology.

If X is a connected finite dimensional reduced Stein space, then by Remmert's embedding theorem there exists an injective holomorphic mapping into some $\mathbb{C}^n$. Since X = M(Hol(X)) by (2.3.2) we have M(Hol(X)) = S(Hol(X)) by theorem (10.1.4) in this case. (We remark for the sake of completeness that this result remains true for not necessarily reduced Stein spaces $(X,\mathcal{O})$ and the algebra $H^0(X,\mathcal{O})$.

For a generalization of this result cf. Ephraim [EPH].)
We give a simple proof - not depending on the deep embedding theorem - for M(Hol(X)) = S(Hol(X)) if (X,p) is a holomorphically separable Riemann domain over $\mathbb{C}^n$. Note that in this case $p_i \in$ Hol(X) (i=1,...,n) if $p=(p_1,...,p_n)$, cf. (2.2.2). Recall from (2.2.5) that (M(Hol(X)),$\hat{p}$) is again a Riemann domain over $\mathbb{C}^n$, where $\hat{p}=(\hat{p}_1,...,\hat{p}_n)$. Since $\hat{p}$ is a local homeomorphism and M(Hol(X)) is a hemicompact space we conclude that for each $z \in \mathbb{C}^n$, $\hat{p}^{-1}(z)$ is either empty or at most countable.(Suppose that this is false. Then $\hat{p}^{-1}(z)$ is not countable for some $z \in \mathbb{C}^n$. Let $(K_n)_n$ be an admissible exhaustion of M(Hol(X)), then there is $k \in \mathbb{N}$ so that

$$K_k \cap \hat{p}^{-1}(z)$$

is not countable, i.e. $\hat{p}^{-1}(z)$ has an accumulation point in M(A), and we get a contradiction.) Now our assertion follows from (10.1.4).

(10.2) The approach of Dixon and Esterle

Next we want to sketch the approach of Dixon and Esterle to Michael's problem. Their work leads to an interesting problem in several complex variables.
Recall that we always consider commutative F-algebras with unit.

(10.2.1) Proposition. Let $f \in$ Hol($\mathbb{C}^n$), and let $a_1,...,a_n$ be elements of an F-algebra A. Let

$$f(z_1,...,z_n) = \sum_{i_1,...,i_n \geq 0} \lambda_{i_1,...,i_n} z_1^{i_1} \cdot ... \cdot z_n^{i_n}$$

be the power series expansion of f (cf. (2.1.5)), then

$$f(a_1,...,a_n) = \sum_{i_1,...,i_n \geq 0} \lambda_{i_1,...,i_n} a_1^{i_1} \cdot ... \cdot a_n^{i_n}$$

defines an element of A and the map

$$A^n \longrightarrow A, \ (a_1,\dots,a_n) \longrightarrow f(a_1,\dots,a_n)$$

is continuous.

Proof. That $f(a_1,\dots,a_n) \in A$ follows from (6.1.2) (use the same argument as in (1.4.4)i)).

Let $(p_l)_l$ be a sequence of defining seminorms for A (cf. (3.1.7)), and let $(a_1^{(k)},\dots,a_n^{(k)})_k$ be a sequence in A^n which converges to $(b_1,\dots,b_n) \in A^n$.

Fix arbitrary $l \in \mathbb{N}$ and $\varepsilon>0$. Choose $r>0$ so that

$$p_l(a_i^{(k)}) < r, \ i=1,\dots,n, \ k \in \mathbb{N}$$

and

$$p_l(b_i) < r, \ i=1,\dots,n.$$

Then

$$p_l(f(a_1^{(k)},\dots,a_n^{(k)})-f(b_1,\dots,b_n)) \le$$

$$\le \sum_{i_1+\dots+i_n>0} |\lambda_{i_1,\dots,i_n}| p_l((a_1^{(k)})^{i_1}\cdot\ldots\cdot(a_n^{(k)})^{i_n}-b_1^{i_1}\cdot\ldots\cdot b_n^{i_n})$$

$$\le \sum_{s\ge i_1+\dots+i_n>0} |\lambda_{i_1,\dots,i_n}| p_l((a_1^{(k)})^{i_1}\cdot\ldots\cdot(a_n^{(k)})^{i_n}-b_1^{i_1}\cdot\ldots\cdot b_n^{i_n}) +$$

$$+ 2\cdot \sum_{i_1+\dots+i_n>s} |\lambda_{i_1,\dots,i_n}| r^{i_1+\dots+i_n}.$$

For s large enough the value of the second sum is less than $\varepsilon/2$ and the first sum tends to zero for $k \longrightarrow \infty$.

So we have proved that $f(a_1^{(k)},\dots,a_n^{(k)})$ converges to $f(b_1,\dots,b_n)$. □

(10.2.2) Remark. If $F = (f_1,\dots,f_k) \in \mathrm{Hol}(\mathbb{C}^n,\mathbb{C}^k)$, i.e. $f_i \in \mathrm{Hol}(\mathbb{C}^n)$ for $i=1,\dots,k$, then the map

$$A^n \longrightarrow A^k, \ (a_1,\dots,a_n) \longrightarrow F(a_1,\dots,a_n) =$$

$$= (f_1(a_1,\dots,a_n),\dots,f_k(a_1,\dots,a_n)),$$

is continuous. This is an easy consequence of the proposition.

(10.2.3) Proposition. Let $F = (f_1,\dots,f_k) \in \mathrm{Hol}(\mathbb{C}^n,\mathbb{C}^k)$, and let A be an F-algebra. Then, for each $a \in A^n$,

$$F(a)\hat{} = F\circ\hat{a},$$

where $F(a)\hat{}$ denotes the Gelfand transform of F(a). More precisely: Let $\varphi \in S(A)$, then

$$\varphi(f_i(a_1,\dots,a_n)) = f_i(\varphi(a_1),\dots,\varphi(a_n)),\ i=1,\dots,k.$$

Proof. Let $\varphi \in S(A)$. Denote by B the closed subalgebra of A generated by $a_1,\dots,a_n$. Since φ restricted to B is continuous by (10.1.6) we get the desired result from the definition of $f_i(a_1,\dots,a_n)$. □

(10.2.4) Theorem. Let $(s_n)_n$ be a sequence of positive numbers. If there is a discontinuous character on an F-algebra A, then the projective limit of every projective system

$$\dots \longrightarrow \mathbb{C}^{s_{n+1}} \xrightarrow{F_n} \mathbb{C}^{s_n} \longrightarrow \dots$$

where $F_n \in \mathrm{Hol}(\mathbb{C}^{s_{n+1}},\mathbb{C}^{s_n})$ is not empty.

Proof. Let φ be a discontinuous character on A.

i) Since the closure of ker φ is an ideal and ker φ is a maximal ideal we see that ker φ is dense in A, cf. (3.2.11).

ii) Set

$$q_1 = 0,\ q_n = s_1+\dots+s_{n-1} \text{ for } n>1.$$

If we endow ker φ with the discrete topology then

$$E_n = A^{s_n} \times (\ker\varphi)^{q_n}$$

is a complete metrizable space for all $n \in \mathbb{N}$. We consider the map

$$T_n: E_{n+1} \longrightarrow E_n,\ (a_1,\dots,a_{s_{n+1}},x_1,\dots,x_{q_{n+1}}) \longrightarrow$$

$$(F_n(a_1,\ldots,a_{s_{n+1}})+(x_{q_n+1},\ldots,x_{q_{n+1}}),x_1,\ldots,x_{q_n}).$$

T_n is continuous for each $n \in \mathbb{N}$ by (10.2.2) and since $\ker \varphi$ is endowed with the discrete topology. Moreover $T_n(E_{n+1})$ is dense in E_n by i). Hence we can choose an element $B = (B_n)_n \in \varprojlim (E_n,T_n)$ by (3.3.2).

For each $n \in \mathbb{N}$ we can write

$$B_n = (b_n,x_n),\ b_n \in A^{s_n},\ x_n \in (\ker \varphi)^{q_n}.$$

Since

$$T_n(b_{n+1},x_{n+1}) = (b_n,x_n)$$

we have

$$F_n(b_{n+1})-b_n \in (\ker \varphi)^{s_n},\ \text{for all } n \in \mathbb{N},$$

hence by (10.2.3)

$$\varphi(b_n) = \varphi(F_n(b_{n+1})) = F_n(\varphi(b_{n+1})).$$

Set

$$z_n = \varphi(b_n),$$

then

$$(z_n)_n \in \varprojlim (\mathbb{C}^{s_n},F_n).$$

□

<u>(10.2.5) Corollary.</u> Let $p \in \mathbb{N}$. If there exists a sequence $(F_n)_n$ of holomorphic mappings from $\mathbb{C}^p$ into itself such that

$$\bigcap_{n\in\mathbb{N}} F_1\circ\ldots\circ F_n(\mathbb{C}^p) = \emptyset ,$$

then $S(A) = M(A)$ for every F-algebra A.

<u>(10.2.6) Remarks.</u> i) Although the statement of the corollary is very simple, the question of the existence of such a sequence $(F_n)_n \in \mathrm{Hol}(\mathbb{C}^p,\mathbb{C}^p)$ seems to be a difficult problem. First note that if one mapping F_i is constant, then

$$\bigcap F_1\circ\ldots\circ F_n(\mathbb{C}^p)$$

is never empty. So we consider only the case when no F_i is constant.

If p=1, the little Picard theorem ([NAR 1],p. 94) asserts that each nonconstant entire function attains each value with one possible exception. It follows that $F_1 \circ \ldots \circ F_n(\mathbb{C})$ is a dense open subset of $\mathbb{C}$ for each $n \in \mathbb{N}$, hence

$$\cap F_1 \circ \ldots \circ F_n(\mathbb{C})$$

is a nonempty subset of $\mathbb{C}$ by Baire's theorem.

If p>1 it is known that there are injective entire functions $F: \mathbb{C}^p \longrightarrow \mathbb{C}^p$, whose jacobian identically equals one but whose range is not dense in $\mathbb{C}^p$. So for p>1 a sequence with the desired property might exist. For a detailed discussion of this question we refer the reader to the paper of Dixon and Esterle [D/E].

ii) Clayton [CLA 1], Schottenloher [SCH], Dixon and Esterle [D/E] and Craw [CRA 2] constructed test algebras for Michael's problem. More precisely they constructed F-algebras A with the property that Michael's problem can be solved iff it can be solved for A.

PART 3

ANALYTIC STRUCTURE IN THE SPECTRUM OF AN F-ALGEBRA

An important subject in the theory of uB-algebras is the question of the existence of analytic structure in spectra, cf. for example chapter III of Stout's book [STO]. One seeks for conditions which ensure that parts of the spectrum of a uB-algebra can be endowed with the structure of a complex space in such a way that all Gelfand transforms become holomorphic functions on it.
As proved in (2.1.13) algebras of holomorphic functions are not uB-algebras but uF-algebras (cf. (2.3.1)). Hence - in the case of uF-algebras - we are moreover interested in the question when a given uF-algebra is topologically and algebraically isomorphic to Hol(X), X a suitable complex space. In some cases we shall obtain a function algebraic characterization of certain classes of holomorphic functions, for instance we shall characterize Hol(X) in the case that $X = \mathbb{C}$, $X \subset \mathbb{C}$ a domain, X a logarithmically convex complete Reinhardt domain, $X \subset \mathbb{C}^n$ a polynomially convex domain etc. .
There are results for uB-algebras, in particular Rossi's maximum principle (1.4.9), which reminds one of the behavior of holomorphic functions. This led to the question whether these results depend on the existence of analytic structure in the spectrum. In general this turned out to be false. In this part of the book we adopt this problem for uF-algebras and ask to what extend properties of Hol(X) (for example the maximum principle, Liouville's theorem, identity theorem, reflexivity, Montel's theorem, etc.) characterize this algebra within the class of uF-algebras.
In our setting we shall prove that a uF-algebra has analytic struc-

ture in every point of its spectrum iff it is a Stein algebra, i.e. iff it is topologically and algebraically isomorphic to the algebra of all holomorphic functions on a (reduced) Stein space.(For the definition of Stein spaces cf. (2.3).) Hence, in connection with the problem of finding analytic structure in the spectrum of a uF-algebra, we are interested in characterizing Stein algebras by intrinsic properties within the class of uF-algebras. This project is also interesting, since Stein algebras play an important role in the theory of several complex variables.

CHAPTER 11

STEIN ALGEBRAS

For the definition and properties of Stein spaces cf. (2.2) and (2.3). As mentioned in the introduction to part 3 of the book we shall prove that a uF-algebra is a Stein algebra iff it has analytic structure in each point of its spectrum

(11.1) Analytic structure in spectra of uF-algebras

(11.1.1) Definition. A uF-algebra A is called Stein algebra if there exists a (reduced) Stein space X such that A is topologically and algebraically isomorphic to Hol(X).

Remark. Let A be a Stein algebra, and let $T: A \longrightarrow Hol(X)$ be an isomorphism, X a Stein space. Recall from (2.3.2) that X = M(Hol(X)), more precisely the structure map $j: X \longrightarrow M(Hol(X))$, $x \longrightarrow \varphi_x$, (cf. (2.3.2)) defines a homeomorphism. So the adjoint spectral map $T^*: X \longrightarrow M(A)$ is a homeomorphism by (3.2.5). Note in particular that the spectrum of a Stein algebra is always locally compact. Furthermore the map

$$(\hat{A}, M(A)) \longrightarrow (Hol(X)^\wedge, M(Hol(X))) = (Hol(X), X),\ \hat{f} \longrightarrow \hat{f} \circ T^*,$$

where $Hol(X)^\wedge$ denotes the algebra of all Gelfand transforms, defines a topological homomorphism by (4.2.4).

(11.1.2) Definition. Let A be a uF-algebra, and let $\varphi \in M(A)$. We say that $\hat{A}$ has analytic structure in φ provided there are an open neighbourhood U of φ in M(A), an analytic subset Y of a domain $G \subset \mathbb{C}^n$

and a homeomorphism $p: Y \longrightarrow U$ such that $\hat{f} \circ p \in Hol(Y)$ for all $\hat{f} \in \hat{A}$.

Remark. Let φ, U, Y, p and G be as above. Note that U is locally compact. Hence there exists an open, hemicompact, relatively compact and $\hat{A}$-convex neighbourhood V of φ such that $V \subset U$ by (4.3.7). Then $p^{-1}(V)$ is an analytic subset of $G\backslash p^{-1}(U\backslash V)$.
Let $K \subset p^{-1}(V)$ be a compact subset. Then

$$L = p^{-1}(\hat{p}(K)_A),$$

where $\hat{p}(K)_A$ denotes the $\hat{A}$-convex hull of p(K), is a compact subset of $p^{-1}(V)$, since p is a homeomorphism and V is $\hat{A}$-convex. Let $x \in p^{-1}(V)\backslash L$, then $p(x) \in M(A)\backslash\hat{p}(K)_A$, i.e. we find $\hat{f} \in \hat{A}$ so that

$$|\hat{f} \circ p(x)| > \|\hat{f}\|_{\hat{p}(K)_A} = \|\hat{f} \circ p\|_L .$$

It follows that

$$\{y \in p^{-1}(V): |g(y)| \le \|g\|_K \text{ for all } g \in Hol(p^{-1}(V))\} \subset L,$$

since $\hat{f} \circ p \in Hol(p^{-1}(V))$. So $p^{-1}(V)$ is holomorphically convex, hence Stein, since the coordinate functions separate the points of $p^{-1}(V)$. Thus we can w.l.o.g. assume in the definition above that Y is a Stein analytic subset and that U is a hemicompact, relatively compact and $\hat{A}$-convex subset.

(11.1.3) We need a result of Rossi [ROI], which can be stated in the following form:

Theorem. Let X be a reduced complex space, and let (A,X) be a uF-algebra such that X is A-morphically convex (cf. (4.3.2)) and A is a subalgebra of Hol(X). Then M(A) can be given the structure of a Stein space such that $\hat{A} = Hol(M(A))$.

<u>(11.1.4) Theorem [KRA 6].</u> Let A be a uF-algebra. Then the following statements are equivalent:

i) A is a Stein algebra;

ii) $\hat{A}$ has analytic structure in each point $\varphi \in M(A)$;

iii) There exists an open cover $(U_i)_{i \in I}$ of M(A) consisting of hemicompact, relatively compact and $\hat{A}$-convex subsets such that each algebra $\hat{A}_{U_i}$ is a Stein algebra.

<u>Proof.</u> i) $\Rightarrow$ iii) Let A be a Stein algebra, and let $T: A \longrightarrow Hol(X)$ be an isomorphism, X a Stein space. Let $\varphi \in M(A)$ be an arbitrary point. Since M(A) is locally compact there exists an open, relatively compact, hemicompact and $\hat{A}$-convex neighbourhood $U \subset M(A)$ of φ. Then $(T^*)^{-1}(U)$ is an open and Hol(X)-convex subset of X, hence a Stein space. The map

$$(\hat{A}_U, U) \longrightarrow (Hol(X)_{(T^*)^{-1}(U)}, (T^*)^{-1}(U)),\ f \longrightarrow f \circ T^*,$$

is well-defined and defines a topological and algebraical isomorphism. By (2.3.6) and since $Hol((T^*)^{-1}(U))$ is complete we have

$$Hol(X)_{(T^*)^{-1}(U)} = Hol((T^*)^{-1}(U)).$$

iii) $\Rightarrow$ ii) Let $\varphi \in M(A)$ be an arbitrary point, and let $\varphi \in U_i$. By definition there exist a Stein space X_i and a topological algebra homomorphism

$$T_i: \hat{A}_{U_i} \longrightarrow Hol(X_i).$$

We can identify $M(Hol(X_i))$ with X_i by (2.3.2). By (4.3.4)iii) we can identify $M(\hat{A}_{U_i})$ with U_i as topological spaces. Thus $T_i^*: X_i \longrightarrow U_i$ defines a homeomorphism. By the definition of a reduced complex space (cf. (2.3.1)) there exist an analytic subset Y in a domain G of some $\mathbb{C}^n$, an open neighbourhood V of $(T_i^*)^{-1}(\varphi)$ in X_i and a homeomorphism $p: Y \longrightarrow V$ so that $f \circ p \in Hol(Y)$ for all $f \in Hol(X_i)$. Then

$$T_i^* \circ p: Y \longrightarrow T_i^*(V)$$

is a homeomorphism onto the open neighbourhood $T_i^*(V)$ of φ. Let $\hat{g} \in \hat{A}$ be an arbitrary element. Then $\hat{g}|_{U_i} \in \hat{A}_{U_i}$ and

$$\hat{g}|_{U_i} \circ T_i^* \in \mathrm{Hol}(X_i) \text{ by remark (11.1.1).}$$

Thus $\hat{g} \circ T_i^* \circ p \in \mathrm{Hol}(Y)$, i.e. $\hat{A}$ has analytic structure in φ.
ii) ⇒ i) For each $\varphi \in M(A)$ choose an analytic subset Y_φ in a domain of some $\mathbb{C}^{n_\varphi}$, an open neighbourhood U_φ of φ in $M(A)$ and a homeomorphism $p_\varphi : Y_\varphi \longrightarrow U_\varphi$ so that $\hat{f} \circ p_\varphi \in \mathrm{Hol}(Y_\varphi)$ for all $\hat{f} \in \hat{A}$. By remark (11.1.2) we can assume that each U_φ is a hemicompact and $\hat{A}$-convex set. Endow U_φ with the complex structure so that p_φ becomes a biholomorphic mapping, i.e.

$$\mathrm{Hol}(U_\varphi) = \{g : U_\varphi \longrightarrow \mathbb{C} : g \circ p_\varphi \in \mathrm{Hol}(Y_\varphi)\}.$$

Then $\hat{A}|_{U_\varphi} \subset \mathrm{Hol}(U_\varphi)$, thus $\hat{A}_{U_\varphi} \subset \mathrm{Hol}(U_\varphi)$, since $\mathrm{Hol}(U_\varphi)$ is complete with respect to the compact open topology.
U_φ is $\hat{A}_{U_\varphi}$-morphically convex, since it is $\hat{A}$-convex and by (4.3.3), so $(\hat{A}_{U_\varphi}, U_\varphi)$ satisfies the hypothesis of theorem (11.1.3). Hence $U_\varphi = M(\hat{A}_{U_\varphi})$ (cf. (4.3.4)iii)) can be given the structure of a Stein space so that $\hat{A}_{U_\varphi} = \mathrm{Hol}(U_\varphi, \mathcal{O})$, where $\mathrm{Hol}(U_\varphi, \mathcal{O})$ denotes the algebra of all holomorphic functions with respect to this new structure.
<u>Claim.</u> These new structures coincide on their intersection.

Let $\varphi, \psi \in M(A)$ such that $U_\varphi \cap U_\psi \neq \emptyset$, and let $\varphi_0 \in U_\varphi \cap U_\psi$. Let V be an open,hemicompact and $\hat{A}$-convex neighbourhood of φ_0 so that $V \subset U_\varphi \cap U_\psi$. Denote by $\mathrm{Hol}(V)_\varphi$ resp. $\mathrm{Hol}(V)_\psi$ the algebra of all holomorphic functions on V with respect to the complex structure which is induced from $(U_\varphi, \mathcal{O})$ resp. $(U_\psi, \mathcal{O})$. Then V is $\hat{A}_{U_\varphi}$-convex, thus $\mathrm{Hol}(U_\varphi, \mathcal{O})$-convex, and the restriction mapping $\mathrm{Hol}(U_\varphi, \mathcal{O}) \longrightarrow \mathrm{Hol}(V)_\varphi$ has dense range by (2.3.6). So

$$\text{Hol}(V)_\varphi = \text{Hol}(U_\varphi,\mathcal{O})_V = (\hat{A}_{U_\varphi})_V.$$

Repeating these arguments for $\text{Hol}(U_\psi,\mathcal{O})$ we get

$$\text{Hol}(V)_\varphi = \text{Hol}(V)_\psi,$$

since $(\hat{A}_{U_\varphi})_V = \hat{A}_V = (\hat{A}_{U_\psi})_V$ by (4.1.5).

Since φ_o was an arbitrary point of $U_\varphi \cap U_\psi$ and since every point of M(A) has a neighbourhood consisting of open, hemicompact and $\hat{A}$-convex subsets (cf. (4.3.7)ii)), we get our assertion.

Hence we have equipped M(A) with the structure of a complex space so that $\hat{A} \subset \text{Hol}(M(A))$. Since M(A) is $\hat{A}$-morphically convex we get our result from (11.1.3). □

<u>(11.1.5)</u> The next example shows that subalgebras of Stein algebras need not be Stein algebras.

<u>Example.</u> Set

$$A = \{f \in \text{Hol}(\mathbb{C}) : f(0) = f(i) \text{ for all } i=1,2,\ldots\}.$$

Then A - as a closed subalgebra of the Stein algebra $\text{Hol}(\mathbb{C})$ - is a uF-algebra. Set

$$\varphi_o : A \longrightarrow \mathbb{C},\ f \longrightarrow f(0),$$

then $\varphi_o \in M(A)$.

<u>Claim.</u> $M(A) = \mathbb{C}\backslash\{1,2,\ldots\}$ as sets.

i) For each $x \in \mathbb{C}\backslash\{1,2,\ldots\}$ the evaluation homomorphism at x defines an element of M(A). If $x, y \in \mathbb{C}\backslash\{1,2,\ldots\}$ are distinct points there is $f \in A$, by the theorem of Weierstraß, such that $f(x) \neq f(y)$, i.e. $\mathbb{C}\backslash\{1,2,\ldots\} \subset M(A)$.

ii) Let $\varphi \neq \varphi_o$ be an arbitrary element of M(A). We show that φ can be extended to an element of $M(\text{Hol}(\mathbb{C}))$.

Choose $f \in A$ such that $\varphi(f-f(0)) = 1$ and define

$$\tilde{\varphi} : \text{Hol}(\mathbb{C}) \longrightarrow \mathbb{C},\ g \longrightarrow \varphi(g(f-f(0))).$$

Then $\tilde{\varphi}$ is well-defined, since $f(i) = f(0)$ for $i=1,2,\dots$, i.e. $g(f-f(0)) \in A$. Let $g, h \in Hol(\mathbb{C})$ be arbitrary elements, then

$$\tilde{\varphi}(gh) = \varphi(gh(f-f(0))) = \varphi(gh(f-f(0)))\varphi(f-f(0))$$
$$= \varphi(gh(f-f(0))^2) = \varphi(g(f-f(0)))\varphi(h(f-f(0))) = \tilde{\varphi}(g)\tilde{\varphi}(h).$$

It follows that $\tilde{\varphi}$ defines an algebra homomorphism. Moreover $\tilde{\varphi}|_A=\varphi$. Since φ is continuous there is a compact set $K \subset \mathbb{C}$ such that

$$|\varphi(h)| \le \|h\|_K, \text{ for all } h \in A,$$

hence

$$|\tilde{\varphi}(g)| \le \|g\|_K \|f-f(0)\|_K, \text{ for all } g \in Hol(\mathbb{C}),$$

so $\tilde{\varphi} \in M(Hol(\mathbb{C}))$, i.e. $\tilde{\varphi}$ is the evaluation homomorphism at a point $x \in \mathbb{C}$. We have $x \neq 0,1,\dots$, since $\varphi \neq \varphi_o$, and our claim is proved.

Claim. φ_o has no compact neighbourhood, hence A is not a Stein algebra.

iii) For $n \in \mathbb{N}$ set $K_n = \{|x| \le n\}$. The seminorms $(\|\cdot\|_{K_n})_n$ generate the topology of A, hence

$$\tilde{K}_n = \{\varphi \in M(A): |\varphi(f)| \le \|f\|_{K_n} \text{ for all } f \in A\},\ n \in \mathbb{N},$$

is an admissible exhaustion of M(A) by (3.2.6)(*), (3.2.8).

Let $x \in \mathbb{C}\setminus\{0,1,\dots\}$, $|x|>n$. We claim that x is not contained in $\tilde{K}_n$. To prove this choose $h \in Hol(\mathbb{C})$, such that

$$\{y \in \mathbb{C}: h(y) = 0\} = \{0,1\dots\}.$$

Choose $k \in \mathbb{N}$, so that

$$(|x|/n)^k > \|h\|_{K_n}/|h(x)|.$$

Then $g = z^k \cdot h \in A$ and $|g(x)| > \|g\|_{K_n}$.

iv) Now suppose that φ_o has a compact neighbourhood $K \subset M(A)$. Choose $n \in \mathbb{N}$, so that $K \subset \tilde{K}_n$ and choose $f_1,\dots,f_r \in A$ so that

$$U = \{\varphi \in M(A): |\varphi(f_i)-\varphi_o(f_i)| < 1,\ i=1,\dots,r\} \subset \tilde{K}_n$$

(cf. (3.2.2)i)). Since all f_i are continuous functions on the plane and $f_i(0) = f_i(j)$ for $j=1,2,\dots$, we find $x \in \mathbb{C}\setminus\{1,2,\dots\}$, $|x|>n$, so that

$x \in U$, i.e. $x \in \tilde{K}_n$, and we get a contradiction to iii).

(11.1.6) Using (11.1.3) we get a criterion when a subalgebra of a Stein algebra is itself a Stein algebra.

Lemma. Let X be a Stein space, and let (A,X) be a closed subalgebra of Hol(X) such that the adjoint spectral map of the inclusion homomorphism $i: A \longrightarrow \mathrm{Hol}(X)$, $f \longrightarrow f$, is a proper mapping. Then A is a Stein algebra.

Proof. By (2.3.2) the structure map

$$j: X \longrightarrow M(\mathrm{Hol}(X)),\ x \longrightarrow \varphi_x,$$

where $\varphi_x(f) = f(x)$ for all $f \in \mathrm{Hol}(X)$, is a homeomorphism. It follows that

$$i^* \circ j: X \longrightarrow M(A)$$

is a proper mapping. Let $x \in X$. Then

$$i^* \circ j(x)(f) = f(x) \text{ for all } f \in A,$$

i.e. the structure map

$$j_A: X \longrightarrow M(A),\ x \longrightarrow \varphi_x,$$

is a proper mapping, hence X is A-morphically convex by (4.3.5) and our assertion follows from (11.1.3). □

CHAPTER 12

CHARACTERIZING SOME PARTICULAR STEIN ALGEBRAS

We start with a function algebraic characterization of Hol(G) in the case that G is

- an open subset of the plane,
- a polynomially convex open subset of $\mathbb{C}^n$,
- a logarithmically convex complete Reinhardt domain,
- a domain of holomorphy in some $\mathbb{C}^n$.

(12.1) Polynomially convex analytic subsets

(12.1.1) Many results of this chapter are based on the following theorem, which has been used in a similar form by various authors.

Theorem. Let A be a rationally n-generated resp. n-generated semisimple F-algebra with generating elements $f_1,\ldots,f_n$. Suppose that the map

$$\hat{F}: M(A) \longrightarrow \mathbb{C}^n,\ \varphi \longrightarrow (\hat{f}_1(\varphi),\ldots,\hat{f}_n(\varphi)),$$

is a homeomorphism onto an open subset $G \subset \mathbb{C}^n$. Then G is a rationally convex resp. polynomially convex open subset and A is topologically and algebraically isomorphic to Hol(G).

Proof. Recall that a subset $S \subset \mathbb{C}^n$ is called rationally convex resp. polynomially convex if the rationally convex hull resp. the polynomially convex hull of each compact subset of S is contained in S. Since $\hat{F}$ is a homeomorphism, G is rationally resp. polynomially convex by (5.2.4)i) resp. (5.1.5).

Denote by ϑ the (unique) continuous homomorphism

$$\vartheta\colon \mathrm{Hol}(G) \longrightarrow A$$

such that

i) $\vartheta(z_i) = f_i$, $i=1,\dots,n$,

ii) $\hat{\vartheta}(h) = h\circ\hat{F}$ for all $h \in \mathrm{Hol}(G)$,

where $\hat{\vartheta}(h)$ denotes the Gelfand transform of $\vartheta(h)$ (cf. (6.1.2)).

We see that ϑ is injective by ii). If we can show that ϑ is surjective, our assertion will follow from the open mapping theorem for Fréchet spaces.

Let $g \in A$ be an arbitrary element. Since A is rationally n-generated there exist polynomials p_k, q_k ($k \in \mathbb{N}$) so that $q_k(f_1,\dots,f_n)$ is invertible in A and $g_k = p_k(f_1,\dots,f_n)/q_k(f_1,\dots,f_n)$ converges to g as $k \longrightarrow \infty$. It follows from (6.1.4) that $\hat{q}_k(f_1,\dots,f_n)$ – the Gelfand transform of $q_k(f_1,\dots,f_n)$ – has no zero on M(A), so $\hat{q}_k(f_1,\dots,f_n)\circ\hat{F}^{-1}$ has no zero on G, i.e.

$$\hat{g}_k\circ\hat{F}^{-1} \in \mathrm{Hol}(G) \text{ for all } k \in \mathbb{N},$$

in fact $\hat{g}_k\circ\hat{F}^{-1}$ is a rational function which is analytic on G.

By (3.2.12) $\hat{g}_k$ converges to $\hat{g}$, hence $\hat{g}_k\circ\hat{F}^{-1}$ converges compactly to $\hat{g}\circ\hat{F}^{-1}$ on G, since $\hat{F}$ is a homeomorphism. So $\hat{g}\circ\hat{F}^{-1} \in \mathrm{Hol}(G)$ by (2.1.7). Set $h = \hat{g}\circ\hat{F}^{-1}$, then $\hat{\vartheta}(h) = \hat{g}$ by ii). Hence $\vartheta(h) = g$, since A is semisimple and we have proved that ϑ is surjective. □

(12.1.2) Corollary. Let A be a rationally n-generated semisimple F-algebra with generating elements $f_1,\dots,f_n$. Suppose that the map $\hat{F}$ defines a homeomorphism onto an analytic subset of a domain of holomorphy $G \subset \mathbb{C}^n$. Then A is topologically and algebraically isomorphic to Hol(X).

Proof. α) Denote by ϑ the continuous homomorphism from Hol(G) to A such that

i) $\vartheta(z_i) = f_i$ (i=1,...,n),

ii) $\hat{\vartheta}(h) = h \circ \hat{F}$ for all $h \in \mathrm{Hol}(G)$ (cf. (6.1.2)).

Let $g \in \mathrm{Hol}(X)$. Choose $h \in \mathrm{Hol}(G)$ so that $h|_X = g$ by (2.3.3) and set

$$\vartheta_X: \mathrm{Hol}(X) \longrightarrow A,\ g \longrightarrow \vartheta(h).$$

We show that ϑ_X is well-defined. Let $h \in \mathrm{Hol}(G)$ so that $h|_X \equiv 0$. Then $\hat{\vartheta}(h) = h \circ \hat{F} = 0$ by ii), hence $\vartheta(h) = 0$, since A is semisimple. It follows that $\vartheta_X(g)$, $g \in \mathrm{Hol}(X)$, is independent of the choice of the extension $h \in \mathrm{Hol}(G)$. Moreover we see that ϑ_X is an injective algebra homomorphism.

Let $g \in A$ be an arbitrary element. Repeating the proof of (12.1.1) we get a sequence of rational functions $(r_k)_k$, analytic on X, such that r_k converges compactly on X to $h = \hat{g} \circ \hat{F}^{-1}$. Since $r_k \in \mathrm{Hol}(X)$ for all $k \in \mathbb{N}$, we have $h \in \mathrm{Hol}(X)$, since Hol(X) is complete with respect to the compact open topology. Moreover

$$\hat{\vartheta}_X(h) = h \circ \hat{F} = \hat{g},$$

i.e. $\vartheta_X(h) = g$ since A is assumed to be semisimple. Hence ϑ_X is an algebra isomorphism.

β) Consider the map

$$\tilde{\vartheta}: A \longrightarrow \mathrm{Hol}(X),\ g \longrightarrow \hat{g} \circ \hat{F}^{-1}.$$

Then $\tilde{\vartheta}$ is well-defined by α), and it is easily seen - using ii) and the hypothesis "semisimple" - that $\tilde{\vartheta} = \vartheta_X^{-1}$. The map

$$\Gamma: A \longrightarrow \hat{A},\ f \longrightarrow \hat{f},$$

is continuous by (3.2.12) and the map

$$T: \hat{A} \longrightarrow \mathrm{Hol}(X),\ \hat{g} \longrightarrow \hat{g} \circ \hat{F}^{-1},$$

is continuous since $\hat{F}$ is a homeomorphism. Hence $\tilde{\vartheta} = T \circ \Gamma$ is continuous, and A is topologically isomorphic to Hol(X) by the open mapping theorem.

□

(12.1.3) Corollary. Let A be an n-generated uF-algebra with generating elements $f_1,\dots,f_n$. Suppose that $\hat{F}$ is a homeomorphism onto an analytic subset X of a domain $G \subset \mathbb{C}^n$, then A is topologically and algebraically isomorphic to Hol(X).

Proof. Recall first that A is isomorphic to $\hat{A}$. Since $\hat{F}$ is a homeomorphism we see that X is polynomially convex (cf. (5.1.8)). Using similar arguments as above we show that

$$T: \hat{A} \longrightarrow \mathrm{Hol}(X),\ \hat{f} \longrightarrow \hat{f} \circ \hat{F}^{-1},$$

defines an injective and continuous algebra homomorphism.
We show that T is surjective.
Let $f \in \mathrm{Hol}(X)$ be an arbitrary element. We shall show that there is a sequence of polynomials $(p_k)_k$ so that p_k converges compactly to f. Then $p_k \circ \hat{F}$ converges to an element $\hat{g} \in \hat{A}$, since $\hat{A}$ is complete, and $T(\hat{g}) = f$.
Now let $K \subset X$ be a polynomially convex compact subset. Choose an open relatively compact subset $U \subset G$ which contains K, and a closed polycylinder P which contains U. For each $x \in P \backslash U$ we find a polynomial p_x such that $|p_x(x)| > 1 > \|p_x\|_K$. By the compactness of $P\backslash U$ we find polynomials $p_1,\dots,p_r$ such that

$$K \subset Q = \{x \in P: |p_i(x)| < 1,\ i=1,\dots,r\},$$

and $Q \subset U$. Hence $Q \cap X$ is an analytic subset of the polynomially convex open set Q. Extend $f|_{Q \cap X}$ to a holomorphic function $\tilde{f}$ on Q (cf. (2.3.3)). By (1.4.5) $\tilde{f}$ - and hence f - can be approximated uniformly on K by polynomials. □

(12.2) Polynomially convex open subsets of $\mathbb{C}^n$

(12.2.1) Let $G \subset \mathbb{C}$ be a polynomially convex open subset. Then Hol(G) is a singly generated uF-algebra with generating element

$$z: G \longrightarrow \mathbb{C},\ x \longrightarrow x,$$

by (5.1.2)i). The map

$$D: \mathrm{Hol}(G) \longrightarrow \mathrm{Hol}(G),\ f \longrightarrow f',$$

where f' denotes the derivative of f, defines a derivation on Hol(G) (cf. (8.2)) such that $D(z) = 1$ is invertible in Hol(G).
Our first theorem - due to Carpenter [CAR 6] - shows that these properties characterize in fact the algebra Hol(G), $G \subset \mathbb{C}$ a polynomially convex open set.

Theorem [CAR 6]. Let A be a singly generated uF-algebra with generating element f. Suppose there is a derivation $D: A \longrightarrow A$ such that D(f) is invertible in A. Then $\hat{f}(M(A))$ is a polynomially convex open subset of the plane and A is topologically and algebraically isomorphic to $\mathrm{Hol}(\hat{f}(M(A)))$.

Proof. By (5.1.8) it suffices to consider the algebra $A = \varprojlim P(K_n)$, where $\ldots \subset K_n \subset K_{n+1} \subset \ldots$ is a sequence of polynomially convex compact subsets of the plane.
Recall from (5.1.6) that $(K_n)_n$ is an admissible exhaustion of M(A).
Moreover A is singly generated by the polynomial $z: \cup K_n \longrightarrow \mathbb{C}$, $x \longrightarrow x$. By our hypothesis there exists a derivation D on A so that D(z) is invertible in A, i.e. $\widehat{D}(z)$, the Gelfand transform of D(z), has no zero on $\cup K_n = M(A)$ by (6.1.4).
By (5.1.4) $\hat{z}$ is an injective and continuous mapping. We want to show that $\hat{z}$ is an open mapping, i.e. $\hat{z}$ is a homeomorphism.
Suppose the contrary. Let U be an open subset of M(A) (Gelfand topology) such that $\hat{z}(U) = U$ is not an open subset of the plane.
Let y be a boundary point of $\hat{z}(U)$ such that $y \in \hat{z}(U)$. W.l.o.g. $y \in K_1$.
Since D is continuous by (8.2.3), and since A is topologically and algebraically isomorphic to $\widehat{A}$ there is $l \in \mathbb{N}$ and $c>0$ such that

$$\|\widehat{D}(f)\|_{K_1} \le c\|\hat{f}\|_{K_l} \text{ for all } \hat{f} \in \widehat{A},$$

where $\widehat{D}(f)$ denotes the Gelfand transform of D(f), hence

(*) $\qquad |\widehat{D}(f)(y)| \leq c\|\hat{f}\|_{K_l}$ for all $\hat{f} \in \hat{A}$.

Consider the map

$$d: \hat{A}|_{K_l} \longrightarrow \mathbb{C},\ \hat{h}|_{K_l} \longrightarrow \widehat{D}(h)(y).$$

Note that

$$d(\hat{f}\hat{g}) = d(\hat{f})\hat{g}(y) + d(\hat{g})\hat{f}(y) \text{ for all } \hat{f},\ \hat{g} \in \hat{A}|_{K_l}.$$

Since d is continuous by (*) it can be extended to a continuous point derivation on $\hat{A}_{K_l} = P(K_l)$ at y (cf. (1.3.14)). The point y is a boundary point of K_l (euclidean topology) since on K_l the (relative) Gelfand topology and the euclidean topology coincide by (5.1.7)ii). Hence $d \equiv 0$ by (1.3.14). This is a contradiction, since

$$d(\hat{z}|_{K_l}) = \widehat{D}(z)(y),$$

and $\widehat{D}(z)$ has no zero on $\cup K_n$.
Our assertion follows from (12.1.1). □

<u>(12.2.2) Remarks.</u> i) Theorem (12.2.1) becomes false if we replace "uF-algebra" by "semisimple F-algebra".
The algebra $C^\infty([0,1])$ is a semisimple F-algebra by (8.1.3)ii). It is singly generated by the map

$$z: [0,1] \longrightarrow \mathbb{C},\ x \longrightarrow x,$$

(cf. (5.1.2)iii)), and

$$D: C^\infty([0,1]) \longrightarrow C^\infty([0,1]),\ f \longrightarrow f',$$

where f' denotes the derivative of f, is a derivation on $C^\infty([0,1])$. Clearly D(z) is invertible in $C^\infty([0,1])$.
ii) We can sharpen theorem (12.2.1) in the following way (see [CAR 6]): Let A be a uF-algebra. Suppose there are $f \in A$, $x_1,\dots,x_n \in \mathbb{C}$ and a derivation $D: A \longrightarrow A$ such that

α) $f-x_i\cdot 1$ is invertible in A for $i=1,\dots,n$,

β) The polynomials in f, $(f-x_1\cdot 1)^{-1},\dots,(f-x_n\cdot 1)^{-1}$ are dense in A,

γ) D(f) is invertible in A.

Then $\hat{f}(M(A))$ is a finitely connected open set in $\mathbb{C}$ and A is algebraically and topologically isomorphic to $\mathrm{Hol}(\hat{f}(M(A)))$.

The proof of this theorem runs along the lines of the proof of (12.2.1), using the fact, that for each $\hat{A}$-convex compact subset K of M(A) our assumptions imply that every boundary point of $\hat{f}(K)$ is a peak point for $R(\hat{f}(K))$.

In general it is false that each boundary point of a compact subset $K \subset \mathbb{C}$ is a peak point for the algebra R(K). Hence we cannot copy the proof of (12.2.1) for rationally singly generated uF-algebras, and it is unknown whether an analogous result holds for these algebras.

<u>(12.2.3)</u> Let $G \subset \mathbb{C}^n$ be a polynomially convex open subset. Then Hol(G) is n-generated by the coordinate functions $z_1,\dots,z_n$ and

$$\hat{z}: M(\mathrm{Hol}(G)) \longrightarrow \mathbb{C}^n,\ \varphi \longrightarrow (\hat{z}_1(\varphi),\dots,\hat{z}_n(\varphi)),$$

is a homeomorphism onto G by example (5.1.7). Hence Hol(G) is a n-generated semisimple F-algebra and M(Hol(G)) is a 2n-dimensional topological manifold.

We use this last observation for a function algebraic characterization of Hol(G), G as above.

<u>Theorem [H/W].</u> Let A be an n-generated semisimple F-algebra with generating elements $f_1,\dots,f_n$. Let

$$\hat{F}: M(A) \longrightarrow \mathbb{C}^n,\ \varphi \longrightarrow (\hat{f}_1(\varphi),\dots,\hat{f}_n(\varphi)).$$

Suppose that M(A) is a topological 2n-dimensional manifold (without boundary), then $\hat{F}(M(A))$ is a polynomially convex open subset of $\mathbb{C}^n$ and A is topologically and algebraically isomorphic to $\mathrm{Hol}(\widehat{F}(M(A)))$.

Proof. $\hat{F}$ is an injective and continuous map by (5.1.4). If we can show that $\hat{F}$ is an open mapping, hence a homeomorphism, our assertion will follow from theorem (12.1.1).

Let V be an arbitrary open subset of M(A). Let $\varphi \in V$. Choose a chart (U,h,W) around φ (i.e. U is an open neighbourhood of φ, W is an open subset of $\mathbb{R}^{2n}$ and $h: W \longrightarrow U$ is a homeomorphism) such that $U \subset V$ and $\overline{U}$ - the closure of U - is a compact subset of M(A). Since $\hat{F}|_{\overline{U}}$ is a homeomorphism we see that $\hat{F} \circ h$ is a homeomorphism from W onto the subset $\hat{F}(U) \subset \mathbb{C}^n = \mathbb{R}^{2n}$. By Brouwer's domain of invariance theorem (cf. [DUG]) $\hat{F}(U)$ is an open subset of $\mathbb{C}^n$. Hence $\hat{F}$ is an open mapping. □

(12.2.4) Corollary [H/W]. Let A be a semisimple F-algebra generated by $f_1,\ldots,f_n$. Suppose that M(A) is a topological 2n-dimensional manifold (without boundary) such that for each $c>0$ the set

$$K_c = \{\varphi \in M(A): |\hat{f}_i(\varphi)| \le c,\ i=1,\ldots,n\}$$

is compact in M(A). Then A is topologically and algebraically isomorphic to $\mathrm{Hol}(\mathbb{C}^n)$.

Proof. By (12.2.3) $\hat{F}(M(A))$ is an open subset of $\mathbb{C}^n$ and A is topologically and algebraically isomorphic to $\mathrm{Hol}(\hat{F}(M(A)))$. We show that $\hat{F}(M(A))$ is a closed subset of $\mathbb{C}^n$. Then $\hat{F}(M(A)) = \mathbb{C}^n$ and we are done.

Let $(x_k)_k$ be a sequence in $\hat{F}(M(A))$ which converges to y. Choose $c>0$ so that

$$\{y,x_1,x_2,\ldots\} \subset \{(w_1,\ldots,w_n) \in \mathbb{C}^n: |w_i|<c,\ i=1,\ldots,n\}.$$

Then $(\hat{F}^{-1}(x_k))_k$ is a sequence in the compact set K_c. Hence a subsequence converges to an element $\varphi \in M(A)$. Clearly $\hat{F}(\varphi) = y$. □

Remark. Originally Heal and Windham used the stronger assumption "uF-algebra" in (12.2.3) and (12.2.4).

(12.3) Open subsets of the plane

(12.3.1) Corollary. i) Let A be a rationally singly generated semisimple F-algebra with generating element f. Suppose that M(A) is a topological two dimensional manifold (without boundary). Then $\hat{f}(M(A))$ is an open subset of $\mathbb{C}$ and A is topologically and algebraically isomorphic to $Hol(\hat{f}(M(A)))$.
ii) Let $G \subset \mathbb{C}$ be an open set. Then Hol(G) is a rationally singly generated semisimple F-algebra and M(Hol(G)) is a topological two dimensional manifold.

Proof. i) Copy the proof of (12.2.3).
ii) Hol(G) is a rationally singly generated uF-algebra by (5.2.2) and M(Hol(G)) is homeomorphic to G by (4.1.7)ii) or (2.2.7). □

(12.3.2) Another characterization of Hol(G), G an open subset of $\mathbb{C}$, is due to Brooks [BRO 7].

Theorem [BRO 7]. Let A be a rationally singly generated semisimple F-algebra with generating element f. Suppose that $\gamma A = \emptyset$. Then $U = \hat{f}(M(A))$ is an open subset of the plane and A is algebraically and topologically isomorphic to Hol(U).
(Here γA denotes the Shilov boundary of A, cf. chapter 9.)

For the proof we need a result on the Shilov boundary of rationally singly generated B-algebras.

(12.3.3) Proposition. Let A be a rationally singly generated B-algebra with generating element f. Then

$$\hat{f}(\gamma A) = \partial \hat{f}(M(A)),$$

where $\partial \hat{f}(M(A))$ denotes the boundary of $\hat{f}(M(A))$.

Proof. i) The inclusion "$\supset$" follows from (1.3.7).

ii) Recall that $\hat{f}: M(A) \longrightarrow \hat{f}(M(A))$ is a homeomorphism by (5.2.3).

Let $g \in A$ be an arbitrary element. Choose polynomials p_k and q_k ($k \in \mathbb{N}$), so that $q_k(f)$ is invertible in A and $g_k = p_k(f)/q_k(f)$ converges to g as $k \longrightarrow \infty$. It follows from (1.2.9) that $\hat{q}_k(f)$, the Gelfand transform of $q_k(f)$, has no zero on M(A), i.e. q_k has no zero on $\hat{f}(M(A))$, so $p_k/q_k \in R(\hat{f}(M(A)))$ for all $k \in \mathbb{N}$.

We have $\hat{g}_k \longrightarrow \hat{g}$ as $k \longrightarrow \infty$ by (1.2.3), hence p_k/q_k converges uniformly to $\hat{g} \circ \hat{f}^{-1}$ on $\hat{f}(M(A))$ and $\hat{g} \circ \hat{f}^{-1} \in R(\hat{f}(M(A)))$.

So

$$\|\hat{g} \circ \hat{f}^{-1}\|_{\hat{f}(M(A))} = \|\hat{g} \circ \hat{f}^{-1}\|_{\partial \hat{f}(M(A))},$$

i.e.

$$\|\hat{g}\|_{M(A)} = \|\hat{g}\|_{\hat{f}^{-1}(\partial \hat{f}(M(A)))},$$

and we see that

$$\gamma A \subset \hat{f}^{-1}(\partial \hat{f}(M(A))).$$

□

Proof of (12.3.2). We claim that $U = \hat{f}(M(A))$ is an open subset of the plane and $\hat{f}: M(A) \longrightarrow U$ is a homeomorphism.

Let $\varphi \in M(A)$ be an arbitrary point. Recall from (9.1.2) the definition of the seminorms P_φ. Since $\varphi \in M(A)\backslash\gamma A$ there exists $p \in P_\varphi$ so that $\varphi \in M(A_p)\backslash\gamma(A_p)$. The algebra A_p is rationally singly generated since the natural map

$$A \longrightarrow A_p,\ g \longrightarrow g + \{h \in A: p(h) = 0\},$$

has dense range. This implies that $\hat{f}(\varphi) \in \operatorname{int} \hat{f}(M(A_p))$ (interior of $\hat{f}(M(A_p))$) by (12.3.3), so U is an open subset of the plane. Moreover

$$\hat{f}|_{\hat{f}^{-1}(\operatorname{int} \hat{f}(M(A_p)))}: \hat{f}^{-1}(\operatorname{int} \hat{f}(M(A_p))) \longrightarrow \operatorname{int} \hat{f}(M(A_p))$$

is a homeomorphism, since $M(A_p)$ is compact and $\hat{f}$ is an injective and continuous map by (5.2.3). We now conclude that $\hat{f}$ is in fact a homeomorphism.

The rest of our assertion follows from (12.1.1). □

(12.4) Domains of holomorphy

(12.4.1) We recall some notations. We set $\mathbb{N}_0 = \mathbb{N} \cup \{0\}$. For $v = (v_1,\dots,v_n) \in \mathbb{N}_0^n$, $t = (t_1,\dots,t_n) \in \mathbb{C}^n$ resp. $\mathbb{R}^n$ set $v! = v_1!\cdot\dots\cdot v_n!$, $t^v = t_1^{v_1}\cdot\dots\cdot t_n^{v_n}$. Furthermore

$$\mathbb{R}_+^n = \{(r_1,\dots,r_n) \in \mathbb{R}^n : r_i > 0,\ i=1,\dots,n\}.$$

If $t=(t_1,\dots,t_n) \in \mathbb{C}^n$, $r=(r_1,\dots,r_n) \in \mathbb{R}_+^n$, define

$$D_{t,r} = \{(w_1,\dots,w_n) \in \mathbb{C}^n : |w_i - t_i| < r_i,\ i=1,\dots,n\}.$$

Let A be an F-algebra and let $D_1,\dots,D_n$ be derivations on A (cf. (8.2)). If $v = (v_1,\dots,v_n) \in \mathbb{N}_0^n$ and $f \in A$ set

$$D^v(f) = \underbrace{D_n \circ \dots \circ D_n}_{v_n\text{-times}} \circ \dots \circ \underbrace{D_1 \circ \dots \circ D_1}_{v_1\text{-times}}(f),$$

$$D^0(f) = f \text{ for } 0 = (0,\dots,0).$$

Theorem. Let A be a semisimple F-algebra with connected spectrum M(A), and let $(p_n)_n$ be a defining sequence of seminorms for A (cf. (3.1.7)). Suppose there are elements $f_1,\dots,f_n \in A$ and derivations $D_1,\dots,D_n$ on A such that

i) $\hat{F} : M(A) \longrightarrow \mathbb{C}^n$, $\varphi \longrightarrow (\hat{f}_1(\varphi),\dots,\hat{f}_n(\varphi))$, is an injective map,

ii) $D_i(f_j) = \delta_{ij}\cdot 1$, for $i,j=1,\dots,n$, where δ_{ij} denotes the Kronecker symbol,

iii) there is a sequence $r_1, r_2, \dots$ in $\mathbb{R}_+^n$ such that

$$p_k(D^v(f)) \le p_{k+1}(f)v!/r_k^v \quad \text{for all } f \in A,\ k \in \mathbb{N},\ v \in \mathbb{N}_0^n.$$

Then $G = \hat{F}(M(A))$ is a domain of holomorphy in $\mathbb{C}^n$ and A is topologically and algebraically isomorphic to Hol(G).

Proof. As usual denote by A_k the completion of the algebra $A/\{p_k=0\}$ with respect to the norm

$$p'_k(g+\{p_k=0\}) = p_k(g).$$

By (3.2.8) $(M(A_k))_k$ is an admissible exhaustion of $M(A)$.

α) Claim. $\hat{F}(M(A_k))$ is contained in int $\hat{F}(M(A_{k+1}))$ - the interior of $\hat{F}(M(A_{k+1}))$ - for each $k \in \mathbb{N}$.

Let $r_k = (r_1^{(k)},\dots,r_n^{(k)})$. Choose an arbitrary $t=(t_1,\dots,t_n) \in D_{o,r_k}$.

Let $g \in A$ be an arbitrary element. Then

$$(*) \qquad \sum_{v\in\mathbb{N}_o^n} p_k(D^v(g)\cdot t^v/v!) \le p_{k+1}(g)\cdot\prod_{i=1}^{n} 1/(1-(|t_i|/r_i^{(k)})),$$

by iii). It follows that

$$\sum_{v\in\mathbb{N}_o^n} D^v(g)t^v/v! + \{p_k=0\},$$

defines an element in A_k, and that the series converges in any order of summation. We note that

$$T_t\colon A/\{p_{k+1}=0\} \longrightarrow A_k,$$

$$g + \{p_{k+1}=0\} \longrightarrow \sum_{v\in\mathbb{N}_o^n} D^v(g)t^v/v! + \{p_k=0\},$$

is a linear map, since all D_i are linear maps. In fact - using Leibniz' rule - we see that T_t is even an algebra homomorphism. This homomorphism is continuous by (*), hence can be extended to A_{k+1}. We denote this map again by T_t. We remark that

$$(**) \qquad T_t(f_i + \{p_{k+1}=0\}) = f_i + t_i\cdot 1 + \{p_k=0\} \text{ for } i=1,\dots,n,$$

by ii) and since $D_j(1)=0$ $(j=1,\dots,n)$.

Consider the spectral map

$$T_t^*\colon M(A_k) \longrightarrow M(A_{k+1}),$$

cf. (3.2.5). By (**) we have for $\varphi \in M(A_k)$

$$\hat{f}_i(T_t^*(\varphi)) = \varphi(f_i) + t_i,$$

i.e.

(***) $\qquad \hat{F}(T_t^*(\varphi)) = \hat{F}(\varphi) + t.$

Hence

$$\hat{F}(M(A_k)) \subset \bigcup_{\varphi \in M(A_k)} D_{\hat{F}(\varphi), r_k} \subset \operatorname{int} \hat{F}(M(A_{k+1})),$$

and our claim is proved.

β) Since $\hat{F}|_{M(A_k)}$ is a homeomorphism for each $k \in \mathbb{N}$, we learn from α) that $\hat{F}: M(A) \longrightarrow G$ is a homeomorphism onto an open subset $G \subset \mathbb{C}^n$. Since $M(A)$ is connected the same is true for G.

γ) Claim. $\hat{g} \circ \hat{F}^{-1} \in \mathrm{Hol}(G)$ for all $g \in A$.

W.l.o.g. $0 \in G$. Let $\varphi_o \in M(A)$ such that $\hat{F}(\varphi_o) = 0$, w.l.o.g. $\varphi_o \in M(A_1)$.

Let $t \in D_{o,r_1}$ be an arbitrary point and define $T_t: A_2 \longrightarrow A_1$ as in α).

Then

$$\hat{F}^{-1}(t) = T_t^*(\varphi_o)$$

by (***), so we have for $g \in A$

$$\hat{g}(\hat{F}^{-1}(t)) = \hat{g}(T_t^*(\varphi_o)) =$$

$$= \varphi_o\Big(\sum_{\nu \in \mathbb{N}_o^n} D^\nu(g)t^\nu/\nu! + \{p_1=0\}\Big) = \sum_{\nu \in \mathbb{N}_o^n} \varphi_o(D^\nu(g))t^\nu/\nu!,$$

i.e. $\hat{g} \circ \hat{F}^{-1}|_{D_{o,r_1}} \in \mathrm{Hol}(D_{o,r_1})$ by (2.1.5). In the same way we can show that $\hat{g} \circ \hat{F}^{-1}$ is holomorphic in a neighbourhood of any point $x \in G$ and our claim is proved.

δ) Denote by ϑ the continuous homomorphism from $\mathrm{Hol}(G)$ to A such that

$\qquad$ i) $\vartheta(z_i) = f_i$ for $i=1,\dots,n$,

and

$\qquad$ ii) $\hat{\vartheta}(h) = h \circ \hat{F}$ for all $h \in \mathrm{Hol}(G)$,

cf. (6.1.2). Let $\vartheta(h) = 0$. Then $0 = \hat{\vartheta}(h) = h \circ \hat{F}$, hence $h \equiv 0$, i.e. ϑ is injective. Now let $g \in A$ be an arbitrary element. Then

$\hat{g} \circ \hat{F}^{-1} \in Hol(G)$ by γ), hence

$$\hat{\vartheta}(\hat{g} \circ \hat{F}^{-1}) = \hat{g},$$

where $\hat{\vartheta}(\hat{g} \circ \hat{F}^{-1})$ denotes the Gelfand transform of $\vartheta(\hat{g} \circ \hat{F}^{-1})$, and it follows that $\vartheta(\hat{g} \circ \hat{F}^{-1}) = g$ since A is semisimple, i.e. ϑ is surjective. So ϑ is a topological isomorphism by the open mapping theorem.

ε) Let $x \in G\backslash\hat{F}(M(A_k))$ be an arbitrary point. Then

$$\varphi = \hat{F}^{-1}(x) \in M(A)\backslash M(A_k).$$

By (3.2.6)(*)

$$M(A_k) = \{\varphi \in M(A): |\varphi(f)| \le p_k(f) \text{ for all } f \in A\}.$$

Hence there is $g \in A$ so that

$$|\hat{g}(\varphi)| = |\varphi(g)| > p_k(g) \ge \|\hat{g}\|_{M(A_k)}.$$

The last inequality follows by (1.2.3). So

$$|h(x)| > \|h\|_{\hat{F}(M(A_k))}$$

for the holomorphic function $h = \hat{g} \circ \hat{F}^{-1}$. Hence G is holomorphically convex, since $(\hat{F}(M(A_k)))_k$ is an admissible exhaustion of G, thus G is a domain of holomorphy by (2.2.7). □

<u>(12.4.2) Remarks.</u> i) Theorem (12.4.1) is based on an idea of Arens [ARE 1] who considered rationally singly generated (semisimple) F-algebras which satisfy hypothesis ii) and iii).

ii) Let $G \subset \mathbb{C}^n$ be a domain of holomorphy. Denote by $z_1,\dots,z_n$ the coordinate functions. By (2.2.7) we can identify G and M(Hol(G)), hence

$$\hat{z}: M(Hol(G)) \longrightarrow \mathbb{C}^n,\ \varphi \longrightarrow (\hat{z}_1(\varphi),\dots,\hat{z}_n(\varphi)),$$

is injective. Set for $i=1,\dots,n$

$$D_i: Hol(G) \longrightarrow Hol(G),\ f \longrightarrow \frac{\partial f}{\partial z_i}.$$

Then $D_1,\dots,D_n$ are derivations on Hol(G) which satisfy ii) and iii) (cf.

(2.1.4)). Hence theorem (12.4.1) characterizes Hol(G), G a domain of holomorphy among the semisimple F-algebras.

(12.5) Logarithmically convex complete Reinhardt domains

(12.5.1) Finally we consider F-algebras A which have a special basis. We say that $f_1,\dots,f_n$ generate a basis in A, provided each $g \in A$ has a unique representation

$$g = \sum_{(v_1,\dots,v_n)\in\mathbb{N}_o^n} a_{v_1,\dots,v_n} f_1^{v_1}\cdot\ldots\cdot f_n^{v_n},$$

$a_{v_1,\dots,v_n} \in \mathbb{C}$ for all $(v_1,\dots,v_n) \in \mathbb{N}_o^n$.

This means that the series converges with respect to a fixed order of summation in the topology of A. We use the notation

$$g = \sum a_v f^v.$$

Theorem [S/W]. Let A be an F-algebra with basis generated by $f_1,\dots,f_n$. Suppose that

$$G = \{(\hat{f}_1(\varphi),\dots,\hat{f}_n(\varphi)) \in \mathbb{C}^n : \varphi \in M(A)\}$$

is an open subset of $\mathbb{C}^n$. Then G is a logarithmically convex complete Reinhardt domain (cf. example (2.2.7)ii)) and A is topologically and algebraically isomorphic to Hol(G).

Proof. We first note that A is n-generated by $f_1,\dots,f_n$ in the usual sense (cf. (5.1.1)). So

$$\hat{F}: M(A) \longrightarrow \mathbb{C}^n,\ \varphi \longrightarrow (\hat{f}_1(\varphi),\dots,\hat{f}_n(\varphi)),$$

is an injective and continuous map by (5.1.4).

We claim that $\hat{F}$ is a homeomorphism and that A is semisimple.

Let $g = \sum a_v f^v$ be an arbitrary element of A. Let $t = (t_1,\dots,t_n) \in G$ such that $|t_i|>0$ $(i=1,\dots,n)$ and set $|t| = (|t_1|,\dots,|t_n|)$. Let $\varphi = \hat{F}^{-1}(t)$. Then

$$\hat{g}\circ\hat{F}^{-1}(t) = \varphi(\sum a_\nu f^\nu) = \sum a_\nu t^\nu .$$

It follows from (2.1.5) that the series $\sum a_\nu z^\nu$ defines a holomorphic function on the polycylinder $D_{o,|t|}$. Moreover

$$\hat{g}\circ\hat{F}^{-1}(x) = \sum a_\nu x^\nu \quad \text{for all } x \in D_{o,|t|}\cap G.$$

Hence

$$h_g(z) = \sum a_\nu z^\nu$$

defines a holomorphic function on

$$\tilde{G} = \{(z_1,\ldots,z_n) \in \mathbb{C}^n\text{: There exists } (t_1,\ldots,t_n) \in G \text{ so that } |z_i| < |t_i| \text{ for } i=1,\ldots,n\},$$

and

$$h_g|_G = \hat{g}\circ\hat{F}^{-1}.$$

Note that $\tilde{G}$ is a complete Reinhardt domain.

Now choose an arbitrary element $g \in \operatorname{rad} A$, i.e. $\hat{g} \equiv 0$ by (8.1.2). Let $g = \sum a_\nu f^\nu$. Then $h_g = 0$ by the identity theorem (2.1.6). Hence $a_\nu=0$ for all $\nu \in \mathbb{N}_o^n$ and $g = 0$. Thus A is semisimple.

Let $U \subset M(A)$ be an open subset, and let $\varphi \in U$ be an arbitrary element. Set $z_o = \hat{F}(\varphi)$. Choose $g_1,\ldots,g_r \in A$ such that

$$W = \{\psi \in M(A)\text{: } |\hat{g}_i(\psi)-\hat{g}_i(\varphi)| < 1,\ i=1,\ldots,r\} \subset U,$$

cf. (3.2.2)i). Then

$$\hat{F}(W) = \{z \in G\text{: } |\hat{g}_i\circ\hat{F}^{-1}(z)-\hat{g}_i\circ\hat{F}^{-1}(z_o)| < 1,\ i=1,\ldots,r\}.$$

Hence $\hat{F}(W)$ is an open subset of G, since $\hat{g}_i\circ\hat{F}^{-1} \in \operatorname{Hol}(G)$ for $i=1,\ldots,r$. So $\hat{F}$ is an open mapping, hence a homeomorphism.

It follows from (12.1.1) that G is polynomially convex and that A is topologically and algebraically isomorphic to Hol(G).

We see from the proof of (12.1.1) that the map

$$A \longrightarrow \operatorname{Hol}(G),\ g \longrightarrow \hat{g}\circ\hat{F}^{-1},$$

defines a topological algebra isomorphism. By the proof above each element $\hat{g}\circ\hat{F}^{-1}$ of Hol(G) can be extended to a holomorphic function

h_g on $\tilde{G}$. Since G is polynomially convex this implies that $G = \tilde{G}$, hence G is a logarithmically convex complete Reinhardt domain (cf. example (2.2.7)ii)). □

(12.5.2) Remarks. i) There are B-algebras A with a basis generated by n elements (see [S/W]). In this case

$$G = \{(\hat{f}_1(\varphi),\dots,\hat{f}_n(\varphi)) \in \mathbb{C}^n : \varphi \in M(A)\}$$

is of course not an open subset. Hence we cannot drop the hypothesis "G is open" in (12.5.1).

ii) Let $G \subset \mathbb{C}^n$ be a logarithmically convex complete Reinhardt domain. The coordinate functions $z_1,\dots,z_n$ generate a basis for Hol(G) by example (2.2.7)ii). Since G is a domain of holomorphy we can naturally identify M(Hol(G)) and G by (2.2.7). Hence

$$\{(\hat{z}_1(\varphi),\dots,\hat{z}_n(\varphi)) : \varphi \in M(\mathrm{Hol}(G))\} = G$$

is an open subset of $\mathbb{C}^n$. So (12.5.1) is again a characterization theorem.

CHAPTER 13

LIOUVILLE ALGEBRAS

(13.1) Liouville algebras

Liouville's theorem for entire functions states that each bounded function $f \in \mathrm{Hol}(\mathbb{C})$ is constant. Since we can naturally identify $M(\mathrm{Hol}(\mathbb{C}))$ and $\mathbb{C}$ (cf. (4.1.7)ii)) resp. $\mathrm{Hol}(\mathbb{C})$ and $\mathrm{Hol}(\mathbb{C})^{\wedge}$, the algebra of all Gelfand transforms, we can rephrase Liouville's theorem in the following way: If $f \in \mathrm{Hol}(\mathbb{C})$ is an element so that $\hat{f}(M(\mathrm{Hol}(\mathbb{C}))$ is a bounded subset of $\mathbb{C}$, then f must be a scalar multiple of $1 \in \mathrm{Hol}(\mathbb{C})$. The last formulation motivates the following definition.

(13.1.1) Definition. An F-algebra A is called Liouville algebra if $\hat{f}(M(A))$ is an unbounded subset of $\mathbb{C}$ for each element $f \in A\backslash\{c\cdot 1 : c \in \mathbb{C}\}$.

(13.1.2) Remarks. i) Note that $C(\{x\})$, the algebra of all (continuous) complex-valued functions on a point x is a Liouville algebra.
ii) A nontrivial Liouville algebra was constructed in example (9.2.2).
iii) Each Liouville algebra is automatically semisimple, i.e. the map $A \longrightarrow \hat{A}$, $f \longrightarrow \hat{f}$, is injective (cf. (8.1.2)). Otherwise there would be an element $g \neq 0$ in A such that $\hat{g}(M(A)) = \{0\}$, but $g \in A\backslash\{c\cdot 1 : c \in \mathbb{C}\}$.
iv) Using Shilov's idempotent theorem (6.1.5) one sees that the spectrum of each Liouville algebra is connected.
v) Let A be a Liouville algebra, $f \in A\backslash\{c\cdot 1 : c \in \mathbb{C}\}$.
Suppose that $\mathbb{C}\backslash\hat{f}(M(A))$ contains an open disc $\{z : |z-z_o| < \varepsilon\}$.
Then $f-z_o\cdot 1$ is invertible by (6.1.4). We have $1/(f-z_o\cdot 1) \in A\backslash\{c\cdot 1 : c \in \mathbb{C}\}$ since the same is true for f, but

$$(1/(f-z_o\cdot 1))^\wedge (M(A)) \subset \{z\colon |z| \le 1/\varepsilon\},$$

where $(1/(f-z_o\cdot 1))^\wedge$ denotes the Gelfand transform of $1/(f-z_o\cdot 1)$, and we get a contradiction.
It follows that $\hat{f}(M(A))$ is a dense subset of $\mathbb{C}$.

<u>(13.1.3)</u> We can give a more detailed description of the spectrum of an unbounded element of a Liouville algebra.

<u>Proposition [DAL].</u> Let A be a Liouville algebra, $f \in A\backslash\{c\cdot 1\colon c \in \mathbb{C}\}$. Then $\mathbb{C}\backslash\hat{f}(M(A))$ contains no closed connected subsets other than single points.

<u>Proof.</u> Let K be a closed connected subset of $\mathbb{C}$ which is contained in $\mathbb{C}\backslash\hat{f}(M(A))$.
i) Assume that K is unbounded.
Consider K as a subset of S, the extended plane, and set $U = S \setminus (K \cup \{\infty\})$. U is an open subset of the plane. U is connected, since $\hat{f}(M(A))$ is a dense subset of U by (13.1.2)v) and since $\hat{f}(M(A))$ is connected by (13.1.2)iv).
$S\backslash U = K \cup \{\infty\}$ is connected, hence U is a simply connected domain of the plane (cf. [NAR 1],p. 151). Since $K \neq \emptyset$ we have $U \neq \mathbb{C}$, hence there exists a biholomorphic map $g \in \mathrm{Hol}(U)$ which maps U onto the open unit disc D by the Riemann mapping theorem. Denote by ϑ the continuous homomorphism from Hol(U) to A such that $\vartheta(z) = f$ and $\vartheta(h)^\wedge = h\circ\hat{f}$ for all $h \in \mathrm{Hol}(U)$, cf. (6.1.2). Then $\vartheta(g) \in A$,

$$\vartheta(g)^\wedge (M(A)) = g\circ\hat{f}(M(A)) \subset D,$$

i.e. $\vartheta(g)^\wedge (M(A))$ is bounded but $\vartheta(g) \in A\backslash\{c\cdot 1\colon c \in \mathbb{C}\}$, since $\vartheta(g)^\wedge (M(A))$ is a dense subset of D. Hence we get a contradiction.
ii) Assume that K is bounded.
Let $x \in K$. Consider the map

$$h\colon S \longrightarrow S,\ z \longrightarrow 1/(z-x).$$

Then h is a homeomorphism, thus h(K) is a connected compact subset of S. Note that $h(K)\setminus\{\infty\}$ is a closed subset of $\mathbb{C}$. Suppose that there is $y \in K\setminus\{x\}$. Let L be the component of $h(K)\setminus\{\infty\}$ which contains h(y). Then L is unbounded, since otherwise L would be a component of h(K) such that $L \neq h(K)$.
Now we get a contradiction as in i), replacing f by $1/(f-x\cdot 1)$, and K by L.

□

<u>(13.1.4)</u> Let $G \subset \mathbb{C}$ be a polynomially convex domain, i.e. G is a simply connected domain by (1.2.4)ii). If $G \neq \mathbb{C}$ then there exists a biholomorphic map from G onto the unit disc D by the Riemann mapping theorem. This shows in particular that the Liouville property characterizes $\mathrm{Hol}(\mathbb{C})$ among the algebras Hol(G), G a polynomially convex domain of the plane.
These algebras are singly generated nontrivial uF-algebras and one is inclined to ask whether $\mathrm{Hol}(\mathbb{C})$ is characterized by the Liouville property among the singly generated nontrivial uF-algebras. The next example shows that this is not the case.

<u>(13.1.5) Example [B/L].</u> We construct a singly generated nontrivial Liouville algebra A, which is not isomorphic to $\mathrm{Hol}(\mathbb{C})$.
For $z \in \mathbb{C}$ denote by Re z resp. Im z the real resp. imaginary part of z. For $n \in \mathbb{N}$ set

$$L_n = \{z \in \mathbb{C}: -n \leq \mathrm{Re}\, z \leq n,\ -n \leq \mathrm{Im}\, z \leq n\},$$

$$E_n = \{z \in \mathbb{C}: \mathrm{Im}\, z > 0,\ 1/(n+1) < \mathrm{Re}\, z < 1/n\}$$

and

$$K_n = L_n \setminus \bigcup_{k \geq n} E_k.$$

Each K_n is a polynomially convex compact set, since $\mathbb{C}\setminus K_n$ is connected, cf. (1.2.4)ii). Set $A = \varprojlim P(K_n)$, then A is a singly generated uF-algebra.

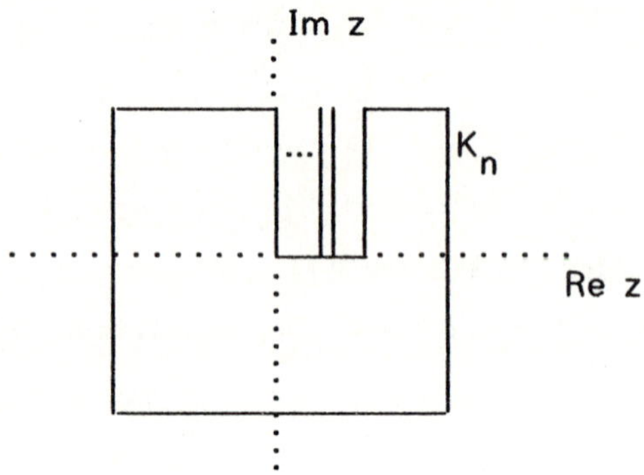

By (5.1.6) $M(A) = \bigcup K_n = \mathbb{C}$ as sets and $(K_n)_n$ is an admissible exhaustion of M(A) with respect to the Gelfand topology. Note that $(K_n)_n$ is not an admissible exhaustion of $\mathbb{C}$ with respect to the euclidean topology, since for example the set $\{z: 0 \le \text{Re } z \le 1, \text{ Im } z=1\}$ is not contained in any K_n. In the same way as in example (5.1.12) we can show that M(A) is not locally compact. More precisely, no point of the set $\{z: 0=\text{Re } z, \text{ Im } z \ge 0\}$ has a compact neighbourhood (Gelfand topology). Hence A is not a Stein algebra (cf. (11.1.1)), in particular A is not isomorphic to $\text{Hol}(\mathbb{C})$.

Claim. A is a Liouville algebra.

Let $f \in A$ be an arbitrary element. Interpret $\hat{f}$ as a complex-valued function on $\mathbb{C}$ which can be approximated uniformly on every K_n by polynomials, cf. (5.1.7)i). Then $\hat{f}|_{\mathbb{C}\setminus\{\text{Im } z \ge 0, \text{Re } z=0\}}$ is holomorphic, since every point of $\mathbb{C}\setminus\{\text{Im } z\ge 0, \text{Re } z=0\}$ is an interior point (euclidean topology) of some K_n.

Now let $\hat{f} \in \hat{A}$ be an element so that $\hat{f}(\mathbb{C})$ is bounded. We shall show that $\hat{f}$ under this assumption is continuous on $\mathbb{C}$ with respect to the euclidean topology. Since a continuous function on $\mathbb{C}$, which is holomorphic on $\mathbb{C}\setminus\{\text{Im } z\ge 0, \text{ Re } z=0\}$, is in fact holomorphic on $\mathbb{C}$, our assertion will follow from Liouville's theorem.

It remains to prove that $\hat{f}$ is continuous, i.e. we must show that $\hat{f}$ is continuous at every point $x \in \{\text{Im } z\ge 0, \text{ Re } z=0\}$.

Denote by D the open and by $\overline{D}$ the closed unit disc. Let $(x_n)_n$ be an arbitrary sequence in $\{\mathrm{Re}\, z>0\}$ which converges to $x \in \{\mathrm{Im}\, z\geq 0, \mathrm{Re}\, z=0\}$. Choose $s > \max\{1, |x|\}$, and set

$$H = \{|z| \leq s,\ \mathrm{Re}\, z \geq 0\}.$$

Choose a homeomorphism

$$g: \overline{D} \longrightarrow H$$

such that $g|_D$ is biholomorphic. (This choice is possible by [B/S], p. 364.)

By our hypothesis $\hat{f}\circ g|_D$ is a bounded holomorphic function on D, so for almost all $\vartheta \in [0,2\pi]$ the radial limit

$$\hat{f}\circ g|_D^*(\exp(i\vartheta)) = \lim_{r\to 1} \hat{f}\circ g|_D(r\exp(i\vartheta))$$

exists (see appendix A.12). By our construction we find for each $\vartheta \in [0,2\pi]$ a sequence $(y_k)_k$ in some suitable K_m so that

$$g^{-1}(y_k) = r_k\exp(i\vartheta),\ r_k \to 1.$$

Since $\hat{f}|_{K_m}$ is continuous it follows that

$$\hat{f}\circ g|_D^*(\exp(i\vartheta)) = \hat{f}\circ g(\exp(i\vartheta))$$

for almost all $\vartheta \in [0,2\pi]$, hence for $r<1$

$$\hat{f}\circ g|_D(r\exp(i\vartheta)) = 1/2\pi \int_0^{2\pi} P(r\exp(i(\vartheta-t)))\hat{f}\circ g(\exp(it))dt$$

by A.13. Since $\hat{f}\circ g|_{\{|z|=1\}}$ is continuous with the only possible exception at $g^{-1}(is)$ we see by A.11 that

$$\lim_{z\to\exp(it)} \hat{f}\circ g|_D(z) = \hat{f}\circ g(\exp(it)),$$

if $\exp(it) \neq g^{-1}(is)$. Consequently

$$\lim_{n\to\infty} \hat{f}(x_n) = \hat{f}(x).$$

□

<u>(13.1.6) Remarks.</u> i) Let A be as in the example, and let $f \in \mathrm{Hol}(\mathbb{C})$ be an arbitrary element. Then $f|_{K_n} \in P(K_n)$ by Mergelyan's theorem (1.3.13). It follows that the map

$$T: Hol(\mathbb{C}) \longrightarrow A,\ f \longrightarrow (f|_{K_n})_{n\in\mathbb{N}}$$

defines an injective continuous algebra homomorphism. Moreover T has dense range.

ii) Daly [DAL] - modifying the example above - constructed a singly generated Liouville algebra A with generating element f, such that $\hat{f}(M(A)) = \mathbb{C}\setminus\{0\}$.

(13.1.7) We now give a characterization of Hol($\mathbb{C}$), combining Brooke's result (12.3.2) and the Liouville property.

Theorem [BIR 1]. Let A be a singly generated Liouville algebra with generating element f. Suppose that the Shilov boundary of A is empty. Then A is topologically and algebraically isomorphic to Hol($\mathbb{C}$).

Proof. A is semisimple by (13.1.2)iii), hence all assumptions of theorem (12.3.2) are satisfied. By the proof of (12.3.2) $\hat{f}$ is a homeomorphism from M(A) onto an open subset G of the plane, and A is topologically and algebraically isomorphic to Hol(G). Note that G is a domain by (13.1.2)iv).
By (12.1.1) G is a polynomially convex domain, hence G = $\mathbb{C}$ as remarked in (13.1.4).

□

(13.1.8) The situation for n-generated F-algebras is much more complicated. We first prove that (13.1.7) cannot be extended to the case $n>1$.
We give an example of a doubly generated Liouville algebra A with locally compact spectrum, which has an empty Shilov boundary but which is not a Stein algebra. Since this example is a slight modification of example (9.2.2) we omit most of the proofs of the assertions.
Define the following sets in $\mathbb{C}^2$

$X_1 = \{1\}\times\mathbb{C}$, $X_2 = \mathbb{C}\times\{0\}$,

$X_n = \{1-1/n\}\times\mathbb{C}$ for $n>2$,

$X = \bigcup_{n\geq 1} X_n$,

$K_l = X \cap \{|z|\leq l,\ |w|\leq l\}$ for $l \in \mathbb{N}$.

Each K_l is a polynomially convex compact set. So $A = \varprojlim P(K_n)$ is a doubly generated uF-algebra. We have $M(A) = X$, Gelfand and euclidean topology coincide on X, in particular M(A) is a locally compact space. A is a Liouville algebra and for each $\hat{f} \in \hat{A}$ and each $n\geq 1$ we can interpret $\hat{f}|_{X_n}$ as a holomorphic function on the plane. It follows from this that $\gamma A = \emptyset$.

Claim. A is not a Stein algebra.

Consider the sequence $(z_1^n \cdot z_2)_n$ in $\hat{A}$. Then

$$\|z_1^n \cdot z_2\|_{K_l} \leq l \text{ for all } l,\ n \in \mathbb{N},$$

i.e. $(z_1^n \cdot z_2)_n$ is a bounded sequence in $\hat{A}$. Each subsequence converges pointwise to the function

$$g(x,y) = \begin{cases} 0 & \text{if } (x,y) \in X\setminus X_1 \\ y & \text{if } (x,y) \in X_1 \end{cases}$$

It follows that g is not an element of A, since g is not continuous with respect to the euclidean topology. Hence A is not a Montel space (cf. (2.1.14)), thus A is not a Stein algebra by the remarks in (2.3.1).

So Liouville's property and the maximum principle do not characterize $Hol(\mathbb{C}^n)$ among the n-generated uF-algebras, in fact these properties do not characterize $Hol(\mathbb{C}^n)$ among the algebras Hol(G), G a polynomially convex domain in $\mathbb{C}^n$. Diederich and Sibony [D/S] constructed polynomially convex domains $G \subset \mathbb{C}^2$ such that Hol(G) is a Liouville algebra, but G is not biholomorphic equivalent to $\mathbb{C}^2$, although G is diffeomorphic to $\mathbb{R}^4$. Consequently Hol(G) is not isomorphic to $Hol(\mathbb{C}^2)$, cf. (2.3.5).

CHAPTER 14

MAXIMUM MODULUS PRINCIPLE

It was believed for some time that an analytic type property like Rossi's maximum principle for B-algebras (1.4.9) is based on the existence of analytic structure in the spectrum, i.e. it was conjectured that these properties only occur if parts of the spectrum can be given the structure of a complex space such that all Gelfand transforms become holomorphic functions on it. In general this turned out to be false (cf. [STO] for examples of Stolzenberg and Garnett, and the example of Wermer in (14.1.3)v)).
In this section we consider the maximum principle in the context of uF-algebras. As already mentioned it is in general not sufficient for a characterization of the algebra of holomorphic functions on a suitable complex space. In special cases however the maximum principle helps to introduce analytic structure in the spectrum of a uF-algebra. A key observation is a subharmonicity theorem, due to Wermer and Senichkin, which brought potential theory to bear on the study of uF-algebras.

(14.1) Maximum modulus algebras

(14.1.1) Definition. A uF-algebra (A,X) is called maximum modulus algebra (m.m.a.) provided that

i) there exists no nonempty compact subset $K \subset X$ such that $\partial K = \emptyset$;

ii) $\|f\|_K = \|f\|_{\partial K}$ for each compact set $K \subset X$ and each $f \in A$.

(We denote the boundary of a set K in a space X by ∂K.)

(14.1.2) Remarks. i) Assuming i) of (14.1.1) we avoid the difficulty of defining $\|f\|_\emptyset$ for $f \in A$. Note in particular that if X is a compact space, i.e. if (A,X) is a uB-algebra, then (A,X) is not a m.m.a..
ii) The maximum modulus property is not preserved under topological algebra isomorphisms. For example set

$$X = \{0\} \cup \{z \in \mathbb{C}: |z| \geq 1\} \text{ and } A = \mathrm{Hol}(\mathbb{C})|_X.$$

Then (A,X) is a uF-algebra which is isomorphic to the m.m.a. $(\mathrm{Hol}(\mathbb{C}),\mathbb{C})$, but which is not a m.m.a., since $K= \{0\}$ is a compact subset of X with empty boundary.
iii) The term "maximum modulus algebra" is not only used for uF-algebras. Some authors call (A,X) a maximum modulus algebra if X is a Hausdorff space and A is a pointseparating subalgebra of C(X)- not necessarily complete in the compact open topology - which contains the constants and satisfies i) and ii) of definition (14.1.1). Most of the results will be valid for these algebras, which we shall denote by m.m.a.(*).

(14.1.3) Examples. i) Let X be a reduced holomorphically separable complex space with a countable topology, then (Hol(X),X) is a m.m.a. (cf. (2.1.9) and remark (2.3.1)).
ii) For each $n \in \mathbb{N}$ set

$$K_n = \{(z,w) \in \mathbb{C}^2: |z| \leq n, \mathrm{Im}\, z = 0, |w| \leq n\}.$$

Then K_n is polynomially convex, since it is the product of the interval $[-n,n]$ and the closed disc with radius n. Moreover $(K_n)_n$ is an admissible exhaustion of $\bigcup K_n$ with respect to the euclidean topology. Set $A = \varprojlim P(K_n)$, then

$$M(A) = \bigcup K_n = \{(z,w): \mathrm{Im}\, z = 0\}$$

and the Gelfand topology and the euclidean topology are equivalent on $\bigcup K_n$ by (5.1.7)ii). Let $f \in A$ be an arbitrary element. As in (5.1.7)i) interpret each $\hat{f} \in \hat{A}$ as a complex-valued function on $\{(z,w): \mathrm{Im}\, z=0\}$

which can be approximated uniformly on every K_n by polynomials. Then

$$f_t: \mathbb{C} \longrightarrow \mathbb{C},\ w \longrightarrow \hat{f}(t,w)$$

is a holomorphic function for every $t \in \mathbb{R}$. Hence $(\hat{A},M(A))$ is a m.m.a..

iii) Let A be an F-algebra and denote by γA the Shilov boundary of A (cf. (9.1)). Then $(\hat{A},M(A))$ is a m.m.a.(*) by (9.1.7)i) and (9.1.6) if $\gamma A \neq \emptyset$.

iv) It follows from Rossi's maximum principle (1.4.9) that $(\hat{B},M(B)\setminus\gamma B)$ is a m.m.a.(*) for each B-algebra B.

v) We shortly sketch an example due to Wermer [WER 1]. We shall use it to show that there exists a m.m.a. (A,X) such that X does not admit analytic structure, in fact X does not even contain an analytic disc. (We say that X contains an analytic disc if there are a disc D in the plane and a nonconstant continuous function $f: D \longrightarrow X$ so that $g \circ f \in Hol(D)$ for all $g \in A$.)

We denote by $a_1,a_2,\ldots$ the points in the disc $\{|z| < 1/2\}$ with rational real and imaginary part. For each $j \in \mathbb{N}$ denote by B_j the algebraic function

$$B_j(z) = (z-a_1)(z-a_2)\ldots(z-a_{j-1})\sqrt{(z-a_j)}$$

and set

$$g_j(z) = \sum_{k=1}^{j} c_k B_k(z),$$

where $c_1,\ldots,c_j$ are positive constants. Denote by $\Sigma(c_1,\ldots,c_n)$ the subset of the Riemann surface of g_n which lies in $\{|x|\leq 1/2\}$, i.e.

$$\Sigma(c_1,\ldots,c_n) = \{(z,w): |z|\leq 1/2,\ w=w_j^{(n)}(z),\ j=1,\ldots,2^n\}$$

where $w_j^{(n)}(z)$, $j=1,\ldots,2^n$ are the values of g_n at z.

For a sequence of positive numbers $(c_n)_n$ define $X_{(c_n)}$ to be the set of all points $(z,w) \in \mathbb{C}^2$ such that

i) $|z| \leq 1/2$ and

ii) there is a sequence $(z,w_n) \in \Sigma(c_1,\ldots c_n)$ such that

$$w_n \longrightarrow w \text{ as } n \longrightarrow \infty.$$

It was proved by Wermer that there exists a sequence $(c_n)_n$ of positive constants so that

α) $X = X_{(c_n)}$ is a polynomially convex compact set in $\mathbb{C}^2$ which contains no analytic disc,

β) X is the polynomially convex hull of the compact set $Y = X \cap \{(z,w): |z| = 1/2\}$,

γ) $z_1(X) = \{|z| \le 1/2\}$, where z_1 denotes the first coordinate function.

For later use we remark that

δ) $\Sigma(c_1,\dots,c_n) \cap (\{a_1\} \times \mathbb{C}) = \{(a_1,0)\}$ for all $n \in \mathbb{N}$, hence $X \cap (\{a_1\} \times \mathbb{C}) = \{(a_1,0)\}$ by the construction.

We now use Wermer's example to construct a m.m.a.. For $n>2$ set

$$K_n = X \cap \{(z,w): |z| \le (1/2)-(1/n)\}.$$

Each K_n is a polynomially convex compact subset of $\mathbb{C}^2$, since the same is true for X. Note that K_n is not empty by γ).

Set $A = \varprojlim P(K_n)$. Then $(\hat{A}, M(A))$ is a uF-algebra. Since $(K_n)_n$ is an admissible exhaustion of $\bigcup K_n = X \setminus Y$ with respect to the euclidean topology, we see that $M(A) = X \setminus Y$ and that Gelfand and euclidean topology coincide on $X \setminus Y$ (cf. (5.1.7)ii)).

We claim that $(\hat{A}, M(A))$ is a m.m.a..

Suppose that there exists a compact subset $K \subset (X \setminus Y)$ so that the boundary of K (relative to $X \setminus Y$) is empty. Then the same is true for the boundary of K relative to X, i.e. K is a closed and open subset of X. Since $M(P(X)) = X$ and $P(X) = P(X)^\wedge$ by (1.2.4)ii), where $P(X)^\wedge$ denotes the algebra of all Gelfand transforms, it follows from Shilov's idempotent theorem (1.4.6) that there is $f \in P(X)$ so that $f|_K = 1$ and $f|_{X \setminus K} = 0$. If we approximate f sufficiently close by a polynomial p we get the inequalities

$$\|p\|_X \ge \|p\|_K > \|p\|_{X \setminus K} \ge \|p\|_Y,$$

a contraction to β). In fact β) implies that the Shilov boundary $\gamma P(X)$ is contained in Y. Hence we have

(*) $\quad \|p\|_K = \|p\|_{\partial K}$

for all polynomials p and all compact subsets $K \subset (X\backslash Y)$ by Rossi's maximum modulus principle (1.4.9), since $X\backslash Y \subset M(P(X))\backslash \gamma P(X)$. Since the polynomials lie dense in $\hat{A}$ the equality (*) remains valid for all elements $\hat{f} \in \hat{A}$ and our claim is proved.

Remark. The first example of a compact subset $K \subset \mathbb{C}^2$ such that K is a proper subset of the polynomially convex hull $\hat{K}$ and $\hat{K}$ contains no analytic disc was given by Stolzenberg (cf. [STO]). In the same way as above we could have used this example to construct a m.m.a. (A,X) such that X contains no analytic disc. In some sense Wermer's construction is more explicit and allows concrete calculations (cf. [LEV], [GOL 4]). We shall consider it later in (14.2.5) under a different aspect.

(14.1.4) In this point we prove that elements of maximum modulus algebras are open mappings under rather mild assumptions.

Lemma. Let (A,X) be a m.m.a.(*). Then

i) $\partial f(K) \subset f(\partial K)$ for every $f \in A$ and every compact subset $K \subset X$;

ii) $f \in A$ is an open mapping if X is locally compact and if $f^{-1}(f(x))$ is a discrete subset of X for each $x \in X$.

Proof. Set $D_{y,\varepsilon} = \{z : |z-y| < \varepsilon\}$ and $\overline{D}_{y,\varepsilon} = \{z : |z-y| \le \varepsilon\}$, $\varepsilon>0$, $y \in \mathbb{C}$.

i) Let $f \in A$, and let $K \subset X$ be a compact subset. Assume there is

$$y \in \partial f(K)\backslash f(\partial K).$$

Since $f(\partial K)$ is compact there exists $\varepsilon>0$ such that

$$\overline{D}_{y,\varepsilon} \cap f(\partial K) = \emptyset.$$

For each positive constant $\delta<\varepsilon$ we have

$$\partial D_{y,\delta} \cap f(K) \neq \emptyset$$

otherwise

$$W = f^{-1}(\overline{D}_{y,\delta}) \cap K = f^{-1}(D_{y,\delta}) \cap K = f^{-1}(D_{y,\delta}) \cap \text{int } K,$$

where int K denotes the interior of K, would be a nonempty compact subset of X with $\partial W = \emptyset$. This cannot happen by the definition of a m.m.a.(*).

Choose a positive constant $\delta < \varepsilon$ such that

(*) $$\partial D_{y,\delta} \cap f(K) \neq \partial D_{y,\delta} .$$

Such a δ exists, since otherwise

$$D_{y,\varepsilon} \subset f(K),$$

a contradiction to $y \in \partial f(K)$. It follows from (*) that

$$\partial D_{y,\delta} \cap f(K)$$

is a nonempty polynomially convex compact subset of the plane, cf. (1.2.4)ii). Hence there is a polynomial p so that

$$|p(y)| > \|p\|_{\partial D_{y,\delta} \cap f(K)}.$$

Clearly $p \circ f \in A$. Choose a point $x \in f^{-1}(y) \cap K$. Then

$$S = f^{-1}(\overline{D}_{y,\delta}) \cap K$$

is a compact neighbourhood of x in X, since $x \in \text{int } K$, and

$$\partial S \subset f^{-1}(\partial D_{y,\delta} \cap f(K)),$$

but

$$|p \circ f(x)| > \|p \circ f\|_{\partial S}$$

by construction. This is a contradiction to the definition of a m.m.a.(*).

ii) Let $U \subset X$ be an arbitrary open subset. Let $x \in U$. Choose a compact neighbourhood K of x such that $K \subset U$ and

$$K \cap f^{-1}(f(x)) = \{x\}.$$

Then

$$f(x) \in f(K) \setminus f(\partial K),$$

so

$$f(x) \in f(K) \setminus \partial f(K)$$

by i), i.e. $f(x) \in \operatorname{int} f(K)$ resp. $f(x) \in \operatorname{int} f(U)$. □

Remarks. i) Part i) - in the case that X is locally compact - was first proved by Järvi [JÄR].

ii) In general nonconstant functions in a m.m.a. may not be open mappings. Consider for instance example (14.1.3)ii). The function

$$\hat{z}_1 : M(A) = \{(z,w) : \operatorname{Im} z = 0\} \longrightarrow \mathbb{C},\ (z,w) \longrightarrow z,$$

defines an element of $\hat{A}$, but $\hat{z}_1(M(A)) = \mathbb{R}$.

(14.2) Maximum algebras and subharmonicity

(14.2.1) We now prove a subharmonicity theorem due to Wermer [WER 2] and Senichkin [SEN 1] which is basic for many of the following results. We first recall two definitions.

Let X be a topological space and let $u: X \longrightarrow \mathbb{R} \cup \{-\infty\}$ be a mapping. Then u is called upper semi-continuous if for any $\lambda \in \mathbb{R}$ the set $\{x \in X : u(x) < \lambda\}$ is open in X.

Let X,Y be topological spaces and let $g: X \longrightarrow Y$ be a continuous map. Then g is called proper provided that $g^{-1}(K)$ is a compact subset of X for any compact subset $K \subset Y$.

Proposition. Let X, Y be locally compact spaces, and let f be a proper map from X onto Y. Then

$$Z_h : Y \longrightarrow \mathbb{R},\ y \longrightarrow \|h\|_{f^{-1}(y)},$$

is an upper semi-continuous function on Y for every continuous function $h: X \longrightarrow \mathbb{C}$.

(Note that $f^{-1}(y)$ is a compact subset of Y.)

Proof. Fix $\lambda \in \mathbb{R}$. Then

$$\{y \in Y: Z_h(y) < \lambda\} = Y\backslash\{Z_h(y) \geq \lambda\} = Y\backslash f(|h|^{-1}\{t\geq\lambda\}).$$

Since h is continuous the set $|h|^{-1}\{t\geq\lambda\}$ is closed. Since proper mappings between locally compact spaces are closed mappings, the set $f(|h|^{-1}\{t\geq\lambda\})$ is closed, too. So $\{y \in Y: Z_h(y) < \lambda\}$ is an open subset of Y.

□

(14.2.2) For the definition and properties of subharmonic functions cf. appendix A.

Theorem [WER 3]. Let X be a Hausdorff space and let A be a point-separating subalgebra of C(X) - not necessarily a uF-algebra - which contains the constants. Suppose that there are an open set $U \subset \mathbb{C}$ and $f \in A$ so that

i) $f: X \longrightarrow U$ is a proper surjective mapping, and

ii) for each closed disc $D(\lambda_o) \subset U$ with center λ_o and for each $x_o \in f^{-1}(\lambda_o)$ we have
$|g(x_o)| \leq \|g\|_{f^{-1}(\partial D(\lambda_o))}$ for all $g \in A$.

Then

$$\log Z_g: U \longrightarrow \mathbb{R} \cup \{-\infty\},\ \lambda \longrightarrow \log \|g\|_{f^{-1}(\lambda)},$$

is a subharmonic function for each $g \in A$.
(Note that $f^{-1}(\partial D(\lambda_o))$ and $f^{-1}(\lambda)$ are compact subsets of X by i).)

Proof. α) By i) X is a locally compact space. Hence

$$Z_g: U \longrightarrow \mathbb{R},\ \lambda \longrightarrow \|g\|_{f^{-1}(\lambda)},$$

is upper semi-continuous by (14.2.1) and the same is true for $\log Z_g$.
β) Let $D \subset U$ be a closed disc centered at λ_o. Fix $x_o \in f^{-1}(\lambda_o)$. Since $f^{-1}(D)$ is a compact subset of X we see that $\log Z_g$ is bounded from above on D, hence we can find a sequence of continuous functions $(u_l)_l$ on D such that $u_l \downarrow \log Z_g$, i.e. $u_1 \geq u_2 \geq \ldots \geq \log Z_g$ and

$u_n(x) \longrightarrow \log Z_g(x)$ as $n \to \infty$ for each $x \in D$, cf. appendix A.2. For each $k \in \mathbb{N}$ choose a polynomial p_k such that

$$|\mathrm{Re}(p_k)-u_k| < 1/k \text{ on } \partial D.$$

The choice of p_k is possible by A.5. Then

$$(*) \qquad \log Z_g \leq \mathrm{Re}(p_k) + 1/k \text{ on } \partial D.$$

Fix $k \in \mathbb{N}$. Let $(q_n)_n$ be a sequence of polynomials which converges uniformly on D to $\exp(-p_k-1/k)$. Since $g \cdot q_n(f) \in A$ there is $y_n \in f^{-1}(\partial D)$ so that

$$|g(x_o)q_n(f(x_o))| \leq |g(y_n)q_n(f(y_n))|$$

by hypothesis ii). Hence we find $x_k \in f^{-1}(\partial D)$ such that

$$|g(x_o)\exp(-(p_k(f)(x_o)+1/k))| \leq |g(x_k)\exp(-(p_k(f)(x_k)+1/k))|.$$

So

$$|g(x_o)|\exp(\mathrm{Re}(-(p_k(\lambda_o)+1/k)))$$
$$\leq Z_g(f(x_k))\exp(\mathrm{Re}(-(p_k(f)(x_k)+1/k))) \leq 1.$$

The last inequality follows from (*). Hence

$$|g(x_o)| \leq \exp(\mathrm{Re}(p_k(\lambda_o))+1/k)$$

and

$$\log Z_g(\lambda_o) \leq \mathrm{Re}(p_k(\lambda_o))+1/k = 1/2\pi \int_{\partial D} (\mathrm{Re}(p_k)(t)+1/k)dt$$
$$\leq 1/2\pi \int_{\partial D} (u_k(t)+2/k)dt.$$

For the equality cf. A.4. For $k \longrightarrow \infty$ we get

$$\log Z_g(\lambda_o) \leq 1/2\pi \int_{\partial D} (\log Z_g)(t)dt .$$

Since this inequality holds for each closed disc $D \subset U$, $\log Z_g$ is subharmonic on U by A.7.

□

(14.2.3)Corollary. Let (A,X) be a m.m.a.(*) resp. a m.m.a.. Suppose there are an open subset $U \subset \mathbb{C}$ and an element $f \in A$ so that $f: X \longrightarrow U$ is a surjective and proper mapping. Then

$$\log Z_g: U \longrightarrow \mathbb{R} \cup \{-\infty\},\ \lambda \longrightarrow \log \|g\|_{f^{-1}(\lambda)},$$

is a subharmonic function for each $g \in A$.

Proof. Let $D \subset U$ be a closed disc centered at λ_o, and let $x_o \in f^{-1}(\lambda_o)$. Then $f^{-1}(D) \subset X$ is a compact neighbourhood of x_o and $\partial f^{-1}(D) \subset f^{-1}(\partial D)$. Hence we have

$$|g(x_o)| \leq \|g\|_{\partial f^{-1}(D)} \leq \|g\|_{f^{-1}(\partial D)}$$

for every $g \in A$ by the definition of a m.m.a.(*), and our assertion follows from (14.2.2).

□

Remark. This theorem was also obtained by Senichkin [SEN 1].

(14.2.4) We give a first application of this theorem.

Definition. Let $U \subset \mathbb{C}$ be an open set. A subset $E \subset U$ is called polar set if there exists a subharmonic function $p \not\equiv -\infty$ on U such that

$$E \subset \{z \in U: p(z) = -\infty\}.$$

We remark that a polar set has Lebesgue measure zero (cf. A.14), in particular an open subset $W \subset U$ is not a polar set. We are now able to prove an identity theorem for maximum modulus algebras.

Theorem. Let A,X,f,U be as in theorem (14.2.2). Suppose that X is connected. Let $E \subset U$ be a nonpolar set and let $g \in A$ such that g vanishes on $f^{-1}(E)$. Then g vanishes on X.

Proof. First note that U is connected. By (14.2.2)

$$\log Z_g\colon U \longrightarrow \mathbb{R} \cup \{-\infty\},\ \lambda \longrightarrow \|g\|_{f^{-1}(\lambda)},$$

is a subharmonic function and $\log Z_g(\lambda) = -\infty$ for all $\lambda \in E$. Since E is a nonpolar set we see that $\log Z_g \equiv -\infty$ on U, i.e. $g \equiv 0$ on X. □

<u>(14.2.5) Corollary.</u> Let X,U,f,A be as in (14.2.4). Let $x \in U$. Suppose $g \in A$ vanishes in an open neighbourhood V of $f^{-1}(x)$, then g vanishes identically on X.

<u>Proof.</u> We show that there exists an open neighbourhood $W \subset U$ of x so that g vanishes on $f^{-1}(W)$. Then our assertion follows by (14.2.4).

Suppose W does not exist. Then there exists a sequence $(x_n)_n$ in $X\setminus V$ such that $f(x_n)$ converges to x. Since

$$f^{-1}(\{f(x_n)\colon n\in\mathbb{N}\} \cup \{x\})\setminus V$$

is a compact subset of X a subsequence of $(x_n)_n$ converges to a point $y \in X\setminus V$. But $f(y) = x$, and we get a contradiction. □

<u>Example.</u> Let A,X,Y be as in example (14.1.3)v). We remarked that (using the notations of (14.1.3)v)):

i) $M(A) = X\setminus Y$ contains no analytic disc,

ii) $\hat{z}_1(M(A)) = \{z \in \mathbb{C}\colon |z|<1/2\}$, where $\hat{z}_1 \in \hat{A}$ denotes the Gelfand transform of the first coordinate function,

iii) $M(A) \cap (\{a_1\}\times\mathbb{C}) = (X\setminus Y) \cap (\{a_1\}\times\mathbb{C}) = \{(a_1,0)\}$.

Let $K \subset \{|z|<1/2\}$ be an arbitrary compact subset. Then

$$\hat{z}_1^{-1}(K) \cap X = (K\times\mathbb{C}) \cap X$$

is a compact set. Since $\hat{z}_1^{-1}(K) \cap Y = \emptyset$ we see that $\hat{z}_1^{-1}(K)$ is a compact subset of M(A), i.e.

$$\hat{z}_1\colon M(A) \longrightarrow \{|z|<1/2\}$$

is a proper surjective mapping.

Let $\hat{g} \in \hat{A}$ vanish in an arbitrary open neighbourhood of $(a_1,0)$. Then -

applying (14.2.5) to $\hat{A}$, $\hat{z}_1$, M(A), $\{|z|<1/2\}$ - we see that $\hat{g}$ vanishes on M(A).

<u>Remark.</u> Using the same idea one can in fact show that the following is true:
If $\hat{f} \in \hat{A}$ vanishes on an arbitrary nonempty open subset of M(A) then $\hat{f}$ vanishes identically, cf. [GOL 4].
We remark a striking consequence for our "characterization program" for the algebra of holomorphic functions: Although the uF-algebra $(\hat{A}, M(A))$ has the maximum modulus property and satisfies the identity theorem, M(A) does not contain an analytic disc, in particular A is not a Stein algebra.

CHAPTER 15

MAXIMUM MODULUS ALGEBRAS AND ANALYTIC STRUCTURE

Using the maximum modulus property we give a characterization of Hol(X) in the case that X

- is a domain in $\mathbb{C}$,
- is a Riemann domain over $\mathbb{C}^n$.

In the second part of this chapter we consider algebras (A,X) which satisfy the maximum principle on certain subsets of X. We state a deep result on the existence of analytic structure in X for these algebras due to Wermer, Senichkin , Rusek, Kumagai and others.

(15.1) Riemann domains over $\mathbb{C}^n$

(15.1.1) Theorem. Let A be a semisimple rationally singly generated F-algebra with generating element f. Suppose that M(A) is locally compact and that $(\hat{A},M(A))$ is a m.m.a.(*). Then U = $\hat{f}$ (M(A)) is an open subset of the plane and A is topologically and algebraically isomorphic to Hol(U).

Proof. By (5.2.3) $\hat{f}$ is an injective and continuous mapping, by (14.1.4)ii) $\hat{f}$ is an open mapping, hence a homeomorphism and our assertion follows from (12.1.1).

□

Remarks. i) The assumption "locally compact spectrum" cannot be dropped in theorem (15.1.1). Consider for instance example (9.1.8). There we constructed a singly generated uF-algebra such that $(\hat{A},M(A))$ is a m.m.a. but M(A) is not locally compact, hence the conclusion of (15.1.1) does not hold, since M(Hol(U)) = U is locally compact.

ii) There is no analogous result for finitely generated maximum modulus algebras. Consider for instance example (14.1.3)v). In this example A is a doubly generated uF-algebra, M(A) is locally compact, $(\hat{A}, M(A))$ is a m.m.a., but M(A) does not contain an analytic disc. Hence A is not a Stein algebra.

(15.1.2) With regard to the last example we sharpen the maximum modulus assumption to characterize Hol(G), G a Riemann domain over $\mathbb{C}^n$.

Notation. Let (A,X) be a uF-algebra. Let $f_1,\dots,f_n \in A$, then set

$$F: X \longrightarrow \mathbb{C}^n,\ x \longrightarrow (f_1(x),\dots,f_n(x)).$$

Let L be an affine complex line in $\mathbb{C}^n$ such that $L \cap F(X) \neq \emptyset$. Then define $X_L = F^{-1}(L)$.

Theorem [GOL 2]. i) Let (A,X) be a uF-algebra on a locally compact and connected space X, and let $f_1,\dots,f_n \in A$ such that the map F is locally injective, i.e. for every $x \in X$ there exists a neighbourhood of x on which F is injective. Suppose that $(A|_{X_L}, X_L)$ is a m.m.a.(*) for all affine complex lines $L \subset \mathbb{C}^n$ which are parallel to a coordinate axis and for which $L \cap F(X) \neq \emptyset$.

Then F is an open mapping and X can be equipped with the uniquely determined structure of a Riemann domain over $\mathbb{C}^n$ such that $A \subset \mathrm{Hol}(X)$.

ii) If moreover X = M(A), then X is Stein and A = Hol(X).

(For the definition of Riemann domains and Stein spaces cf. (2.2.2) and (2.2.6).)

Proof. i) Let $x \in X$ be an arbitrary point, and let K be a compact neighbourhood of x such that F is injective on an open set U containing K.

Since ∂K is a compact set and $F|_K$ is an injective and continuous map, there is $r>0$ such that

$$(*) \qquad F^{-1}(D_{F(x),r}) \cap K \subset \text{int } K,$$

where $D_{F(x),r}$ denotes the open polycylinder centered at F(x) with polyradius r.

<u>Assertion.</u> $D_{F(x),r} \subset F(K)$.

We assume the contrary. Hence there is a complex line L parallel to a coordinate axis such that

$$\text{a) } L \cap (D_{F(x),r} \setminus F(K)) \neq \emptyset,$$

and

$$\text{b) } L \cap (D_{F(x),r} \cap F(K)) \neq \emptyset.$$

W.l.o.g. we can assume that $L = \{(z_1,\ldots,z_n): z_2=\ldots=z_n=0\}$. Note that $(A|_{X_L\cap U}, X_L\cap U)$ is a m.m.a.(*), $X_L\cap U$ is locally compact and $f_1|_{X_L\cap U}$ is injective. Hence

$$(**) \qquad f_1: X_L\cap U \longrightarrow \mathbb{C}$$

is an open mapping by (14.1.4)ii). Since X_L is closed in X

$$f_1(X_L\cap K) \subset \mathbb{C}$$

is a compact set and we may choose a boundary point y of $f_1(X_L\cap K)$ such that $|y|<r$ by a).

Denote by $\tilde{y}$ the uniquely determined point in $X_L\cap U$ so that $f_1(\tilde{y}) = y$. We have

$$\tilde{y} \in F^{-1}(D_{F(x),r})\cap K$$

thus $\tilde{y} \in \text{int } K$ by (*). This is a contradiction since

$$f_1(X_L\cap \text{int } K) \subset \mathbb{C}$$

is an open set by (**) and y is a boundary point of $f_1(X_L\cap K)$. So the assertion is proved and it follows that F is an open mapping. Since F is a local homeomorphism we get a uniquely determined structure

of a Riemann domain over $\mathbb{C}^n$ such that F is a holomorphic map. Now let $g \in A$ be an arbitrary element. We claim that $g \in Hol(X)$, X endowed with the above introduced complex structure.
Let $U \subset X$ be an open set such that $\overline{U}$ - the closure of U - is a compact subset of X, F(U) is an open polycylinder with polyradius $r>0$ and $F|_U: U \longrightarrow F(U)$ is a homeomorphism. W.l.o.g. $F(U) = D_{0,r}$. Let

$$L = \{(z_1,\ldots,z_n): z_2=\ldots=z_n=0\},$$

then

$$L \cap F(U) = D\times\{0\},$$

where $D = \{z \in \mathbb{C}: |z|<r\}$ and $0 \in \mathbb{C}^{n-1}$. By Wermer's subharmonicity theorem (14.2.3), applied to the m.m.a.(*) $(A|_{X_L\cap U}, X_L\cap U)$ and to $f_1|_{X_L\cap U} \in A|_{X_L\cap U}$, the function

$$D \longrightarrow \mathbb{R} \cup \{-\infty\},\ \lambda \longrightarrow \log|h\circ(f_1|_{X_L\cap U})^{-1}(\lambda)|,$$

is subharmonic for all $h \in A|_{X_L\cap U}$ (note that $f_1: X_L\cap U \longrightarrow D$ is a homeomorphism), in particular for $g|_{X_L\cap U}-a$, $a \in \mathbb{C}$.

Then - by a theorem of Hartogs (cf. A.8) - either

$$(*) \qquad h: D \longrightarrow \mathbb{C},\ \lambda \longrightarrow g\circ(f_1|_{X_L\cap U})^{-1}(\lambda)$$

or the conjugate of this function, is holomorphic. Repeating the arguments for the function $g\cdot f_1|_{X_L\cap U} \in A|_{X_L\cap U}$ we see that either

$$(**) \qquad D \longrightarrow \mathbb{C},\ \lambda \longrightarrow \lambda\cdot g\circ(f_1|_{X_L\cap U})^{-1}(\lambda)$$

or the conjugate of this function is holomorphic. We conclude easily from (*) and (**) that h is holomorphic.
Repeating these arguments for an arbitrary complex line L which is parallel to a coordinate axis and for which $L \cap F(U) \neq \emptyset$ we see that $g\circ F^{-1}|_{L\cap F(U)}$ is holomorphic. By Hartogs' theorem on separate analyticity (2.1.2)

$$g\circ F^{-1}|_{F(U)}: F(U) \longrightarrow \mathbb{C}$$

is a holomorphic map and thus $g \in Hol(X)$.

ii) We have $\hat{A} \subset Hol(X)$ by i). The holomorphically convex hull $\hat{K}_{Hol(X)}$ (cf. (2.2.6)) is compact for every compact subset $K \subset X$, since the $\hat{A}$-convex hull of K is already compact, cf. (4.3.3). Hol(X) separates the points of X, since $\hat{A}$ does. Hence X is a Stein manifold (cf. (2.2.6)). Moreover F forms a coordinate system at every point $z \in X$. Since $\hat{f}_1,\dots,\hat{f}_n \in \hat{A}$ we see that $\hat{A}$ is a dense subalgebra of Hol(X) by (2.3.7), thus $\hat{A} = Hol(X)$, because $\hat{A}$ is complete. □

Remarks. i) Theorem (15.1.2) is based on an idea of Rusek [RUS].

ii) Let (X,π) be a Riemann domain over $\mathbb{C}^n$. Then $\pi \in Hol(X)^n$, cf. (2.2.2). If L is a complex line in $\mathbb{C}^n$ such that $L \cap \pi(X) = \emptyset$, then X_L is an analytic subset of (X,π). It follows that $(Hol(X)|_{X_L}, X_L)$ is a m.m.a.(*), cf. (2.1.9) and remark (2.3.1). Thus theorem (15.1.2) gives a characterization of the algebra of all holomorphic functions on a Riemann domain over $\mathbb{C}^n$ within the class of uF-algebras.

(15.1.3) Corollary. i) Let $(\hat{A},M(A))$ be a rationally n-generated uF-algebra with locally compact spectrum M(A) and generating elements $f_1,\dots,f_n$ such that $(\hat{A}|_{M(A)_L}, M(A)_L)$ is a m.m.a.(*) for all affine complex lines $L \subset \mathbb{C}^n$ which are parallel to a coordinate axis and for which $L \cap \hat{F}(M(A)) \neq \emptyset$ $(\hat{F}=(\hat{f}_1,\dots,\hat{f}_n))$.

Then $\hat{F}(M(A))$ is a rationally convex open set in $\mathbb{C}^n$ and A is topologically and algebraically isomorphic to $Hol(\hat{F}(M(A)))$.

ii) If $(\hat{A},M(A))$ is n-generated by $f_1,\dots,f_n$ then - under the assumptions of part i) - $\hat{F}(M(A))$ is moreover a polynomially convex open set.

Proof. By (15.1.2) $\hat{F}$ is an open mapping, hence a homeomorphism by (5.1.4) resp. (5.2.3) and our assertion follows from (12.1.1). □

Example. Consider example (14.1.3)ii). In this example A is doubly

generated, M(A) is locally compact and $(\hat{A}|_{M(A)_L}, M(A)_L)$ is a m.m.a. for all affine complex lines which are parallel to the second coordinate axis. But A is not a Stein algebra.
Suppose the contrary. Then there exists a Stein space X and a topological algebra homomorphism $T: A \longrightarrow Hol(X)$. By remark (11.1.1) $\hat{z}_1 \circ T^*$ is a nonconstant holomorphic function on X, hence $\hat{z}_1 \circ T^*$ is an open subset of the plane. But

$$\hat{z}_1 \circ T^*(X) = \hat{z}_1(M(A)) = \mathbb{R}$$

by (14.1.3)ii).
In particular A is not isomorphic to the algebra of all holomorphic functions on a polynomially convex domain, although $\hat{A}$ is a m.m.a.(*) "on many lines".

(15.2) Maximum modulus algebras and finite mappings

Let (A,X) be a uB-algebra, $f \in A$ and W be a component of $\hat{f}(M(A)) \setminus f(X)$. Suppose that there exists a set of positive plane measure $G \subset W$ so that for all $\lambda \in G$ the set $\hat{f}^{-1}(\lambda) \subset M(A)$ is finite. Then - due to a classical result of Bishop (cf. [WER 4],p. 65) - each point $p \in \hat{f}^{-1}(W)$ has a neighbourhood in M(A) which is a finite union of analytic discs.
Basener [BAS] and independently Sibony [SIB], introducing the concept of higher Shilov boundaries, obtained a result on analytic structure of dimension $n>1$ in the spectrum of uB-algebras if certain functions have finite fibers.
Among others Wermer [WER 3], Senichkin [SEN 2], Rusek [RUS] and Kumagai [KUM 1] put these results in the setting of maximum modulus algebras. We follow the proof of Rusek and shall obtain the results of Basener and Sibony as a corollary in the next chapter.

(15.2.1) Definition. We say that a triple (X,F,Y) is an analytic cover with critical set S if

i) X is a locally compact Hausdorff space, $Y \subset \mathbb{C}^n$ is a domain, and F is a proper, discrete and continuous mapping of X onto Y;

ii) S is an negligible set in Y, i.e. S is nowhere dense and for every domain $D \subset Y$ and every $f \in \mathrm{Hol}(D\backslash S)$ which is locally bounded on D there exists a unique holomorphic extension $\tilde{f}$ to all of D;

iii) there is $k \in \mathbb{N}$ such that F is a k-sheeted covering map from $X\backslash F^{-1}(S)$ onto $Y\backslash S$, i.e. $F|_{X\backslash F^{-1}(S)}$ is a local homeomorphism and $\# F^{-1}(y) = k$ for all $y \in Y\backslash S$;

iv) $X\backslash F^{-1}(S)$ is dense in X.

(15.2.2) We shall deal with algebras of the following type:
Let X be a locally compact Hausdorff space, and let A be a subalgebra of $C(X)$ - not necessarily closed - which contains the constants and separates the points of X. Let $f_1,\ldots,f_n \in A$ such that

i) $F: X \longrightarrow W \subset \mathbb{C}^n$, $x \longrightarrow (f_1(x),\ldots,f_n(x))$

is a proper mapping onto a domain $W \subset \mathbb{C}^n$;

ii) $(A|_{X_L}, X_L)$ is a m.m.a.(*) for all affine complex lines for which $L \cap W \neq \emptyset$.

(X_L is defined as in (15.1.2), for the definition of a m.m.a.(*) cf. (14.1.2)iii).)

(15.2.3) Theorem [RUS]. Let A, X, F, W, as in (15.2.2). Suppose that there exists a nonpluripolar subset E of W such that $\# F^{-1}(\lambda)$ - the number of points in $F^{-1}(\lambda)$ - is finite for every $\lambda \in E$. Set for $k \in \mathbb{N}$

$$W_k = \{\lambda \in W: \# F^{-1}(\lambda) = k\}.$$

Then

i) there is $k \in \mathbb{N}$ such that $W = W_1 \cup \ldots \cup W_k$ and $W_k \neq \emptyset$,

ii) the set $S = W_1 \cup \ldots \cup W_{k-1}$ is a proper analytic subset of W,

iii) the triple (X,F,W) is an analytic cover with critical set S,

iv) there exists an analytic space structure of pure dimension n on X such that $A \subset Hol(X)$, if we endow X with this structure.

(For the definition of a pluripolar set cf. A.14, for the definition of the dimension of a complex space cf. (7.1).)

(15.2.4) We divide the proof of theorem (15.2.3) into several steps. We first prove a lemma due to Senichkin [SEN 2].

Notation. Let (A,X) be a m.m.a.(*), and let $K \subset X$ be a compact set, then A_K denotes the uB-algebra on K which is obtained by completing $A|_K$ with respect to the norm $\|\cdot\|_K$.

Lemma [SEN 2]. Let (A,X) be a m.m.a.(*). Let $f \in A$, and let $G \subset \mathbb{C}$ be an open set such that $f: X \longrightarrow G$ is a proper mapping onto G. Let $a \in G$, and let D be an open disc centered at a such that $\overline{D}$ - the closure of D - is contained in G.

Let $m \in f^{-1}(a)$, and let μ be a representing measure for m concentrated on $f^{-1}(\partial D)$ with respect to the algebra $A_{f^{-1}(\overline{D})}$. For $g \in A$ define

$$\eta_g = \eta: D \longrightarrow \mathbb{C},\ z \longrightarrow \int_{f^{-1}(\partial D)} (f-a)g/(f-z)\, d\mu.$$

Then

i) η is a bounded analytic function and $\max\{|\eta(z)|: z \in D\} \le \|g\|_{f^{-1}(\partial D)}$,

ii) if $\zeta \in \partial D$ and if a nontangential limit $\eta(\zeta) = \lim_{t \to \zeta} \eta(t)$ exists, then $|\eta(\zeta)| \le \|g\|_{f^{-1}(\zeta)}$,

iii) $\eta(a) = g(m)$.

(Here ∂D denotes the boundary of D.)

Before we prove this lemma we give some explanations. Note that $f^{-1}(\overline{D})$ is a compact subset of X since f is a proper mapping. Denote by $\gamma A_{f^{-1}(\overline{D})}$ the Shilov boundary of $A_{f^{-1}(\overline{D})}$. Then $\gamma A_{f^{-1}(\overline{D})}$ is contained in $f^{-1}(\partial D)$ since (A,X) is a m.m.a.(*). By (1.3.10) there exists a representing measure μ for m concentrated on $\gamma A_{f^{-1}(\overline{D})}$, i.e. a positive measure μ such that

$$\int_{\gamma A_{f^{-1}(\overline{D})}} d\mu = 1$$

and

$$\int_{\gamma A_{f^{-1}(\overline{D})}} g\, d\mu = g(m), \text{ for all } g \in A_{f^{-1}(\overline{D})}.$$

Hence the existence of a measure μ in lemma (15.2.4) is always guaranteed.

Proof of (15.2.4). i) For simplicity we assume that D is the open unit disc, hence a = 0 and

$$(*) \qquad \eta(t) = \int_{f^{-1}(\partial D)} fg/(f-t)\, d\mu.$$

It is easy to show that η is complex differentiable and hence analytic on D.

Let $\varphi \in M(A_{f^{-1}(\overline{D})})$, then by (1.2.1)

$$|\hat{f}(\varphi)| \le \|f\|_{f^{-1}(\overline{D})} \le 1.$$

Hence $(1-\bar{t}f)^\wedge$ - the Gelfand transform of $1-\bar{t}f$ - has no zero for $t \in D$,where $\bar{t}$ denotes the complex conjugate of t, i.e. $1-\bar{t}f$ is invertible in $A_{f^{-1}(\overline{D})}$ by (1.2.9). We have

$$\int_{f^{-1}(\partial D)} fg/(1-\bar{t}f)\, d\mu = fg/(1-\bar{t}f)(m) = 0,$$

since f(m) = 0. We multiply the last equation by $\bar{t}$ and add it to (*), then

$$(**) \qquad \eta(t) = \int_{f^{-1}(\partial D)} \left(\frac{1}{f-t} + \frac{\bar{t}}{f-\bar{t}f}\right) fg\, d\mu =$$

$$\int_{f^{-1}(\partial D)} fg(1-|t|^2)/((f-t)(1-\bar{t}f))\, d\mu = \int_{f^{-1}(\partial D)} g(1-|t|^2)/|f-t|^2\, d\mu.$$

For the last equation note that for $\zeta \in \partial D$

$$\zeta/((\zeta-t)(1-\bar{t}\zeta)) = \zeta/((\zeta-t)(\zeta\bar{\zeta}-\zeta\bar{t})) = 1/|\zeta-t|^2 .$$

Denote by υ the projection of μ on ∂D. Then υ is a positive measure on ∂D and $\int_{\partial D} d\upsilon = 1$. Let g be an arbitrary element of the disc algebra $P(\bar{D})$. Then $g\circ f|_{f^{-1}(\bar{D})} \in A_{f^{-1}(\bar{D})}$, since $p(f) \in A_{f^{-1}(\bar{D})}$ for each polynomial p. Then

$$g(0) = g(f(m)) = \int_{f^{-1}(\partial D)} g\circ f \, d\mu = \int_{\partial D} g \, d\upsilon,$$

since μ is a representing measure and f is constant on the fibers $f^{-1}(t)$. It follows that υ is a representing measure for the point evaluation at 0 with respect to the disc algebra, concentrated on ∂D. Thus υ coincides with the normalized Lebesgue measure $1/2\pi \, d\vartheta$, cf. [STO], p. 46, [GAM], p.34. Therefore

$$|\eta(t)| \le \left(\int_{f^{-1}(\partial D)} (1-|t|^2)/|f-t|^2 \, d\mu\right)\|g\|_{f^{-1}(\partial D)} =$$

$$(1/2\pi)\int_{\partial D} (1-|t|^2)/|\vartheta-t|^2 \, d\vartheta \, \|g\|_{f^{-1}(\partial D)} = \|g\|_{f^{-1}(\partial D)}$$

using (**) and A.9i).

ii) Since η is a bounded holomorphic function, there exists a non-tangential limit $\eta(\zeta) = \lim_{t \to \zeta} \eta(t)$ for almost all $\zeta \in \partial D$, cf. A.12. Let $\xi_o \in \partial D$ be a point such that $\eta(\zeta_o)$ exists, w.l.o.g. $\zeta_o=1$. Fix $\varepsilon>0$. It follows from (14.2.1) that the map

$$\bar{D} \longrightarrow \mathbb{R}_+ \cup \{0\}, \; z \longrightarrow \|g\|_{f^{-1}(z)}$$

is upper semi-continuous, hence there exists a neighbourhood U of 1 in $\bar{D}$ so that

(***) $\qquad \|g\|_{f^{-1}(t)} < \|g\|_{f^{-1}(1)}+\varepsilon$ for all $t \in U$.

Let $\delta>0$ such that $\{\exp(it)\colon \delta \ge t \ge -\delta\} \subset U$. Let $r<1$ be a positive number, then

$$\left|\int_{f^{-1}(\partial D\setminus U)} g(1-r^2)/|f-r|^2 \, d\mu\right| \le$$

$$\|g\|_{f^{-1}(\partial D)}(1/2\pi)\int_{\partial D\setminus U}(1-r^2)/(1-(\cos(\delta))^2)\,d\vartheta$$

by A.9ii). The last expression tends to zero as $r \longrightarrow 1$. Using (**) and the last observation we get

$$|\eta(1)| = \lim_{r\to 1}|\eta(r)| = \lim_{r\to 1}\left|\int_{f^{-1}(\partial D\cap U)} g(1-r^2)/|f-r|^2 d\mu\right| \le$$

$$(\|g\|_{f^{-1}(1)}+\varepsilon)\lim_{r\to 1}(1/2\pi)\int_{\partial D\cap U}(1-r^2)/|\vartheta-r|^2\,d\vartheta \le \|g\|_{f^{-1}(1)}+\varepsilon.$$

The last inequalities follow from (***) and A.10.

iii) This is obvious. □

<u>(15.2.5) Notation.</u> Let (A,X), f and G be as in (15.2.4). For $n\ge 2$ denote by Σ_n the subset of the n-fold Cartesian product of X which consists of all points $(x_1,\dots x_n)$ such that $f(x_1) = \dots = f(x_n)$. We give Σ_n the subspace topology. Set

$$\pi: \Sigma_n \longrightarrow G,\ (x_1,\dots,x_n) \longrightarrow f(x_1).$$

Note that π is a proper mapping. We denote by A'_n the subalgebra of $C(\Sigma_n)$ consisting of all functions of the form

$$\sum_{i=1}^{k}\prod_{j=1}^{n} g_{ij}(x_j),\ g_{ij} \in A.$$

<u>Lemma [SEN 2].</u> Let $h \in A'_n$ and let D be an open disc centered at $a \in G$ such that $\overline{D}$, the closure of D, is contained in G. Let $m \in \pi^{-1}(a)$. Then

$$|h(m)| \le \|h\|_{\pi^{-1}(\partial D)}.$$

<u>Proof.</u> Let $h(x_1,\dots,x_n) = \sum_{i=1}^{k}\prod_{j=1}^{n} g_{ij}(x_j)$.

Let $m = (m_1,\dots,m_n)$. According to (15.2.4) we associate to each g_{ij} a bounded holomorphic function η_{ij} on D such that $\eta_{ij}(a) = g_{ij}(m_j)$. Choose $\zeta \in \partial D$ such that all nontangential limits

$$\lim_{t\to\zeta}\eta_{ij}(t) = c_{ij},\ 1\le i\le k,\ 1\le j\le n,$$

exist. Set

$$h_1 = \sum_{i=1}^{k}\Big(\prod_{j=2}^{n} c_{ij}\Big)g_{i1}.$$

Then $h_1 \in A$. Choose η_1 on D according to (15.2.4), i.e.

$$\eta_1(z) = \int_{f^{-1}(\partial D)} h_1(f-a)/(f-z)\ d\mu_1 =$$

$$\sum_{i=1}^{k}\Big(\prod_{j=2}^{n} c_{ij}\Big)\int_{f^{-1}(\partial D)} g_{i1}(f-a)/(f-z)\ d\mu_1 =$$

$$\sum_{i=1}^{k}\Big(\prod_{j=2}^{n} c_{ij}\Big)\eta_{i1}(z),\ z \in D.$$

Here μ_1 is the representing measure for m_1 associated to $A_{f^{-1}(\overline{D})}$, which was already used in the construction of the functions η_{i1}. We have

$$\lim_{t\to\zeta}|\eta_1(t)| \le \|h_1\|_{f^{-1}(\zeta)}$$

by (15.2.4)ii) applied to h_1 (note that the nontangential limit $\eta_1(\zeta)$ exists). Choose $y_1 \in f^{-1}(\zeta)$ such that $|h_1(y_1)| = \|h_1\|_{f^{-1}(\zeta)}$. (Recall that $f^{-1}(\zeta)$ is compact, since f is a proper mapping.) Hence

$$\lim_{t\to\zeta}|\eta_1(t)| \le |h_1(y_1)|.$$

Set

$$h_2 = \sum_{i=1}^{k}\Big\{\Big(\prod_{j=3}^{n} c_{ij}\Big)g_{i1}(y_1)\Big\}g_{i2}$$

The corresponding function η_2 has the form

$$\eta_2 = \sum_{i=1}^{k}\Big\{\Big(\prod_{j=3}^{n} c_{ij}\Big)g_{i1}(y_1)\Big\}\eta_{i2}.$$

As above choose $y_2 \in f^{-1}(\zeta)$ such that

$$\lim_{t\to\zeta}|\eta_2(t)| \le |h_2(y_2)|.$$

We have

$$\lim_{t\to\zeta}\eta_2(t) = \sum_{i=1}^{k}\{(\prod_{j=3}^{n}c_{ij})g_{i1}(y_1)\}c_{i2} = h_1(y_1),$$

hence

$$\lim_{t\to\zeta}|\eta_1(t)| \le |h_1(y_1)| \le |h_2(y_2)|.$$

Continuing in this way we obtain at the n-th step that

$$(*)\qquad \lim_{t\to\zeta}|\eta_1(t)| \le |h_n(y_n)|,$$

where

$$h_n = \sum_{i=1}^{k}(\prod_{j=1}^{n-1}g_{ij}(y_j))g_{in}$$

and $y = (y_1,\dots,y_n) \in \pi^{-1}(\zeta)$. Set

$$\eta(t) = \sum_{i=1}^{k}\prod_{j=1}^{n}\eta_{ij}(t).$$

Then η is a bounded holomorphic function on D,

$$\lim_{t\to\zeta}\eta(t) = \lim_{t\to\zeta}\eta_1(t),$$

and

$$\eta(a) = \sum_{i=1}^{k}\prod_{j=1}^{n}g_{ij}(m_j) = h(m).$$

Since $h_n(y_n) = h(y)$ we get from (*)

$$\lim_{t\to\zeta}|\eta(t)| = \lim_{t\to\zeta}|\eta_1(t)| \le |h_n(y_n)| = |h(y)| \le \|h\|_{\pi^{-1}(\partial D)}.$$

This inequality holds for almost all $\zeta \in \partial D$. Since η is a bounded holomorphic function on D we have

$$|h(m)| = |\eta(a)| \le \|h\|_{\pi^{-1}(\partial D)}.$$

□

<u>(15.2.6) Remark.</u> Let (A,X), f, G be as in (15.2.4) and use the notations of (15.2.5). Let $g \in A$.
It follows by (14.2.2) and (15.2.5) that the map

$$G \longrightarrow \mathbb{R} \cup \{-\infty\},\ \lambda \longrightarrow \log \|h\|_{\pi^{-1}(\lambda)},$$

is a subharmonic function for each $h \in A'_n$. This observation is of particular interest in the case that

$$h(x_1,\ldots,x_n) = \prod_{i<j} (g(x_i)-g(x_j)).$$

Note that $h \in A'_n$, since h is the sum of terms of the form

$$g^{k_1}(x_1)\cdot\ldots\cdot g^{k_n}(x_n),\ 0\le k_1,\ldots,k_n.$$

(15.2.7) Notation. Let A, X, F, W be as in (15.2.2). Fix $k\ge 2$. Denote by X_k the subset of the k-fold Cartesian product of X consisting of all points $(x_1,\ldots,x_k)$ such that $F(x_1)=\ldots=F(x_k)$. Endow X_k with the subspace topology and set

$$\tilde{\pi}: X \longrightarrow W,\ (x_1,\ldots,x_k) \longrightarrow F(x_1).$$

Then X_k is a locally compact space and $\tilde{\pi}$ is a proper mapping onto W. Fix $g \in A$. Define

$$\tilde{g}: X_k \longrightarrow \mathbb{C},\ (x_1,\ldots,x_k) \longrightarrow \prod_{i<j} (g(x_i)-g(x_j))$$

and

$$\psi_{k,g}: W \longrightarrow \mathbb{R} \cup \{-\infty\},\ \lambda \longrightarrow \log \|\tilde{g}\|_{\tilde{\pi}^{-1}(\lambda)}.$$

We remark that $\psi_{k,g}(\lambda)=-\infty$ if $\#F^{-1}(\lambda)<k$.

Lemma [RUS]. $\psi_{k,g}$ is a plurisubharmonic function.

Proof. i) Clearly $\tilde{g}$ is continuous. hence

$$Z_{\tilde{g}}: W \longrightarrow \mathbb{R},\ \lambda \longrightarrow \|\tilde{g}\|_{\tilde{\pi}^{-1}(\lambda)},$$

is upper semi-continuous by (14.2.1). Hence the same is true for $\log Z_{\tilde{g}} = \psi_{k,g}$.

ii) Let L be an arbitrary affine complex line such that $L \cap W \neq \emptyset$. Recall that $(A|_{X_L}, X_L)$ is a m.m.a.(*) by our assumption. Identify L with $\mathbb{C}$ and denote by p the orthogonal projection from $\mathbb{C}^n$ onto L. Then $p \circ F \in A$ and $p \circ F|_{X_L} \in A|_{X_L}$. Since $p \circ F|_{X_L}(x) = F(x)$ for all $x \in X_L$ we have $F|_{X_L} \in A|_{X_L}$. Then $F|_{X_L}$ is a proper mapping from X_L onto $L \cap W$. Furthermore

$$\tilde{\pi}^{-1}(\lambda) = \pi^{-1}(\lambda), \text{ for } \lambda \in L \cap W,$$

where we use the notation of (15.2.5) replacing f by $F|_{X_L}$, (A,X) by $(A|_{X_L}, X_L)$ and G by $L \cap W$. Denote by Σ_k the intersection of X_k with the k-fold product of X_L and set

$$\tilde{g}_L: \Sigma_k \longrightarrow \mathbb{C},\ (x_1,\dots,x_k) \longrightarrow \tilde{g}|_{\Sigma_k}.$$

Then

$$L \cap W \longrightarrow \mathbb{R} \cup \{-\infty\},\ \lambda \longrightarrow \log \|\tilde{g}_L\|_{\pi^{-1}(\lambda)},$$

is subharmonic by (15.2.6). So $\psi_{k,g}|_{L \cap W}$ is subharmonic, since

$$\psi_{k,g}(\lambda) = \log \|\tilde{g}\|_{\tilde{\pi}^{-1}(\lambda)} = \log \|\tilde{g}_L\|_{\pi^{-1}(\lambda)} \text{ for all } \lambda \in L \cap W.$$

□

<u>(15.2.8) Proof of (15.2.3).</u> i) For $i \in \mathbb{N}$ set

$$E_i = \{\lambda \in E: \#F^{-1}(\lambda) = i\}.$$

Then $E = \bigcup_{i \in \mathbb{N}} E_i$. Since a countable union of pluripolar sets is pluripolar, there is $k \in \mathbb{N}$ such that E_k is nonpluripolar. Assume there is $\lambda \in W$ such that $F^{-1}(\lambda)$ contains k+1 different points $x_1,\dots,x_{k+1}$. Since A separates the points of X and contains the constants, there is $g \in A$ so that $g(x_i) \neq g(x_j)$ for $i \neq j$ (cf. the proof of (3.2.3)), hence

$$\psi_{k+1,g}(\lambda) \neq -\infty.$$

But $\psi_{k+1,g}$ is plurisubharmonic on W by (15.2.7) and $\psi_{k+1,g} \equiv -\infty$ on E_k, and we get a contradiction. Hence

$$W = W_1 \cup \ldots \cup W_k.$$

Suppose that there exists an open set $U \subset W \setminus W_k = W_1 \cup \ldots \cup W_{k-1}$. Then there is $l<k$ so that W_l is nonpluripolar, since sets which have positive Lebesgue measure are nonpluripolar. Repeating the argument above this implies that $W_j = \emptyset$ for $j>l$. But this is a contradiction to $W_k \neq \emptyset$, thus W_k is dense in W.

ii) We first show that F is an open mapping.

Fix $x_1 \in X$ and set $y=F(x_1)$. By i) there is $l \leq k$ such that $F^{-1}(y)=\{x_1,\ldots,x_l\}$.

Let $K \subset X$ be a compact neighbourhood of x_1, and let U be a compact neighbourhood of $\{x_2,\ldots,x_l\}$ such that $K \cap U = \emptyset$.

We claim that there is a closed polycylinder $\overline{D} \subset \mathbb{C}^n$ centered at y so that $F^{-1}(\overline{D}) \subset K \cup U$.

Otherwise there would be a sequence $(z_v)_v$ in $X \setminus (K \cup U)$ so that $F(z_v) \longrightarrow y$ as $v \to \infty$. Since F is proper

$$F^{-1}(\{F(z_v) : v \in \mathbb{N}\} \cup \{y\})$$

is a compact subset of X and a subsequence of $(z_v)_v$ converges to a point $z \in X$. Then $z \in \{x_1,\ldots,x_l\}$ and we get a contradiction to $z_v \in X \setminus (K \cup U)$ for all $v \in \mathbb{N}$.

Next we claim that $D \subset F(K)$.

Suppose the contrary. Then there is an affine complex line L so that

α) $L \cap D \cap F(K) \neq \emptyset$ and

β) $L \cap (D \setminus F(K)) \neq \emptyset$.

Since K is compact it follows by α) and β) that $F(K) \cap L \cap D$ contains a boundary point (in the relative topology of L). But $(A|_{X_L}, X_L)$ is a m.m.a.(*), and $F|_{X_L} \in A|_{X_L}$ by the proof of (15.2.7), hence

$$F(F^{-1}(L \cap D) \cap K) = F(K) \cap L \cap D$$

is an open set by lemma (14.1.4)ii) (in the relative topology of L). Thus we get a contradiction, hence F is an open mapping.

Let k be as in i), and let $x \in W_k$. Then $F^{-1}(x) = \{x_1,\ldots,x_k\}$. Choose disjoint open and relatively compact neighbourhoods $U_1,\dots,U_k$ of $x_1,\dots,x_k$ in X. Then

$$V = \bigcap_{i=1}^{k} F(U_i)$$

is a neighbourhood of x, since F is open, and $V \subset W_k$. Hence W_k is open and

$$S = W_1 \cup \dots \cup W_{k-1} = W \setminus W_k$$

is closed. Let x, $U_1,\dots,U_k$ be as above. Then $U_i \cap F^{-1}(V)$ is an open neighbourhood of x_i and F is injective on $U_i \cap F^{-1}(V)$ for i=1,...,k, i.e.

$$F: F^{-1}(W_k) \longrightarrow W_k$$

is a local homeomorphism. We can now follow the proof of theorem (15.1.2) to show that there is a uniquely determined structure of an n-dimensional manifold on $F^{-1}(W_k)$ such that $g|_{F^{-1}(W_k)}$ is a holomorphic map (with respect to this structure) for all $g \in A$.

Next we show that S is an analytic subset of W.

Fix $g \in A$ and define

$$\tilde{g}: W \longrightarrow \mathbb{C},\ z \longrightarrow \begin{cases} \prod_{i<j} (g(x_i^z)-g(x_j^z))^2 & \text{if } z \in W_k \\ 0 & \text{if } z \in W\setminus W_k \end{cases}$$

where $x_1^z,\dots,x_k^z$ are the points of $F^{-1}(z)$ in some ordering. Note that the value of $\tilde{g}$ is independent of which ordering is chosen. Let U_i be open neighbourhoods of x_i^z so that $F|_{U_i}$ is a homeomorphism (i=1,...,k).

Then

$$g \circ F^{-1}|_{F(U_i)} \in \mathrm{Hol}(F(U_i)) \text{ for } i=1,\dots,k$$

as observed above, and $\tilde{g}$ is locally just the sum of products of functions of this form, thus $\tilde{g}$ is analytic on W_k.

We claim that $\tilde{g}$ is continuous on W.

To prove this let $b \in (\partial W_k) \cap W$ be an arbitrary point. Let $(z_\nu)_\nu$ be a sequence in W_k which converges to b. Suppose there exists $\delta>0$ and a subsequence $(z_{v_i})_i$ such that

$$|g(x_j^{z_{v_i}})-g(x_l^{z_{v_i}})| \geq \delta \text{ if } l \neq j,\ i \in \mathbb{N}.$$

Let y_l be an accumulation point of $(x_l^{z_{v_i}})_i$. (Such a point exists since $\{z_{v_i}: i \in \mathbb{N}\}\cup\{b\}$ is a compact set and F is a proper mapping.) Then $y_l \in F^{-1}(b)$ and

$$|g(y_l)-g(y_j)|\geq\delta \text{ if } j \neq l.$$

But $b \in W\setminus W_k$, so $\#F^{-1}(b)<k$, thus $y_l=y_j$ for some pair l,j with $l \neq j$. This is a contradiction, hence no such δ does exist.

We now show that $\tilde{g}(z_v) \longrightarrow 0$ as $v \longrightarrow \infty$.

Choose a compact neighbourhood K of b in W, then $F^{-1}(K)$ is a compact subset of X. Let $\varepsilon>0$. Choose $m \in \mathbb{N}$ so that for all $v\geq m$

α) $z_v \in K$ and

β) there exists a pair $i(v)<j(v)$ such that

$$|g(x_{j(v)}^{z_v})-g(x_{i(v)}^{z_v})| < \varepsilon.$$

Then

$$|\tilde{g}(z_v)| \leq \varepsilon^2\cdot \prod_{\substack{(i,j)\neq(i(v),j(v))\\ 1\leq i<j\leq k}} |g(x_j^{z_v})-g(x_i^{z_v})|^2 \leq$$

$$\varepsilon^2\cdot \prod_{\substack{(i,j)\neq(i(v),j(v))\\ 1\leq i<j\leq k}} (|g(x_j^{z_v})|+|g(x_i^{z_v})|)^2 \leq$$

$$\varepsilon^2\cdot \prod_{\substack{(i,j)\neq(i(v),j(v))\\ 1\leq i<j\leq k}} 4\|g\|^2_{F^{-1}(K)}, \text{ for all } v\geq m.$$

Hence $\tilde{g}$ is continuous on W and analytic on $W\setminus\{x \in W: \tilde{g}(x)=0\}$. By a theorem of Rado (for the n=1 case cf. [WER 4], p. 60) $\tilde{g}$ is in fact holomorphic on W.

Since A separates the points of X, we find for each $x \in W_k$ an element $g_x \in A$ such that $\tilde{g}_x(x) \neq 0$. Thus

$$S = \bigcap_{g\in A} \{x \in W: \tilde{g}(x) = 0\},$$

and ii) is proved.
iii) We show that $X\setminus F^{-1}(S)$ is dense in X. $F^{-1}(S)$ is closed, since S is closed. Recall that F is an open mapping. Suppose there exists an open set $U \subset F^{-1}(S)$, then F(U) is an open subset of S, and we get a contradiction since $W\setminus S$ is dense in W.
The rest of the assertion has already been proved in i) and ii). (To show that S is negligible use ii) and remark (7.1.1)vii).)

iv) By a theorem of Grauert and Remmert ([GR/R 1],Theorem 32) analytic covers are analytic spaces, i.e. X can be given the structure of a reduced complex space of pure dimension n such that F is holomorphic with respect to this structure.
Let $g \in A$. Then g is continuous on X and holomorphic on

$$F^{-1}(W_k) = X\setminus F^{-1}(S),$$

where $F^{-1}(S)$ is a nowhere dense analytic subset of X, hence g is holomorphic on X by the Riemann extension theorem, cf. [GR/R 1], Theorem 13.

□

Remark. Theorem (15.2.3) remains valid if $(A|_{X_L}, X_L)$ is a m.m.a.(*) only for those affine complex lines L which are parallel to a coordinate axis and for which $L \cap W \neq \emptyset$. This was proved by Kumagai [KUM 1], [KUM 2] in a similar way.

Example. Let

$$A = \{f \in C(\mathbb{C}^2): f_{(t_1,t_2)}: \mathbb{C} \longrightarrow \mathbb{C},\ z \longrightarrow f(t_1z,t_2z) \text{ is holomorphic for all } (t_1,t_2) \in \mathbb{C}^2\setminus\{0\}\}.$$

Then $(A,\mathbb{C}^2)$ is a uF-algebra which contains $\mathrm{Hol}(\mathbb{C}^2)$. The map

$$F: \mathbb{C}^2 \longrightarrow \mathbb{C}^2,\ x \longrightarrow (z_1(x),z_2(x)),$$

where z_1, z_2 denotes the coordinate functions, is a homeomorphism, in particular F is a proper mapping onto $\mathbb{C}^2$ such that $\#F^{-1}(\lambda)=1$ for all $\lambda \in \mathbb{C}^2$, i.e. $W_1=\mathbb{C}^2$ in the notation of (15.2.3). Moreover $(A|_{X_L}, X_L)$ is a m.m.a.(*) for all complex lines (which contain the origin).

Suppose that X can be given the structure of a two dimensional analytic space such that $A \subset \mathrm{Hol}(X)$. Then, since $F \in A^2$, this structure coincides with the usual complex structure on $\mathbb{C}^2$. Choose a bounded function $h \in C(\mathbb{C})\backslash \mathrm{Hol}(\mathbb{C})$ and set

$$f(z,w) = \begin{cases} zh(w/z) & \text{if } z \neq 0 \\ 0 & \text{if } z = 0. \end{cases}$$

Then $f \in C(\mathbb{C}^2)$ and

$$f_{(t_1,t_2)}(z) = \begin{cases} t_1 z h(t_2/t_1) & \text{if } t_1 \neq 0 \\ 0 & \text{if } t_1 = 0 \end{cases}$$

is holomorphic for all $(t_1,t_2) \in \mathbb{C}^2\backslash\{0\}$, i.e. $f \in A\backslash \mathrm{Hol}(\mathbb{C}^2)$ and we get a contradiction.

Hence although $(A|_{X_L}, X_L)$ is a m.m.a.(*) on "many" complex lines the conclusion of (15.2.3) does not hold.

CHAPTER 16

HIGHER SHILOV BOUNDARIES

We introduce a generalization of the Shilov boundary for uB-algebras due to Basener and Sibony. We use the higher Shilov boundaries to derive results on the analytic structure in the spectrum of a uB-algebra. Later we define higher Shilov boundaries for uF-algebras.

(16.1) Higher Shilov boundaries for uB-algebras

(16.1.1) Let A be a uB-algebra, and let $K \subset M(A)$ be a compact subset. Denote by $\hat{A}_K$ the closure of $\{\hat{f}|_K : f \in A\}$ in $C(K)$.

Let $F = (f_1,\dots,f_n) \in A^n$ and set

$$V(\hat{F}) = \{x \in M(A) : \hat{f}_1(x)=\dots=\hat{f}_n(x)=0\}.$$

Then $V(\hat{F})$ is a compact $\hat{A}$-convex subset of $M(A)$, hence we can identify $M(\hat{A}_{V(\hat{F})})$ with $V(\hat{F})$ by (4.3.3). Denote by $\gamma_0 A$ resp. $\gamma_0 A_{V(\hat{F})}$ the usual Shilov boundary of A resp. $\hat{A}_{V(\hat{F})}$, cf. (1.3.5).

Definition. For $n \geq 1$ set

$$\gamma_n A = \text{closure}\Big(\bigcup_{F \in A^n} \gamma_0 \hat{A}_{V(\hat{F})}\Big)$$

We call $\gamma_n A$ the n-th Shilov boundary of A.

(16.1.2) Examples. i) Let A be a uB-algebra, $k \in \mathbb{N}$ and $F = (f_1,\dots,f_k) \in A^k$. Assume that for $z = (z_1,\dots,z_k) \in \mathbb{C}^k$

$$\hat{F}^{-1}(z) = \{x \in M(A) : \hat{f}_i(x) = z_i,\ i=1,\dots,k\}$$

is totally disconnected. Then $\hat{F}^{-1}(z) \subset \gamma_n A$ for all $n \geq k$.

To see this note first that $\gamma_l A \subset \gamma_m A$ if $m \geq l$. We have $\hat{F}^{-1}(z) = V(\hat{G})$ for $G = (f_1 - z_1, \ldots, f_k - z_k) \in A^k$, and $\hat{A}_{V(\hat{G})} = C(V(\hat{G}))$ by remark (1.4.6).

ii) Denote by $\overline{D}$ the closed unit dicylinder. Consider the algebra $P(\overline{D})$. Since $\overline{D}$ is polynomially convex we have $M(P(\overline{D})) = \overline{D}$ by (1.2.4)ii).
It is easy to prove that $\gamma_o P(\overline{D}) = \{(z,w): |z|=|w|=1\}$.
Let $g \in P(\overline{D})$. Since $g|_D \in Hol(D)$, $V(g) \cap D$ is an analytic subset of D, hence $V(g) \cap D$ is not compact by (7.1.1)ix). It follows from the maximum principle for holomorphic functions that

$$\|f\|_{V(g)} = \|f\|_{V(g) \cap \partial D} \text{ for all } f \in P(\overline{D}).$$

So $\gamma_1 A \subset \partial D$. Let $x = (x_1, x_2)$ be an arbitrary point of ∂D, w.l.o.g. $|x_1| = 1$. Then $x \in V(z_2 - x_2)$, where z_2 denotes the second coordinate function, and we have

$$|g(x)| > \|g\|_{V(z_2 - x_2) \cap \overline{D} \setminus \{x\}}$$

for $g = z_1 + x_1$. Thus $\gamma_1 P(\overline{D}) = \partial D$.
By i) $\gamma_n P(\overline{D}) = \overline{D}$ for $n \geq 2$.
iii) For other examples cf. [SIB].

<u>(16.1.3)</u> Let (B,Y) be a uB-algebra. Set

$$j: Y \longrightarrow M(B),\ y \longrightarrow \varphi_y,$$

with $\varphi_y(f) = f(y)$ for all $f \in B$, cf. (1.2.4).

<u>Proposition [BAS].</u> Let (B,Y) be a uB-algebra. Let $n > 0$ and suppose that $\gamma_{n-1} A \subset j(Y)$. Let $G = (g_1, \ldots, g_n) \in B^n$, and let W be a component of $\mathbb{C}^n \setminus G(Y)$ such that $\hat{G}(M(B)) \cap W \neq \emptyset$. Then $W \subset \hat{G}(M(B))$.
(Here $\hat{G}: M(B) \longrightarrow \mathbb{C}^n$, $\varphi \longrightarrow (\hat{g}_1(\varphi), \ldots, \hat{g}_n(\varphi))$.)

<u>Proof.</u> For n=1 our assertion follows by (1.3.7).
Let $n > 1$. $\hat{G}(M(B)) \cap W$ is a closed nonempty subset of W. Hence it suf-

fices to show that $\widehat{G}(M(B))\cap W$ is open.

Suppose the contrary. Then there is $x \in \widehat{G}(M(B))\cap W$ and an affine complex line L parallel to a coordinate axis such that $x \in L$ and x is a boundary point (relative to L) of $\widehat{G}(M(B))\cap L$. W.l.o.g. $x=(0,\dots,0)$ and $L = \{(z_1,\dots,z_n): z_2=\dots=z_n=0\}$.

We have $H=(g_2,\dots,g_n) \in A^{n-1}$,

$$\widehat{G}(V(\widehat{H})) = L\cap\widehat{G}(M(B)),$$

$$\hat{g}_1|_{V(\widehat{H})} \in \widehat{A}_{V(\widehat{H})}, \text{ and}$$

$$\partial\hat{g}_1(V(\widehat{H})) = \partial\hat{g}_1(M(\widehat{A}_{V(\widehat{H})})) \subset \hat{g}_1(\gamma_o\widehat{A}_{V(\widehat{H})})$$

by (1.3.7). Since $(0,\dots,0)$ is a boundary point of $\widehat{G}(M(B))\cap L$ (relative to L), 0 is a boundary point of $\hat{g}_1(M(\widehat{A}_{V(\widehat{H})}))$, i.e. there exists $y \in \gamma_o\widehat{A}_{V(\widehat{H})}$ such that $\hat{g}_1(y) = 0$. But then

$$(0,\dots,0) = \widehat{G}(y) \in \widehat{G}(\gamma_{n-1}A) \subset G(Y),$$

and we get a contradiction. □

<u>(16.1.4) Theorem [BAS].</u> Let (B,Y) be a uB-algebra. Let $n>0$ and suppose that $\gamma_{n-1}A \subset j(Y)$. Let $G = (g_1,\dots,g_n) \in B^n$, and let W be a component of $\mathbb{C}^n\backslash G(Y)$ such that $\widehat{G}(M(B))\cap W \neq \emptyset$. Suppose there exists a nonpluripolar set $E \subset W$ such that $\#\widehat{G}^{-1}(z)<\infty$ for all $z \in E$. Set

$$W_k = \{y \in W: \#\widehat{G}^{-1}(y)=k\} \text{ for } k \in \mathbb{N}.$$

Then

i) there is $k \in \mathbb{N}$ such that $W = W_1\cup\dots\cup W_k$, and $W_k \neq \emptyset$,

ii) the set $S = W_1\cup\dots\cup W_{k-1}$ is a proper closed analytic subset of W,

iii) the triple $(\widehat{G}^{-1}(W),\widehat{G},W)$ is an analytic cover with critical set S,

iv) there exists an analytic space structure of pure dimension n on $\widehat{G}^{-1}(W)$ such that $A \subset \mathrm{Hol}(\widehat{G}^{-1}(W))$, if we endow $\widehat{G}^{-1}(W)$ with this structure.

Proof. Set $X = \hat{G}^{-1}(W)$, $F = \hat{G}|_X$, $A = \hat{B}|_X$.
Clearly $F: X \longrightarrow W$ is a proper mapping onto W, so all we have to show - by theorem (15.2.3) - is that $(A|_{X_L}, X_L)$ is a m.m.a.(*) for each affine complex line L in $\mathbb{C}^n$ with $L \cap W \neq \emptyset$.
Let

$$L = \bigcap_{j=1}^{n-1} \{(z_1,\dots,z_n) \in \mathbb{C}^n: \sum_{i=1}^{n} a_{ij} z_i = b_j\},\ a_{ij}, b_j \in \mathbb{C}.$$

Then

$$X_L = F^{-1}(L \cap W) \subset \{x \in M(B): \hat{H}(x)=0\} = V(\hat{H}),$$

where $H = (h_1,\dots,h_{n-1})$,

$$h_j = \sum_{i=1}^{n} a_{ij} g_i - b_j,\quad j=1,\dots,n-1.$$

Hence $H \in B^{n-1}$. We note that X_L is an open subset of $V(\hat{H})$ (relative to $V(\hat{H})$) and that

$$X_L \cap \gamma_o \hat{B}_{V(\hat{H})} = \emptyset$$

by our assumption. So $B_{V(\hat{H})}|_{X_L}$ is a m.m.a.(*) by (14.1.3)iv), and the same is true for $(A|_{X_L}, X_L)$.

□

Remarks. i) In the original version of the theorem the stronger assumption is made that E is of positive 2n-dimensional Lebesgue measure. The proof above is due to Rusek [RUS].
ii) Consider example (14.1.3)v). Set $A = P(X)$, notations as in (14.1.3)v). Then $(P(X),X)$ is a uB-algebra and $M(P(X)) = X$, since X is polynomially convex (1.2.4)ii). We have

$$\gamma_o P(X) \subset Y = X \cap \{(z,w): |z| = 1/2\},$$

since $\|p\|_X = \|p\|_Y$ for all polynomials p, hence $\{|z|<1/2\}$ is a component of

$$\hat{z}_1(M(P(X)) \setminus \hat{z}_1(\gamma_o P(X)).$$

It follows from the construction of X that

$$\#X\cap(\{a_i\}\times\mathbb{C}) = 2^{i-1} \text{ for all } i \in \mathbb{N},$$

i.e. $\#\hat{z}_1^{-1}(x)<\infty$ for all elements of a dense subset of $\{|z|<1/2\}$ but M(P(X)) does not contain an analytic disc.

(16.1.5) Finally we state two technical lemmata, due to Kramm and Brooks which will be needed later. We first prove that for calculating the n-th Shilov boundary it suffices to take a smaller index set than A^n.

Lemma [KRA 2]. Let A be a uB-algebra and let $B \subset A$ be a dense subalgebra. Then for $n \geq 1$

$$\gamma_n A = \text{closure}(\bigcup_{F\in B^n} \gamma_0 A_{V(\hat{F})}).$$

Proof. i) The inclusion "$\supset$" is trivial.

ii) Assertion 1. Let $F=(f_1,\ldots,f_n) \in A^n$, $\varphi \in V(\hat{F})$, and let U be an open neighbourhood of $V(\hat{F})$ in M(A). Then there is $G \in B^n$ such that $\varphi \in V(\hat{G})$ and $V(\hat{G}) \subset U$.

Let $a \in \mathbb{C}\setminus\{0\}$. Obviously $V(\hat{F}) = V(\hat{H})$ for $H=(af_1,\ldots,af_n)$. Therefore - and by the compactness of $M(A)\setminus U$ - we can assume that for each $x \in M(A)\setminus U$ there is $i \in \{1,\ldots,n\}$ such that $|\hat{f}_i(x)|>1$.

Now choose $g_1,\ldots,g_n \in B$ such that $\|\hat{g}_i-\hat{f}_i\|_{M(A)}<1/3$ for $i=1,\ldots,n$. This is possible because $\hat{B}$ is dense in $\hat{A}$. Set

$$G=(g_1-\hat{g}_1(\varphi),\ldots,g_n-\hat{g}_n(\varphi)),$$

and assertion 1 is proved.

For the inverse inclusion it is enough to show that

$$(*) \qquad \bigcup_{F\in A^n} \gamma_0\hat{A}_{V(\hat{F})} \subset \text{closure}(\bigcup_{G\in B^n} \gamma_0\hat{A}_{V(\hat{G})}).$$

Let $\varphi \in M(A)$, $F \in A^n$ such that $\varphi \in V(\hat{F})$ and $\varphi \in \gamma_0\hat{A}_{V(\hat{F})}$. Recall from (1.3.12) that the set of all strong boundary points of a uB-algebra A lies dense in the Shilov boundary $\gamma_0 A$. Since the right side of (*) is a closed subset of M(A) we can w.l.o.g. assume that φ is a strong boundary point of $\hat{A}_{V(\hat{F})}$.

Let $U \subset M(A)$ be an arbitrary open neighbourhood of φ. Then there is $g \in \hat{A}_{V(\hat{F})}$ such that

$$|g(\varphi)| = \|g\|_{V(\hat{F})} = 1$$

and

$$\|g\|_{V(\hat{F})\setminus U} < 1/8.$$

(Recall from (4.3.3) that $M(\hat{A}_{V(\hat{F})}) = V(\hat{F})$.)

Choose $h \in A$ such that

$$\|g-\hat{h}\|_{V(\hat{F})} < 1/8.$$

For each $x \in V(\hat{F})$ set

$$U_x = \{y \in M(A) : |\hat{h}(y)-\hat{h}(x)| < 1/8\}.$$

Then

$$(**) \qquad W = \Big(\bigcup_{x \in U \cap V(\hat{F})} U_x \cap U\Big) \cup \Big(\bigcup_{x \in V(\hat{F})\setminus U} U_x\Big)$$

is an open neighbourhood of $V(\hat{F})$. Thus we find $G \in B^n$ such that $\varphi \in V(\hat{G})$ and $V(\hat{G}) \subset W$ by assertion 1. Then

$$|\hat{h}(\varphi)| \geq |g(\varphi)| - |g(\varphi)-\hat{h}(\varphi)| \geq 7/8.$$

If $\psi \in V(\hat{G})\setminus U$ there exists $x \in V(\hat{F})\setminus U$ such that $\psi \in U_x$ by (**). So

$$|\hat{h}(\psi)| \leq |\hat{h}(\psi)-\hat{h}(x)| + |\hat{h}(x)-g(x)| + |g(x)| \leq 3/8.$$

Hence $U \cap \gamma_0 \hat{A}_{V(\hat{G})} \neq \emptyset$. Since U was taken to be arbitrary we conclude that φ is contained in the right side of (*).

□

(16.2) Higher Shilov boundaries for uF-algebras

(16.2.1) We extend the definition of higher Shilov boundaries to uF-algebras.

Let A be a uF-algebra with locally compact spectrum M(A). Recall from chapter 9 that a point $\varphi \in M(A)$ is independent if $\varphi \in \gamma\hat{A}_K$

for each compact set $K \subset M(A)$ which contains φ. The Shilov boundary γA is defined to be the closure of the set of all independent points of $M(A)$.

Definition. Let A be a uF-algebra with locally compact spectrum $M(A)$. Set $\gamma_o A = \gamma A$, and define the n-th Shilov boundary $\gamma_n A$ for $n \geq 1$ by

$$\gamma_n A = \text{closure}\ (\bigcup_{F \in A^n} \gamma_o \hat{A}_{V(\hat{F})}).$$

Remarks. i) It follows by (9.1.4) that definition (16.2.1) coincides with definition (16.1.1) in the case that A is a uB-algebra.

ii) The definition of $\gamma_n A$ makes also sense for uF-algebras without locally compact spectrum. We restrict our definition to uF-algebras with locally compact spectrum since then $V(\hat{F})$ is a locally compact and hemicompact space, hence $C(V(\hat{F}))$ is a uF-algebra by (4.1.4)ii), and $\hat{A}_{V(\hat{F})}$ is a closed subalgebra of $C(V(\hat{F}))$, hence each element of $\hat{A}_{V(\hat{F})}$ is a continuous function on $V(\hat{F})$, cf. (4.1.5).

Let $(K_n)_n$ be an admissible $\hat{A}$-convex exhaustion of $M(A)$, cf. remark (4.3.4)i). Then $(K_n \cap V(\hat{F}))_n$ is an admissible $\hat{A}$-convex exhaustion of $V(\hat{F})$. Hence

$$M(A_{V(\hat{F})}) = V(\hat{F}) \text{ (as topological spaces)}$$

by remark (4.3.4)iii).

iii) Let $(K_n)_n$ be an admissible exhaustion of $M(A)$. Then

$$\gamma_o A = \text{closure}(\bigcup_{k=1}^{\infty} \bigcap_{n \geq k} \gamma \hat{A}_{K_n})$$

by (9.1.3).

iv) Let $G \subset \mathbb{C}^n$ be a domain of holomorphy. Let $F \in Hol(G)^k$. By (2.2.7) we can identify $M(Hol(G))$ with G and $Hol(G)$ with $Hol(G)^\wedge$, the algebra of all Gelfand transforms. Then $V(\hat{F})$ is an analytic subset of G, and $Hol(G)^\wedge_{V(\hat{F})} \subset Hol(V(\hat{F}))$, since $Hol(V(\hat{F}))$ is complete. It follows from the maximum principle for holomorphic functions that

$x \in \gamma_o \mathrm{Hol}(G)^\wedge_{V(\hat{F})}$ iff x is an isolated point of $V(\hat{F})$. This implies $k \geq n$ by remark (7.1.1)ix). Hence

$$\gamma_k \mathrm{Hol}(G) = \begin{cases} \emptyset & \text{if } k<n \\ G & \text{if } k \geq n \end{cases}$$

since the coordinate functions $z_1,\dots,z_n \in \mathrm{Hol}(G)$.

<u>(16.2.2) Lemma [BRO 5].</u> Let A be a uF-algebra with locally compact spectrum M(A). Let $U \subset M(A)$ be an open, hemicompact and $\hat{A}$-convex neighbourhood of $\varphi \in M(A)$, and let $n \in \{0,1,\dots\}$. Then

$$\varphi \in \gamma_n A \text{ iff } \varphi \in \gamma_n \hat{A}_U.$$

<u>Proof.</u> Note that $M(\hat{A}_U) = U$ as topological spaces by remark (4.3.4)iii). We first assume that $n>0$.

i) Let $\varphi \in \gamma_n A$, and let $W \subset U$ be an arbitrary neighbourhood of φ. There are $F \in A^n$ and $\psi \in W$ such that $\psi \in V(\hat{F})$ and ψ is an independent point with respect to $\hat{A}_{V(\hat{F})}$. Hence $\psi \in \gamma_o \hat{A}_K$ for every compact set $K \subset V(\hat{F})$, in particular $\psi \in \gamma_o \hat{A}_K$ for every compact set $K \subset V(\hat{F}) \cap U$. So $\psi \in \gamma_n \hat{A}_U$, since $\hat{F}|_U \in \hat{A}^n_U$. Since W was taken to be arbitrary it follows that $\varphi \in \gamma_n \hat{A}_U$.

ii) Let $\varphi \in \gamma_n \hat{A}_U$. Choose an arbitrary $\hat{A}$-convex compact neighbourhood $K \subset U$ of φ. Then, repeating the proof of i), we see that $\varphi \in \gamma_n \hat{A}_K$.

$\hat{A}|_K$ is a dense subalgebra of $\hat{A}_K$, so by (16.1.5)

$$\varphi \in \text{closure}\Big(\bigcup_{G \in (\hat{A}|_K)^n} \gamma_o \hat{A}_{V(\hat{G})} \Big).$$

Hence there are $F \in A^n$, $\psi \in \text{int } K$ such that $\psi \in V(\hat{F})$ and ψ is a strong boundary point of $\hat{A}_{V(\hat{F}) \cap K}$.

Let $L \subset \text{int } K$ be a compact neighbourhood of ψ, then we find $g \in A$ and $\tilde{\psi} \in (\text{int } L) \cap V(\hat{F})$ so that

$$\hat{g}(\tilde{\psi}) = \|\hat{g}\|_{V(\hat{F})\cap K} = 1,$$

and

$$\|\hat{g}\|_{V(\hat{F})\cap(K\setminus \mathrm{int}\ L)} < 1/2.$$

The set

$$S = \{\vartheta \in V(\hat{F})\cap K : \hat{g}(\vartheta) = 1\}$$

is a peak set of $A_{V(\hat{F})\cap K}$ (consider $h = (g+1)/2$).

Let $(K_n)_n$ be an admissible $\hat{A}$-convex exhaustion of $M(A)$, w.l.o.g. $K \subset K_1$. For each $n \in \mathbb{N}$ S is a local and hence - by Rossi's maximum principle (1.4.9) - a global peak set of $\hat{A}_{V(\hat{F})\cap K_n}$. (Note that $M(\hat{A}_{V(\hat{F})\cap K_n}) = V(\hat{F})\cap K_n$ by (4.3.3).)

Suppose that for each $\vartheta \in S$ there exists $n(\vartheta) \in \mathbb{N}$ such that ϑ is not contained in $\gamma_o\hat{A}_{V(\hat{F})\cap K_{n(\vartheta)}}$.

Then there exists an open neighbourhood N of ϑ in $V(\hat{F})$ such that

$$N \cap \gamma_o\hat{A}_{V(\hat{F})\cap K_n} = \emptyset \text{ for all } n \geq n(\vartheta).$$

Since S is compact there is $m \in \mathbb{N}$ such that

$$S \cap \gamma_o\hat{A}_{V(\hat{F})\cap K_n} = \emptyset \text{ for all } n \geq m.$$

This is impossible.

By (9.1.3) $S \cap \gamma_o\hat{A}_{V(\hat{F})} \neq \emptyset$, hence $K \cap \gamma_o\hat{A}_{V(\hat{F})} \neq \emptyset$.

Since there exists a neighbourhoodbasis of φ consisting of $\hat{A}$-convex compact subsets by (4.3.7)ii), we are done.

The case n=0 can be proved in a similar way. □

<u>(16.2.3) Corollary.</u> Let A be a uF-algebra with locally compact spectrum $M(A)$, and let $n \in \{0,1,...\}$. Then the following statements are equivalent:

i) $\gamma_n A = \emptyset$;

ii) $\gamma_n\hat{A}_U = \emptyset$, for each hemicompact $\hat{A}$-convex open subset U of $M(A)$;

iii) $\gamma_n \hat{A}_K \subset \partial K$ – the boundary of K – for each $\hat{A}$-convex compact subset K of M(A).

Proof. i)⇔ii) This follows from (16.2.2).

iii)⇒i) Suppose there exists $\varphi \in \gamma_n A$. Choose a compact $\hat{A}$-convex neighbourhood K of φ. Then $\varphi \in \gamma_n \hat{A}_K$ by part i) of the proof of (16.2.2). This is impossible.

ii)⇒iii) Suppose there are a compact $\hat{A}$-convex set $K \subset M(A)$ and

$$\varphi \in (\text{int } K) \cap \gamma_n \hat{A}_K.$$

Choose an open, hemicompact and $\hat{A}$-convex neighbourhood $U \subset K$ of φ by (4.3.7)ii). Then, repeating the proof of part i) in (16.2.2), we see that $\varphi \in \gamma_n \hat{A}_U$. This is a contradiction. □

CHAPTER 17

LOCAL ANALYTIC STRUCTURE IN THE SPECTRUM OF A uF-ALGEBRA

In the theory of several complex variables one is not always interested in the global behavior of a function on its entire domain of definition. There are situations in which one considers the behavior of a function in arbitrary small neighbourhoods of a fixed point. This leads to the notion of germs of holomorphic functions at a point $p \in \mathbb{C}^n$.

This notion can be easily carried over to uF-algebras. Let A be a uF-algebra with locally compact spectrum M(A), and let $\varphi \in M(A)$. We denote by $\hat{A}_\varphi$ the algebra of germs of functions of $\hat{A}$ at φ. In this chapter we show that A has analytic structure in φ, if there are algebraic relations between $\hat{A}_\varphi$ and ${}_n\mathcal{O}_o$ - the algebra of germs of holomorphic functions at $0 \in \mathbb{C}^n$.

Most results of this chapter are due to Brooks.

(17.1) Local rings of functions of uF-algebras

(17.1.1) Throughout this chapter we always consider uF-algebras with locally compact spectrum M(A).

If $U \subset M(A)$ is an open and hemicompact subset, then C(U) is a uF-algebra by (4.1.4)ii), hence $\hat{A}_U \subset C(U)$ by (4.1.5).

Let $\varphi \in M(A)$. Let U and W be open and hemicompact neighbourhoods of φ in M(A). We call $f \in \hat{A}_U$ and $g \in \hat{A}_W$ equivalent at φ, provided there exists an open neighbourhood $V \subset U \cap W$ of φ such that $f|_V = g|_V$. This gives an equivalence relation on the disjoint union $\dot{\bigcup} \hat{A}_U$, where the union is taken over all open and hemicompact neighbourhoods U of φ.

An equivalence class will be called a germ of a function of $\widehat{A}$ at φ. We denote by f_φ the equivalence class of an element $f \in \widehat{A}_U$ and by $\widehat{A}_\varphi$ the set of all equivalence classes.

Let $g_\varphi, f_\varphi \in \widehat{A}_\varphi$. Choose representative functions $f \in \widehat{A}_U$ and $g \in \widehat{A}_W$. Then $f+g|_V$ resp. $fg|_V$ are well defined elements of $\widehat{A}_V$, $V \subset U \cap W$ an open and hemicompact neighbourhood of φ. We set

$$f_\varphi + g_\varphi = (f+g)_\varphi, \; f_\varphi \cdot g_\varphi = (fg)_\varphi.$$

It is easy to see that these operations are well-defined, i.e. they are independent of the choice of representatives of f_φ and g_φ. Hence $\widehat{A}_\varphi$ is a commutative algebra with unit.

Let $f_\varphi \in \widehat{A}_\varphi$. Choose a representative $f \in \widehat{A}_U$ of f_φ. Set $f_\varphi(\varphi) = f(\varphi)$, then $f_\varphi(\varphi)$ is well-defined. $\widehat{A}_\varphi$ is a local algebra, i.e. there exists exactly one maximal ideal M in $\widehat{A}_\varphi$, and

$$M = \{f_\varphi \in \widehat{A}_\varphi : f_\varphi(\varphi) = 0\}$$

(If $f_\varphi(\varphi) \neq 0$, there exist an open, hemicompact and $\widehat{A}$-convex neighbourhood W of φ and a representative $f \in \widehat{A}_W$ of f_φ such that $f(\psi) \neq 0$ for all $\psi \in W$ (use (4.3.7)ii)). Since $M(\widehat{A}_W) = W$ by (4.3.4)iii) we see that $1/f \in \widehat{A}_W$ by (6.1.4), hence f_φ is invertible in $\widehat{A}_\varphi$.)

Let U, W be open neighbourhoods of $p \in \mathbb{C}^n$. Then $f \in \mathrm{Hol}(U)$, $g \in \mathrm{Hol}(W)$ are called equivalent at p if there is an open neighbourhood $V \subset U \cap W$ of p such that $f|_V = g|_V$. The algebra of germs of holomorphic functions at p, defined as above, will be denoted by ${}_n\mathcal{O}_p$.

<u>(17.1.2) Theorem [BRO 5].</u> Let A be a uF-algebra with locally compact spectrum M(A), and let φ be a G_δ-point in $M(A)\setminus\gamma_{n-1}A$. Then the following statements are equivalent:

i) There exists a surjective algebra homomorphism $\eta : {}_n\mathcal{O}_o \longrightarrow \widehat{A}_\varphi$;

ii) $\hat{A}$ has analytic structure in φ. More precisely, there are a neighbourhood W of φ in M(A), an open set $D \subset \mathbb{C}^n$ and a homeomorphism $F : W \longrightarrow D$, such that the map

$\hat{A}_W \longrightarrow \mathrm{Hol}(D)$, $f \longrightarrow f \circ F^{-1}$,

is a (well-defined) topological algebra isomorphism.

(Here $\gamma_{n-1}A$ denotes the (n-1)-th Shilov boundary of A, cf. chapter 16. Recall that φ is a G_δ-point if it is the intersection of countable many open subsets of M(A).)

We divide the proof of the theorem into several steps.

(17.1.3) Lemma [BRO 6], [CLA 2]. Let A be a uF-algebra with locally compact spectrum M(A), and let $\varphi \in M(A)$ be a G_δ-point. Suppose that there are a neighbourhood $U \subset M(A)$ of φ, a topological space Y and a map $F: U \longrightarrow Y$ with the following property:
For each $f \in A$ there exists a complex-valued function G defined in some neighbourhood of $F(\varphi)$ so that $\hat{f} = G \circ F$ in some neighbourhood of φ.
Then F is injective in some neighbourhood of p.

Proof. i) By proposition (4.4.2) there exists a countable descending open neighbourhoodbasis $V_1 \supset V_2 \supset \ldots$ of φ.
ii) We may assume that V_1 is contained in U, the domain of definition of F.
We shall prove the lemma by contradiction. Suppose the lemma is false. Then there is a sequence of pairs $(x_i, y_i)_i$ such that for all $i \in \mathbb{N}$

i) x_i and y_i are elements of $V_i \backslash \{\varphi\}$,

ii) $x_i \neq y_i$ but $F(x_i) = F(y_i)$.

W.l.o.g. we can assume that x_i, y_i, x_j, y_j are pairwise disjoint points for $i \neq j$.
By (8.1.4) we can choose an element $f \in A$ such that $\hat{f}$ is injective

on the union of all points x_i, y_i.
Now choose G defined in a neighbourhood of F(p) such that $G \circ F = \hat{f}$ in a neighbourhood W of p, w.l.o.g. $W = V_1$. Then

$$\hat{f}(x_i) = G \circ F(x_i) = G \circ F(y_i) = \hat{f}(y_i) \quad \text{for all } i \in \mathbb{N},$$

and we get a contradiction. □

(17.1.4) Let A be a uF-algebra with locally compact spectrum M(A), $\varphi \in M(A)$. Let $U \subset M(A)$ be an open, hemicompact neighbourhood of φ, and let $F \in (\hat{A}_U)^n$ such that $F(\varphi) = 0$. Set

$$F^*: {}_n\mathcal{O}_o \longrightarrow \hat{A}_\varphi, \; \tilde{g} \longrightarrow (g \circ F)_\varphi,$$

where g represents $\tilde{g}$ in a neighbourhood W of $0 \in \mathbb{C}^n$. We claim that F^* is a well-defined algebra homomorphism.
Since M(A) is locally compact and $F: U \longrightarrow \mathbb{C}^n$ is continuous there exists an open, hemicompact and $\hat{A}$-convex neighbourhood V of φ in M(A) such that $V \subset U$ and $F(V) \subset W$, cf. (4.3.7)ii). We have $g \circ F \in \hat{A}_V$ by (6.1.2), since $M(\hat{A}_V) = V$ by (4.3.4)iii). Using ii) of (6.1.2) one sees that F^* is independent of the choice of g, and our claim is proved.
Let f be a representative of $\tilde{f} \in {}_n\mathcal{O}_o$. Set $\tilde{f}(0) = f(0)$, then $\tilde{f}(0)$ is well-defined, and

$$\tilde{M} = \{\tilde{f} \in {}_n\mathcal{O}_o : \tilde{f}(0) = 0\}$$

is the only maximal ideal in ${}_n\mathcal{O}_o$.

Lemma [BRO 5], [CLA 2]. Let A be a uF-algebra with locally compact spectrum M(A), and let $\varphi \in M(A)$ be a point such that $\hat{A}_\varphi$ is a finite ${}_n\mathcal{O}_o$-algebra via some algebra homomorphism η, i.e. there are $f_1, \ldots, f_r \in \hat{A}_\varphi$ such that $\hat{A}_\varphi = \eta({}_n\mathcal{O}_o)f_1 + \ldots + \eta({}_n\mathcal{O}_o)f_r$.
Then there exist an open, hemicompact and $\hat{A}$-convex neighbour-

hood U of φ and $F \in (\hat{A}_U)^n$ so that $F(\varphi)=0$ and $\eta = F^*$.

Proof. As above denote by M resp. $\tilde{M}$ the maximal ideal of $\hat{A}_\varphi$ resp. ${}_n\mathcal{O}_o$.

i) We first endow $\hat{A}_\varphi$ and ${}_n\mathcal{O}_o$ with the m-adic topology and show that both spaces are Hausdorff spaces.
A fundamental system of neighbourhoods of $\tilde{h} \in {}_n\mathcal{O}_o$ resp. $g_\varphi \in \hat{A}_\varphi$ is given by sets of the form $\tilde{h}+\tilde{M}^k$ resp. $g_\varphi+M^k$, $k \in \mathbb{N}$, where

$$\tilde{M}^k = \{\tilde{g} \in {}_n\mathcal{O}_o : \text{There exist } l \in \mathbb{N}, \text{ elements } \tilde{f}_{ij} \in \tilde{M},\ 1 \le i \le k,\ 1 \le j \le l, \text{ such that } \tilde{g} = \sum_{j=1}^{l} \prod_{i=1}^{k} \tilde{f}_{ij}\}.$$

We have $\bigcap_{k \in \mathbb{N}} \tilde{M}^k = 0$ (cf. [K/K], p. 70), hence ${}_n\mathcal{O}_o$ is a Hausdorff space.

Since ${}_n\mathcal{O}_o$ is a noetherian ring ([K/K], p. 80), the same is true for $\eta({}_n\mathcal{O}_o)$ and – since it is a finitely generated $\eta({}_n\mathcal{O}_o)$-algebra – for $\hat{A}_\varphi$ ([NAG], p. 9). So $\hat{A}_\varphi$ is a noetherian local ring, hence $\bigcap_{k \in \mathbb{N}} M^k = 0$ by Krull's intersection theorem, and we see that $\hat{A}_\varphi$ is a Hausdorff space, too.

ii) We claim that the germs of the polynomials lie dense in ${}_n\mathcal{O}_o$.
Let $\tilde{g} \in {}_n\mathcal{O}_o$ be an arbitrary element, and let g be a representative of $\tilde{g}$. Considering the power series expansion of g at 0, we may write

$$g = \sum_{i=0}^{\infty} P_i,$$

where the P_i's are homogeneous polynomials of degree i. Set

$$Q_j = \sum_{i \ge j} P_i \ ,\ j \in \mathbb{N},$$

then

$$\tilde{g} - (\tilde{P}_o + \ldots + \tilde{P}_{j-1}) = \tilde{Q}_j \in \tilde{M}^j, \text{ for all } j \in \mathbb{N},$$

and our assertion follows.

iii) Denote by $z_1,\dots,z_n$ the coordinate functions and set

$$(f_i)_\varphi = \eta(\tilde{z}_i)\ ,\ i=1,\dots,n.$$

Choose representatives f_i of $(f_i)_\varphi$ in an open, hemicompact and $\hat{A}$-convex neighbourhood U of φ, i=1,...,n. Set $F = (f_1,\dots,f_n) \in (\hat{A}_U)^n$.

We claim that $\eta(\tilde{M}) \subset M$.

Let $\tilde{g} \in {}_n\mathcal{O}_o$ be an element such that $\eta(\tilde{g}) \subset \hat{A}_\varphi \setminus M$, i.e. $\eta(\tilde{g})(\varphi)=c\neq 0$. Then $\eta(\tilde{g}-c\cdot\tilde{1})(\varphi)=0$, hence $\tilde{g}-c\cdot\tilde{1}$ is not invertible in ${}_n\mathcal{O}_o$, so $\tilde{g}(0)=c$, i.e. $\tilde{g} \in {}_n\mathcal{O}_o\setminus\tilde{M}$.

It follows that $\eta(\tilde{h}+\tilde{M}^k) \subset \eta(\tilde{h})+M^k$ for every $k \in \mathbb{N}$, hence η is continuous.

iv) We have $(f_i)_\varphi(\varphi) = 0$ for i=1,...,n, i.e. $F(\varphi) = 0$, by iii).

Let $P(z_1,\dots,z_n)$ be a polynomial and denote by $\tilde{P}(z_1,\dots,z_n)$ the germ of $P(z_1,\dots,z_n)$ in ${}_n\mathcal{O}_o$. Then

$$\eta(\tilde{P}(z_1,\dots,z_n)) = P(\eta(\tilde{z}_1),\dots,\eta(\tilde{z}_n)) = P((f_1)_\varphi,\dots,(f_n)_\varphi) =$$

$$= (P(f_1,\dots,f_n))_\varphi = F^*(\tilde{P}(z_1,\dots,z_n)).$$

Hence η and F^* coincide on the polynomials.

v) Since η and F^* coincide on a dense subset of ${}_n\mathcal{O}_o$ by ii) and iv) we see that $F^* = \eta$ by i) and iii).

□

(17.1.5) Proof of theorem (17.1.2). Assume i). Since $\Upsilon_{n-1}A$ is a closed subset of M(A) and M(A) is locally compact, there exists an open, hemicompact and $\hat{A}$-convex neighbourhood U of φ in M(A), such that

$$\Upsilon_{n-1}A \cap U = \emptyset,$$

by (4.3.7)ii). It follows from (16.2.2) that

$$\Upsilon_{n-1}\hat{A}_U = \emptyset.$$

Since η is surjective, $\hat{A}_\varphi$ is a finite ${}_n\mathcal{O}_o$ algebra (via η). Hence there exists an open, hemicompact and $\hat{A}$-convex neighbourhood $W \subset U$ of φ such that there exists $F \in (\hat{A}_W)^n$ with $\eta = F^*$ by (17.1.4).

Let $h \in A$ be an arbitrary element. Choose $\tilde{g} \in {}_n\mathcal{O}_o$ such that

$$\hat{h}_\varphi = F^*(\tilde{g}).$$

Hence there exist a neighbourhood V of 0 in $\mathbb{C}^n$ and a representative $g \in Hol(V)$ of $\tilde{g}$ such that $h = g \circ F$ in a neighbourhood of φ. Thus F is injective in a neighbourhood $\tilde{W}$ of φ by (17.1.3), w.l.o.g. $W=\tilde{W}$. Note that $M(A_W)=W$.

Now let L be an affine complex line in $\mathbb{C}^n$, parallel to a coordinate axis, so that $L \cap F(W) \neq \emptyset$. Set

$$W_L = F^{-1}(F(W) \cap L).$$

Then $M(\hat{A}_{W_L}) = W_L$ (cf. (4.3.4)iii)) is locally compact and it follows from $\gamma_{n-1}\hat{A}_W = \emptyset$ (cf. (16.2.2)) that $(\hat{A}_W|_{W_L}, W_L)$ is a m.m.a.(*), using (9.1.6) and (9.1.7). We can now apply theorem (15.1.2) to $(\hat{A}_W, W)$ and F. By (15.1.2)i) F is a homeomorphism onto the open subset $F(W) \in \mathbb{C}^n$, and $g \circ F^{-1} \in Hol(F(W))$ for each $g \in \hat{A}_W$ by the proof of (15.1.2)i) p. 246/247. Using (15.1.2)ii) we see that the map

$$\hat{A}_W \longrightarrow Hol(F(W)),\ g \longrightarrow g \circ F^{-1},$$

defines in fact a topological algebra isomorphism.
The inverse statement is evident. □

(17.1.6) Corollary [BRO 5]. Let A be a separable uF-algebra with locally compact spectrum, and let $\gamma_{n-1}A = \emptyset$. Then the following statements are equivalent:

i) For every $\varphi \in M(A)$ there exists a surjective algebra homomorphism $\eta_\varphi : {}_n\mathcal{O}_o \longrightarrow \hat{A}_\varphi$.

ii) M(A) can be equipped with the structure of an n-dimensional Stein manifold such that $\hat{A} = Hol(M(A))$, if we endow M(A) with this structure.

Proof. Assume i). Let $\varphi \in M(A)$ be an arbitrary point. By (4.4.3) φ is a G_δ-point.

For each $\varphi \in M(A)$ there exist an open, hemicompact and $\hat{A}$-convex neighbourhood U_φ of φ in M(A), an open set $W_\varphi \subset \mathbb{C}^n$ and a homeomorphism $F_\varphi : U_\varphi \longrightarrow W_\varphi$, so that the map

$$\hat{A}_{U_\varphi} \longrightarrow \mathrm{Hol}(W_\varphi),\ g \longrightarrow g\circ F_\varphi^{-1},$$

defines a topological algebra isomorphism (note that $F_\varphi \in (\hat{A}_{U_\varphi})^n$ by (17.1.2)).

Let $U_\varphi \cap U_\psi \neq \emptyset$. We conclude easily that

$$F_\varphi \circ F_\psi^{-1} : F_\psi(U_\varphi \cap U_\psi) \longrightarrow F_\varphi(U_\varphi \cap U_\psi)$$

is a biholomorphic mapping, i.e. M(A) can be equipped with the structure of an n-dimensional complex manifold such that $\hat{A} \subset \mathrm{Hol}(M(A))$. Since M(A) is $\hat{A}$-morphically convex we can endow M(A) with the structure of a Stein space such that $\hat{A} = \mathrm{Hol}(M(A),\mathcal{O})$, where $\mathrm{Hol}(M(A),\mathcal{O})$ denotes the algebra of all holomorphic functions with respect to this structure, cf. (11.1.3).

Let f be a holomorphic map on M(A) with respect to the manifold structure, i.e. $f\circ F_\varphi^{-1} \in \mathrm{Hol}(U_\varphi)$ for all $\varphi \in M(A)$. Then

$$f|_{U_\varphi} \in \hat{A}_{U_\varphi} = \mathrm{Hol}(M(A),\mathcal{O})_{U_\varphi} \quad \text{for all } \varphi \in M(A),$$

and it follows that both structures coincide on M(A), since all elements of $\mathrm{Hol}(M(A),\mathcal{O})_{U_\varphi}$ are holomorphic on U_φ with respect to the new structure, because $\mathrm{Hol}(U_\varphi,\mathcal{O})$ is complete with respect to the compact open topology.

□

(17.1.7) The next theorem shows that $\hat{A}$ has still analytic structure in $\varphi \in M(A)$ if $\hat{A}_\varphi$ is only a finite ${}_n\mathcal{O}_o$-algebra.

Theorem [BRO 5]. Let A be a uF-algebra with locally compact spectrum $M(A)$, and let $\varphi \in M(A)\setminus\gamma_{n-1}A$ be a G_δ-point. Then the following statements are equivalent:

i) $\hat{A}_\varphi$ is a finite ${}_n\mathcal{O}_o$-algebra (via some algebra homomorphism η);

ii) There exists a neighbourhood U of φ which can be endowed with an analytic space structure of pure dimension n such that $\hat{A}_U = \mathrm{Hol}(U)$ with respect to this structure.

Proof. Assume i). Choose a neighbourhood U of φ and $F \in (\hat{A}_U)^n$ such that $\eta = F^*$, according to (17.1.4). Recall that $F(\varphi) = 0$.
Let $\tilde{h}_1,\dots,\tilde{h}_r$ generate $\hat{A}_\varphi$ over $F^*({}_n\mathcal{O}_o)$, and choose representatives $h_1,\dots,h_r$ in $\hat{A}_W$, $W \subset U$ an open neighbourhood of φ. Set

$$\pi: W \longrightarrow \mathbb{C}^{n+r},\ \psi \longrightarrow (F(\psi),H(\psi)),$$

where $H = (h_1,\dots,h_r)$.

α) Claim. π is injective in some neighbourhood of φ.
Set $\pi(\varphi) = p$. Let $g \in A$ be an arbitrary element. We have

$$\hat{g}_\varphi = \sum_{i=1}^{r} F^*(\tilde{s}_i)\tilde{h}_i \quad \text{for some } \tilde{s}_1,\dots,\tilde{s}_r \in {}_n\mathcal{O}_o.$$

Choose a polydisc D centered at $0 \in \mathbb{C}^n$ so that each $\tilde{s}_i$ is represented by a function $s_i \in \mathrm{Hol}(D)$. Choose a neighbourhood $V \subset W$ of φ such that $F(V) \subset D$ and

$$\hat{g} = \sum_{i=1}^{r} (s_i \circ F)h_i \quad \text{on } V.$$

Define

$$G: D\times\mathbb{C}^r \longrightarrow \mathbb{C},\ (z,w) \longrightarrow \sum_{i=1}^{r} s_i(z)w_i.$$

Then $G \in \mathrm{Hol}(D\times\mathbb{C}^r)$ and

$$\hat{g} = G\circ\pi \text{ on } V.$$

Our assertion follows now from lemma (17.1.3). W.l.o.g. we assume that π is injective on V.

β) <u>Claim.</u> There exists a neighbourhood $\tilde{V}$ of φ such that

$$\#F^{-1}(z)\cap\tilde{V} < \infty \text{ for all } z \in F(\tilde{V}).$$

Each $\tilde{h}_i$ is integral over $F^*({}_n\mathcal{O}_o)$, i.e. there are monic polynomials $p_i \in F^*({}_n\mathcal{O}_o)[z]$ so that $p_i(\tilde{h}_i) = 0$ for $1\le i\le r$ (cf. [K/K],p. 87). Let

$$p_i = z^{k_i} + \tilde{a}^{(i)}_{k_i-1}z^{k_i-1} + \ldots + \tilde{a}^{(i)}_o .$$

Choose a neighbourhood $\tilde{V} \subset V$, a neighbourhood $\tilde{D}$ of $0 \in \mathbb{C}^n$ and elements

$$a_o^{(1)},a_1^{(1)},\ldots,a_{k_1-1}^{(1)},a_o^{(2)},\ldots,a_{k_r-1}^{(r)} \in \mathrm{Hol}(\tilde{D})$$

such that

$$a_o^{(1)}\circ F,\ldots,a_{k_r-1}^{(r)}\circ F \text{ are representatives of } \tilde{a}_o^{(1)},\ldots,\tilde{a}_{k_r-1}^{(r)} \text{ on } \tilde{V}$$

and

$$(*)\qquad h_i^{k_i} + (a_{k_i-1}^{(i)}\circ F)h_i^{k_i-1} + \ldots + a_o^{(i)}\circ F = 0 \quad \text{on } \tilde{V},\ 1\le i\le r.$$

Let $\tilde{\psi} \in \tilde{V}$, set $x = F(\tilde{\psi})$. If $\psi \in \tilde{V}\cap F^{-1}(x)$ then

$$h_i^{k_i}(\psi) + a_{k_i-1}^{(i)}(x)h_i^{k_i-1}(\psi) + \ldots + a_o^{(i)}(x) = 0 \text{ by } (*),$$

so

$$\{h_i(\psi)\colon \psi \in F^{-1}(x)\cap\tilde{V}\}$$

is a finite set for each i, hence

$$\{H(\psi)\colon \psi \in F^{-1}(x)\cap\tilde{V}\}$$

is a finite set.

This implies that $F^{-1}(x)\cap\tilde{V}$ is a finite set, since H is injective on this set by α).

γ) Since $\gamma_{n-1}A$ is a closed subset of M(A) we can choose a compact $\hat{A}$-convex neighbourhood K of φ such that

$$\varphi \in M(\hat{A}_K)\backslash\gamma_{n-1}\hat{A}_K \text{ (cf. the proof of (16.2.3)).}$$

W.l.o.g. we can assume by β) that there is $F \in (\hat{A}_K)^n$ such that $\#F^{-1}(z) < \infty$ for all $z \in F(K)$ and

$$\{\varphi\} = F^{-1}(F(\varphi)).$$

Hence $F(\varphi)$ is not contained in $F(\gamma_{n-1}\hat{A}_K)$ and there is an open neighbourhood U of φ, $U \subset K$, which can be endowed with the structure of a pure n-dimensional analytic space such that

$$(*) \qquad \hat{A}|_U \subset \mathrm{Hol}(U,\mathcal{O}),$$

where $\mathrm{Hol}(U,\mathcal{O})$ denotes the algebra of all functions which are holomorphic on U with respect to this structure, cf. (16.1.4) and note that $M(\hat{A}_K) = K$. More precisely there exist an open neighbourhood U of φ in K, an open set $W \subset \mathbb{C}^n$ and an analytic subset S of W such that (U,F,W) is an analytic cover with critical set S. By (4.3.7)ii) we can assume that U is $\hat{A}$-convex. Note that $\hat{A}_U \subset \mathrm{Hol}(U,\mathcal{O})$ by $(*)$.
Let $N \subset U \backslash F^{-1}(S)$ be an open, hemicompact and $\hat{A}$-convex set such that $F|_N$ is a homeomorphism. Then $g \circ (F|_N)^{-1} \in \mathrm{Hol}(F(N))$ for all $g \in A$ by the proof of (15.1.2) p.246/247, hence $g \circ (F|_N)^{-1} \in \mathrm{Hol}(F(N))$ for all $g \in A_N$, since $\mathrm{Hol}(F(N))$ is complete and we conclude as in the proof of (15.1.2)ii) that

$$(**) \qquad A_N = \mathrm{Hol}(N,\mathcal{O}).$$

Since U is $\hat{A}_U$-morphically convex, U can be given the structure of a Stein space so that $\hat{A}_U = \mathrm{Hol}(U,\mathcal{O}')$ by (11.1.3), where $\mathrm{Hol}(U,\mathcal{O}')$ denotes the algebra of all functions which are holomorphic with respect to this structure. Note that $\mathrm{Hol}(U,\mathcal{O}') \subset \mathrm{Hol}(U,\mathcal{O})$.
Let $f \in \mathrm{Hol}(U,\mathcal{O})$ be an arbitrary element, let N be as above, then $f|_N \in \hat{A}_N$ by $(**)$, hence $f|_N \in \mathrm{Hol}(U,\mathcal{O}')_N \subset \mathrm{Hol}(N,\mathcal{O}')$, since $\mathrm{Hol}(N,\mathcal{O}')$ is complete with respect to the compact open topology. So

$$\mathrm{Hol}(U,\mathcal{O})|_{U\backslash F^{-1}(S)} = \mathrm{Hol}(U,\mathcal{O}')|_{U\backslash F^{-1}(S)}.$$

Hence

$$\mathrm{Hol}(U,\mathcal{O}) = \mathrm{Hol}(U,\mathcal{O}')$$

since $F^{-1}(S)$ is nowhere dense in U and both algebras are subalgebras of $C(U)$.
For the inverse statement cf. [NAR 2], chapter III. □

(17.1.8) Corollary. Let A be a separable uF-algebra with locally compact spectrum. Suppose that $\gamma_{n-1}A = \emptyset$ and that $\hat{A}_\varphi$ is a finite ${}_n\mathcal{O}_o$-algebra for each $\varphi \in M(A)$. Then M(A) can be given the structure of an n-dimensional Stein space such that $\hat{A} = \mathrm{Hol}(M(A))$ with respect to this structure.

(17.1.9) Remarks. i) The connection of analytic structure and algebraic relations between $\hat{A}_\varphi$ and ${}_n\mathcal{O}_o$ has been established by various authors (cf. [BRO 6], [CAR 2], [CLA 2], [LOY]).
We state a result due to Brooks [BRO 6]: Let V be an analytic subset of an open set $U \subset \mathbb{C}^n$, $0 \in V$. Denote by ${}_V\mathcal{O}_o$ the algebra of germs of holomorphic functions on V at 0 (analogous definition as in (17.1.1)).
Let A be a uF-algebra with locally compact spectrum, and let $\varphi \in M(A)$ be a G_δ-point. Suppose that $\hat{A}_\varphi$ is algebraically isomorphic to ${}_V\mathcal{O}_o$. Then there exist an open neighbourhood U of φ in M(A), an open neighbourhood W of 0 in $\mathbb{C}^n$ and a homeomorphism
$F: W \cap V \longrightarrow U$ such that the map

$$\hat{A}_U \longrightarrow \mathrm{Hol}(W \cap V),\ f \longrightarrow f \circ F,$$

is a (well-defined) topological algebra isomorphism.
ii) Carpenter [CAR 2] called a uF-algebra homogeneous provided that for each pair $\varphi, \psi \in M(A)$ the algebras $\hat{A}_\varphi$ and $\hat{A}_\psi$ are isomorphic. He proved that if A is a singly generated homogeneous uF-algebra then either A is topologically and algebraically isomorphic to Hol(G), G an open subset of the plane, or A = C(M(A)).

CHAPTER 18

REFLEXIVE uF-ALGEBRAS

In this section we deal with uF-algebras, which are reflexive as a topological vectorspace. Recall that a Fréchet space F is called reflexive if F = F'', where F'' denotes the strong bidual of F. Using Montel's theorem one easily sees that algebras of holomorphic functions are reflexive. On the other hand we shall see that reflexive uF-algebras behave in many aspects like algebras of holomorphic functions. The key step in this context is the observation that reflexive uF-algebras have "big" Gleason parts, more precisely the Gleason parts are exactly the components of the spectrum. (Here we have to extend the definition of Gleason parts from uB-algebras to uF-algebras.) It follows from this result that reflexive uF-algebras are antisymmetric, and that they satisfy the maximum principle on the spectrum. If (A,X) is a reflexive uB-algebra, i.e. X is compact, then X consists of finitely many points $x_1,\ldots,x_n$ and A is topologically and algebraically isomorphic to $C(\{x_1,\ldots,x_n\})$.
Most of the results in this chapter can be found in [GKV].

(18.1) Reflexive uF-algebras

(18.1.1) Definition. i) A uF-algebra (A,X) is called reflexive if it is reflexive as a topological vectorspace.

Remarks. i) A Fréchet space is reflexive if F = F''. We give an equivalent description which is more relevant for our work.
Let (A,X) be a uF-algebra, and let $(K_n)_n$ be an admissible exhaustion

of X. A set $E \subset A$ is called bounded if for each $m \in \mathbb{N}$ there exists $m(n) \geq 0$ such that

$$\|f\|_{K_n} \leq m(n) \text{ for all } f \in E.$$

The weak topology on A is generated by the set of all seminorms of the following form

$$(*) \qquad p(f) = \sup\{|a_j(f)| : j=1,\dots,r\},$$

where $a_1,\dots,a_r$ are elements of the topological dual A'.
Then (A,X) is reflexive if every bounded subset of A is weakly relatively compact ([JAR], p. 227, 228).

ii) We mention that closed subspaces of reflexive Fréchet spaces are again reflexive ([JAR], p. 228).
Unfortunately this permanence property does not hold for quotients, i.e. there exist a reflexive Fréchet space F and a closed subspace L of F such that F/L is not reflexive (F/L endowed with the quotient topology). In fact we can assume that F is even a Montel space , cf. [KÖT], p. 433 . (Recall that a Fréchet space E is called Montel space if each bounded subset of E is relatively compact. Hence a Fréchet Montel space is reflexive ([KÖT],p. 369).)

iii) We shall need the following conclusion:
Let (A,X) be a reflexive uF-algebra, and let $(f_n)_n \subset A$ be a bounded sequence. Then there exists a subsequence of $(f_n)_n$ which converges pointwise to an element $f \in A$.
To see this let $(f_{n_k})_k$ be a subsequence which converges weakly to $f \in A$, according to remark i). Let $x \in X$, then the point evaluation functional φ_x is an element of A', hence

$$A \longrightarrow \mathbb{R},\ f \longrightarrow |f(x)|,$$

defines a continuous seminorm (with respect to the weak topology on A) by (*), i.e. $f_{n_k}(x) \longrightarrow f(x)$ as $k \longrightarrow \infty$.

iv) Let X be a hemicompact reduced complex space. Then Hol(X) is a Montel space (cf. remark (2.3.1)), hence Hol(X) is a reflexive uF-algebra.

(18. 2) Gleason parts

(18.2.1) In (1.5.1) we introduced Gleason parts for uB-algebras. We stated one result which shows that in certain cases Gleason parts provides analytic structure in the spectrum (for other results on Gleason parts cf. [STO] chapter III, [GAM] chapter VI).
We now generalize the notion of Gleason parts to uF-algebras. Recall that if A is a uB-algebra, then two points φ, $\psi \in M(A)$ lie in the same Gleason part of A (we write $\varphi \sim \psi$) if

$$\sup\{|\varphi(f)|: f \in A, \|f\| \leq 1, \psi(f) = 0\} < 1.$$

(*) We note that "$\sim$" is an equivalence relation on M(A).
Let A be a uF-algebra, $K \subset M(A)$ be a compact subset and φ, $\psi \in K$. By (4.3.3) we can identify $M(\hat{A}_K)$ with $\hat{K}_A$, hence we can naturally interpret φ, ψ as elements of $M(\hat{A}_K)$.

Definition. Let A be a uF-algebra. We say that two points $\varphi, \psi \in M(A)$ belong to the same Gleason part (we write $\varphi \sim \psi$) if there is a compact set $K \subset M(A)$ such that φ, ψ belong to the same Gleason part of $\hat{A}_K$.

Remarks. i) Let K, L be compact subsets of M(A), $K \subset L$. If φ, ψ belong to the same Gleason part of $\hat{A}_K$ then it is easy to see that they also lie in the same Gleason part of $\hat{A}_L$.
ii) "$\sim$" is an equivalence relation. The transitivity can be shown by using i) and (*).
iii) Let $(K_n)_n$ be an admissible exhaustion of M(A). Then $\varphi \sim \psi$ iff there exists $n \in \mathbb{N}$ such that φ, ψ lie in the same Gleason part of $\hat{A}_{K_n}$ (we write $\varphi \underset{n}{\sim} \psi$).

Example. Let

$$K_n = \{(z,w): |z| \le n,\ |w| \le n,\ \mathrm{Im}\ z = 0\},$$

and set $A = \varprojlim P(K_n)$, cf. example (14.1.3)ii). We have already proved that

$$M(A) = \{(z,w): \mathrm{Im}\ z = 0\},$$

that $(K_n)_n$ is an admissible exhaustion of $M(A)$, and that the induced euclidean topology and the Gelfand topology coincide on $M(A)$.
We want to determine the Gleason parts of A. Let (x_1,x_2), (y_1,y_2) be two points in $M(A)$ with $x_1 > y_1$.
Choose a continuous real-valued function f on $\mathbb{R}$ such that $0 \le f \le 1$ and $f(t) = 1$ for $t \ge x_1$, $f(t) = 0$ for $t \le y_1$. Set

$$g: M(A) \longrightarrow \mathbb{R},\ (z,w) \longrightarrow f(z).$$

Then $g \in \hat{A}$ by the Stone-Weierstraß theorem, hence (x_1,x_2) and (y_1,y_2) are not contained in the same Gleason part of A.
Now let $x_1 = y_1$. Choose $n \in \mathbb{N}$ so that $\max\{|x_1|, |x_2|, |y_2|\} < n$, hence $(x_1,x_2), (y_1,y_2) \in K_n$. Assume that both points are not contained in the same Gleason part of $\hat{A}_{K_n} = P(K_n)$.

Then there exists a sequence $(f_l)_l$ in $P(K_n)$ such that

i) $\|f_l\|_{K_n} \le 1$,

ii) $f_l(x_1,x_2) = 0$,

iii) $f_l(y_1,y_2) = 1-(1/l)$ for all $l \in \mathbb{N}$.

Since for each $f \in P(K_n)$

$$\tilde{f}: \{w \in \mathbb{C}: |w|<n\} \longrightarrow \mathbb{C},\ w \longrightarrow f(x_1,w),$$

is a holomorphic function it follows from Montel's theorem that a subsequence of $(\tilde{f}_l)_l$ converges to a function $g \in \mathrm{Hol}(\{|z|<n\})$. Since $|g| \le 1$ and $g(x_1,y_2) = g(y_1,y_2) = 1$, g must be constant on $\{|z|<n\}$, and we get a contradiction to $g(x_1,x_2) = 0$, so $(x_1,x_2) \sim (y_1,y_2)$ if $x_1=y_1$.

(18.2.2) Remark. Let A be a uF-algebra, and let V be a connected analytic subset in some domain $G \subset \mathbb{C}^n$. Suppose there exists a non-constant continuous map $F: V \longrightarrow M(A)$ such that $f \circ F \in Hol(V)$ for all $f \in \hat{A}$. Then, in a similar way as in the example, we can prove that $\varphi \sim \psi$ for all $\varphi, \psi \in F(V)$.

(18.2.3) Theorem [GKV]. Let A be a reflexive uF-algebra. Then
i) the Gleason parts of A are open and closed subsets of M(A),
ii) if M(A) is connected and $K \subset M(A)$ is a compact subset, then there is a compact set $L \subset M(A)$ such that all points of K belong to the same Gleason part of $\hat{A}_L$.

Proof. i) Let $(K_n)_n$ be an admissible exhaustion of M(A).
We argue indirectly. Suppose there exists $x \in M(A)$ such that the equivalence class

$$[x] = \{y \in M(A): x \sim y\}$$

is not an open subset of M(A).
Then in every neighbourhood U of x there is y_U such that y_U is not contained in [x]. This means that y_U and x lie in different Gleason parts of $\hat{A}_{K_n}$ if $x, y_U \in K_n$, i.e. for each $n \in \mathbb{N}$ such that $x, y_U \in K_n$ we find $f_n^U \in \hat{A}_{K_n}$ so that

$$f_n^U(y_U) = 0, \ \|f_n^U\|_{K_n} \leq 1 \text{ and } f_n^U(x) = 1-(1/n).$$

Approximating f_n^U by elements of $\hat{A}$ we can w.l.o.g. assume that $f_n^U \in \hat{A}$ for each U of the neighbourhood filter Q of x and each $n \in \mathbb{N}$. The sequence $(f_n^U)_n$ is bounded in $\hat{A}$. Since A - and hence $\hat{A}$ - is reflexive we get a function $f^U \in \hat{A}$ so that

$$\|f^U\|_{K_n} \leq 1 \text{ for all } n \in \mathbb{N}, \ f^U(y_U) = 0 \text{ and } f^U(x) = 1,$$

cf. remark (18.1.1)iii).
Now define for every finite subset $M \subset Q$ the function

$$g_M = \prod_{U \in M} f^U.$$

The g_M's, where the M's are directed by inclusion, are a bounded net in the reflexive space $\hat{A}$ which therefore has a subnet converging weakly to some $g \in \hat{A}$.

Since $g(y_U) = 0$ for all $U \in Q$ and $g(x) = 1$ by remark (18.1.1)iii), g is not continuous. This is a contradiction. Hence $[x]$ is open. On the other hand $[x]$ is closed since

$$[x] = M(A) \setminus \bigcup_{y \in M(A) \setminus [x]} [y].$$

ii) Let $(K_n)_n$ be an admissible exhaustion of M(A). W.l.o.g. $K \subset K_1$. As in i) we argue indirectly. Suppose there is $\varphi \in K$ such that for each $n \in \mathbb{N}$ we can find $\varphi_n \in K$ so that φ and φ_n lie in different Gleason parts of $\hat{A}_{K_n}$. Since K is compact a subsequence of $(\varphi_n)_n$ converges to $\psi \in K$, w.l.o.g. we assume that $(\varphi_n)_n$ converges to ψ. By i) $[\varphi]$ is an open and closed subset of M(A), hence $[\varphi] = M(A)$ by our assumption, and there is $l \in \mathbb{N}$ such that $\varphi \underset{l}{\sim} \psi$. Note that for $n > l$ ψ and φ_n lie in different Gleason parts of $\hat{A}_{K_n}$, since otherwise $\varphi \underset{n}{\sim} \varphi_n$.

We can assume that $l=1$. So for $k \geq 2$ there exists $g_k \in \hat{A}_{K_k}$ so that

α) $\|g_k\|_{K_k} \leq 1$,

β) $g_k(\varphi_k) = 0$,

γ) $1 \geq |g_k(\psi)| \geq a_k > 0$,

where the sequence $(a_k)_k$ is chosen in such a way that $\prod_{k=2}^{\infty} a_k > 0$.

Since $\hat{A}|_{K_n}$ is dense in $\hat{A}_{K_n}$ we can assume that $g_k \in \hat{A}$ for all $k \geq 2$. By α) and the reflexivity of $\hat{A}$ a subsequence of

$$h_m = g_2 \cdot \ldots \cdot g_m, \; m \geq 2,$$

converges weakly to $h \in \hat{A}$. But $|h(\psi)| > 0$ by γ), while $h(\varphi_k) = 0$ for $k \geq 2$, cf. remark (18.1.1)iii). This is a contradiction to the continuity of h.

□

(18.2.4) Corollary. If A is a reflexive uF-algebra, then the Gleason parts of A are the components of M(A).

Proof. i) Let $\varphi \in M(A)$. We claim that $[\varphi]$ is connected. By (18.2.3) $[\varphi]$ is a closed and open subset of M(A). If $[\varphi]$ is not connected, there exists a closed and open nonempty proper subset $M \subset [\varphi]$. By Shilov's idempotent theorem (6.1.5) the characteristic function χ_M is contained in $\hat{A}$. But this implies that the points of M and $[\varphi]\backslash M$ cannot belong to the same Gleason part, a contradiction to the definition of $[\varphi]$.
ii) Let $L \subset M(A)$ be a component, $\varphi \in L$. Then $[\varphi] = L$. □

(18.2.5) Remark. Let A be a reflexive uF-algebra, and let $\varphi \in M(A)$. As already proved $[\varphi]$ is a closed and open subset of M(A), hence $(\hat{A}_{[\varphi]},[\varphi])$ is a reflexive uF-algebra, since it can be regarded as a closed subspace of (the topological vectorspace) A, cf. (6.1.8). So we can sometimes restrict our investigations to reflexive uF-algebras with connected spectrum.

(18.2.6) Holomorphic functions on connected compact complex spaces are constant. We prove that an analogue is true for reflexive uF-algebras.

Theorem. Let (A,X) be a reflexive uF-algebra on a compact space X, i.e. (A,X) is a reflexive uB-algebra. Then X consists of finitely many points $x_1,\dots,x_n$ and A is topologically and algebraically isomorphic to $C(\{x_1,\dots x_n\})$.

Proof. By our assumption M(A) is a compact space. Let $\varphi \in M(A)$ be an arbitrary point, then $[\varphi]$ is a closed and open subset of M(A) by (18.2.3). Hence $\chi_{[\varphi]} \in \hat{A}$ by Shilov's idempotent theorem (6.1.5). Suppose that $[\varphi]$ contains at least two points φ_1, φ_2. Choose $f \in A$

such that

$$\hat{f}(\varphi_1) = 0 \text{ and } \hat{f}(\varphi_2) \neq 0.$$

Then

$$\hat{h} = \hat{f}\cdot\chi_{[\varphi]}/\|\hat{f}\cdot\chi_{[\varphi]}\|_{M(A)} \in \hat{A},$$

and $\hat{h}(\varphi_1) = 0$ while $|\hat{h}(\varphi_3)| = 1$ for some $\varphi_3 \in [\varphi]$, hence φ_3 and φ_1 are not contained in the same Gleason part, and we get a contradiction to the definition of $[\varphi]$.

It follows that each $\varphi \in M(A)$ is an isolated point, hence M(A) consists only of a finite number of points and our claim is proved. □

Remark. There exist nontrivial semisimple reflexive (commutative) B-algebras. Keown [KEO] proved that these algebras are essentially l_p-spaces with a suitable defined multiplication.

(18.2.7) C({p}), the algebra of all (continuous) functions on a single point p, is clearly reflexive. (C({p}), {p}) is not a maximum modulus algebra according to the definition (14.1.1), so not every reflexive uF-algebra is a m.m.a..

The next theorem however shows that besides the trivial case $(\hat{A}, M(A))$ is indeed a m.m.a., if A is reflexive.

Theorem. Let A be a reflexive uF-algebra with connected spectrum M(A). Suppose that M(A) contains more than one point, then $(\hat{A}, M(A))$ is a m.m.a..

Proof. Let $K \subset M(A)$ be a compact set. Since M(A) is connected $\partial K = \emptyset$ would imply that $K = M(A)$, hence $K = \{x_1,\dots,x_n\}$ by (18.2.6), i.e. K would consist of a single point, a contradiction.

Suppose that there exists $f \in \hat{A}$ so that

$$(*) \qquad \|\hat{f}\|_K > \|\hat{f}\|_{\partial K}.$$

By (18.2.3)ii) there is a compact set $L \subset M(A)$, $K \subset L$, such that all points of K belong to the same Gleason part of $\hat{A}_L$. W.l.o.g. we can assume that L is $\hat{A}$-convex, i.e. $M(\hat{A}_L) = L$ by (4.3.3). By Rossi's maximum principle (1.4.9) and (*) int K contains a peak set E for $\hat{A}_L$, i.e. there exists $g \in \hat{A}_L$ so that

i) $\|g\|_L = 1 = g(x)$ for all $x \in E$,

ii) $|g(y)| < 1$ if $y \in L \setminus E$.

Let $y \in \partial K$, i.e. $y \in L\setminus E$, and let $x \in E$. Considering the sequence $(g^n)_n$, we see that x and y do not belong to the same Gleason part of $\hat{A}_L$, a contradiction. □

<u>Remark.</u> Let A be a reflexive uF-algebra, and let $K \subset M(A)$ be a compact subset, such that K contains no isolated point of M(A), then

$$\|\hat{f}\|_K = \|\hat{f}\|_{\partial K} \text{ for all } \hat{f} \in \hat{A}.$$

To prove this let $f \in A$ be an arbitrary element. Choose $\varphi \in K$ such that $\|\hat{f}\|_K = |\hat{f}(\varphi)|$. Since φ is not an isolated point of M(A), $[\varphi]$ consists of more than one point by (18.2.3)i).

Consider the reflexive uF-algebra $(\hat{A}_{[\varphi]}, [\varphi])$, cf. (18.2.5). We have $M(\hat{A}_{[\varphi]}, [\varphi]) = [\varphi]$ by (6.1.8). By (18.2.3)i) $[\varphi]$ is a closed subset of M(A), hence $K \cap [\varphi]$ is a compact set and we get

$$|\hat{f}(\varphi)| \le \|\hat{f}\|_{K\cap[\varphi]} = \|\hat{f}\|_{\partial(K\cap[\varphi])},$$

by (18.2.7), where $\partial(K \cap [\varphi])$ denotes the boundary with respect to $[\varphi]$. Since $[\varphi]$ is an open subset of M(A) by (18.2.3)i) we see that

$$\|\hat{f}\|_{\partial(K\cap[\varphi])} \le \|\hat{f}\|_{\partial K}.$$

<u>(18.2.8)</u> As a corollary of (15.1.1) and (18.2.7) we get a new characterization of the algebra Hol(G), G an open set of the plane.

<u>Lemma.</u> Let A be a singly generated reflexive uF-algebra, then M(A) is locally compact.

Proof. Let f be a generating element, and let $\varphi \in M(A)$ be an arbitrary point. W.l.o.g. we can assume that M(A) is connected, otherwise consider $(\hat{A}_{[\varphi]},[\varphi])$.
If there exists a point $\psi \in M(A)\backslash\{\varphi\}$ then choose a compact set $L \subset M(A)$ such that φ and ψ belong to the same Gleason part of $\hat{A}_L$, cf. (18.2.3)ii). We can assume that L is $\hat{A}$-convex, hence $\hat{f}(L)$ is a polynomially convex compact subset of the plane and $\hat{A}_L$ is topologically and algebraically isomorphic to $P(\hat{f}(L))$ via the mapping

$$P(\hat{f}(L)) \longrightarrow \hat{A}_L,\ g \longrightarrow g \circ \hat{f}|_L,$$

by (5.1.5). Then $\hat{f}(\varphi) \in \operatorname{int} \hat{f}(L)$, since otherwise we find $g \in P(\hat{f}(L))$ such that

$|g(\hat{f}(\varphi))| > |g(\hat{f}(\tilde{\varphi}))|$ for all $\tilde{\varphi} \in L\backslash\{\varphi\}$ by example (1.3.12), hence φ and ψ would belong to different Gleason parts of $\hat{A}_L$ (consider the sequence $(g^n \circ \hat{f}/g^n(\hat{f}(\varphi)))_n$). Since $\hat{f}: M(A) \longrightarrow \mathbb{C}$ is injective and continuous by (5.1.4) we see that L is a compact neighbourhood of φ.
If $\{\varphi\} = M(A)$ our assertion becomes trivial.

□

(18.2.9) Theorem. i) Let A be a rationally singly generated reflexive uF-algebra with locally compact spectrum. Then A is topologically and algebraically isomorphic to Hol(U)xC(N), where
- $U \subset \mathbb{C}$ is an open set (possibly empty) and
- N is an at most countable discrete topological space (possibly empty).

In particular if M(A) is connected, then A is isomorphic either to C({p}) or to Hol(U), $U \subset \mathbb{C}$ a domain.
ii) Let A be a singly generated reflexive uF-algebra. Then A is topologically and algebraically isomorphic to Hol(U)xC(N), where
- $U \subset \mathbb{C}$ is a polynomially convex open set (possibly empty)
- N is an at most countable discrete topological space (possibly empty).

If M(A) is moreover connected, then A is isomorphic to either C({p}), Hol($\mathbb{C}$) or Hol(D), D the open unit disc.

Proof. i) By (18.2.3),(18.2.4) $\varphi \in M(A)$ is an isolated point of M(A) iff $\{\varphi\} = [\varphi]$. Denote by N the union of all isolated points of M(A). Then N is an open and closed subset of M(A) by (18.2.3), since

$$N = \bigcup_{\varphi \in N} [\varphi] \text{ and } M(A)\backslash N = \bigcup_{\varphi \in M(A)\backslash N} [\varphi].$$

N is at most countable since M(A) is hemicompact and N has no accumulation point.

The mapping

$$(*) \qquad (\hat{A},M(A)) \longrightarrow (\hat{A}_N,N)\times(\hat{A}_{M(A)\backslash N},M(A)\backslash N),$$

$$\hat{g} \longrightarrow (\hat{g}\chi_N,\hat{g}\chi_{M(A)\backslash N}),$$

defines a topological algebra isomorphism by Shilov's idempotent theorem (6.1.5).

Clearly $\hat{A}_N \subset C(N)$. Let $h \in C(N)$ be an arbitrary element, then

$$h = \sum_{\varphi \in N} h(\varphi)\chi_{\{\varphi\}},$$

and the right side defines an element of $\hat{A}_N$ by (6.1.5), hence $\hat{A}_N = C(N)$.

The algebra $(\hat{A}_{M(A)\backslash N},M(A)\backslash N)$ is rationally singly generated by $\hat{f}\chi_{M(A)\backslash N}$ if A is rationally singly generated by f. Moreover $\hat{A}_{M(A)\backslash N}$ is reflexive, cf. (6.1.8) and remark (18.1.1)ii), hence $(\hat{A}_{M(A)\backslash N},M(A)\backslash N)$ is a m.m.a. by remark (18.2.7), since $M(\hat{A}_{M(A)\backslash N}) = M(A)\backslash N$ by (4.3.4)iii). It follows from (14.1.4)ii) that $\hat{f}\chi_{M(A)\backslash N}$ is an open mapping, hence a homeomorphism by (5.2.3), and $(\hat{A}_{M(A)\backslash N},M(A)\backslash N)$ is isomorphic to $\mathrm{Hol}(\hat{f}(M(A)\backslash N))$ by (12.1.1).

ii) Let A be singly generated by f, then M(A) is locally compact by (18.2.8). By (12.1.1) $\hat{f}(M(A)\backslash N)$ is a polynomially convex open set in $\mathbb{C}$.

If M(A) is connected and A is not isomorphic to Hol($\mathbb{C}$) or C({p}), then A is isomorphic to Hol(U), U $\neq$ $\mathbb{C}$ a polynomially convex domain in the plane. By the Riemann mapping theorem there exists a biholomorphic map h: U $\longrightarrow$ D, hence Hol(U) is topologically and algebraically isomorphic to Hol(D) via the map

$$\mathrm{Hol}(D) \longrightarrow \mathrm{Hol}(U),\ g \longrightarrow g\circ h.$$

□

<u>Remarks.</u> i) We do not know whether the assumption "locally compact spectrum" is necessary in part i) of the theorem. We cannot copy the proof of lemma (18.2.8) in this case, since there are compact subsets K $\subset$ $\mathbb{C}$ such that not every boundary point of K is a peak point for R(K). We can however sharpen i) in the following way:
Let A $\neq$ C({p}) be a reflexive uF-algebra with connected spectrum. Suppose there are f $\in$ A and $x_1,\dots,x_n \in \mathbb{C}$ so that

α) $f-x_i\cdot 1$ is invertible in A for i=1,...,n;

β) the polynomials in f, $(f-x_1)^{-1}, \dots, (f-x_n)^{-1}$ are dense in A.

Then $\hat{f}(M(A))$ is a domain in $\mathbb{C}$ and A is topologically and algebraically isomorphic to $(\mathrm{Hol}(\hat{f}(M(A))), \hat{f}(M(A)))$.
The proof runs along the lines of (18.2.8), (18.2.9) using the fact that for each $\hat{A}$-convex compact set K $\subset$ M(A) our assumptions imply that every boundary point of $\hat{f}(K)$ is a peak point for $R(\hat{f}(K))$.
ii)It is important to note that for n-generated uF-algebras the assumption "reflexive" does not imply that the spectrum of the algebra is locally compact. We give an example.
For n$\in$ $\mathbb{N}$ consider the following sets in $\mathbb{C}^2$:

$$K_n = \{|z| \le 1-(1/n),\ |w| \le 1-(1/n)\} \cup \{|z| \le n,\ w = 0\}.$$

Then each K_n is polynomially convex. To see this consider the polynomials z_1, z_2, and $z_2\cdot(z_1)^l$ ($l \in \mathbb{N}$), where z_1, z_2 denote the coordinate functions.
Set A = $\varprojlim P(K_n)$. A is a doubly generated uF-algebra, and M(A) can be identified as a set with

$$\cup K_n = D \cup \{(z,w) \in \mathbb{C}^2 : w = 0\},$$

where D denotes the open unit bicylinder, cf. (5.1.6).

α) A is reflexive.

By (5.1.7)i) each $\hat{f} \in \hat{A}$ can be interpreted as a complex-valued function on $\cup K_n$ which can be approximated uniformly on every K_n by polynomials. Hence the map

$$T: \hat{A} \longrightarrow \mathrm{Hol}(D)\times\mathrm{Hol}(\mathbb{C}),\ \hat{f} \longrightarrow (\hat{f}|_D, f|_{\{(z,w):\ w=0\}}),$$

is well-defined. In fact it is easy to show that $\hat{A}$ - via the map T - can be regarded as a closed subspace of the reflexive space $\mathrm{Hol}(D)\times\mathrm{Hol}(\mathbb{C})$. So $\hat{A}$ is reflexive by remark (18.1.1)ii).(The same argument can be used to show that $\hat{A}$ is a Schwartz space, cf. next chapter.)

β) In the same way as in example (5.1.12) we can show that M(A) is not locally compact. More precisely no point of the set $\{|z|=1, w=0\}$ has a compact neighbourhood in M(A) (with respect to the Gelfand topology).

iii) We do not know whether an analogue to (18.2.9) is valid for n-generated reflexive uF-algebras with locally compact spectrum. Of course we cannot expect that n-generated reflexive uF-algebras are always of the type Hol(G), $G \subset \mathbb{C}^n$ a domain. For instance let

$$X = \{(z,w) \in \mathbb{C}^2 : z\cdot w=0\}.$$

Then X is an analytic subset of $\mathbb{C}^2$, hence (Hol(X),X) is a reflexive uF-algebra, since Hol(X) is a Montel space, in fact Hol(X) is a Schwartz space by (2.3.8).

We claim that Hol(X) is doubly generated by the coordinate functions. Let $f \in \mathrm{Hol}(X)$ be an arbitrary element. By (2.3.3) there exists $F \in \mathrm{Hol}(\mathbb{C}^2)$ such that $F|_X = f$. Using the power series expansion of F (cf. (2.1.5)) one sees easily that F can be approximated uniformly on each compact set $K \subset \mathbb{C}^2$ by polynomials, in particular f can be approximated uniformly on each compact set $K \subset X$ by polynomials.

Suppose there is a domain $G \subset \mathbb{C}^n$ such that Hol(X) is topologically and algebraically isomorphic to Hol(G). Then Hol(X) is isomorphic to (Hol(G)^ ,M(Hol(G))), where Hol(G)^ denotes the algebra of the Gelfand transforms and M(Hol(G)) can be given the structure of a holomorphically convex Riemann domain over $\mathbb{C}^n$ such that all elements of Hol(G)^ are holomorphic with respect to this structure, cf. (2.2.5). By (2.3.5) there exists a biholomorphic map φ: M(Hol(G)) $\longrightarrow$ X. This cannot happen since for example $X\setminus\{(0,0)\}$ is not connected while $M(Hol(G))\setminus\varphi^{-1}\{(0,0)\}$ is connected.

So the question left open should be formulated in the following way: Are there (rationally) n-generated reflexive uF-algebras with locally compact spectrum, which are not Stein algebras?

Note that ii) shows that the assumption "locally compact spectrum" is necessary, since Stein algebras always have locally compact spectrum.

<u>(18.2.10)</u> Let (A,X) be a m.m.a.. We proved in (14.1.4) that $\partial f(K) \subset f(\partial K)$ for every $f \in A$ and each compact set $K \subset X$. By (18.2.7) this result holds also for reflexive uF-algebras $(\hat{A},M(A))$ with connected spectrum, $A \neq C(\{p\})$.

In fact we can improve this result for reflexive uF-algebras. Denote by λ the Lebesgue measure in the plane and put for $x \in \mathbb{C}$, $\delta>0$ $D_{x,\delta} = \{z \in \mathbb{C}: |z-x| < \delta\}$.

<u>Theorem.</u> Let A be a reflexive uF-algebra with connected spectrum M(A). Let $\hat{f} \in \hat{A}$ be a nonconstant function, and let $y \in M(A)$. Then there is a compact set $K \subset \hat{f}(M(A))$, such that $f(y) \in K$ and

(*) $$\lim_{\delta \to 0} \lambda(K \cap D_{f(y),\delta})/\lambda(D_{f(y),\delta}) = 1.$$

In particular $K \cap D_{f(y),\delta}$ has positive Lebesgue measure for all $\delta>0$.

<u>Proof.</u> Let $(K_n)_n$ be an admissible $\hat{A}$-convex exhaustion of M(A), cf. (4.3.4)i).

For every $n \in \mathbb{N}$ denote by B_n the algebra of all continuous functions on $\hat{f}(K_n)$ which can be approximated uniformly on $\hat{f}(K_n)$ by rational functions with poles off $\hat{f}(M(A))$. Clearly B_n is a uB-algebra. The restriction mapping

$$B_{n+1} \longrightarrow B_n,\ g \longrightarrow g|_{\hat{f}(K_n)},$$

has dense range. Hence $B = \varprojlim B_n$ is a uF-algebra by (4.1.3). Consider the mapping

$$T\colon B = \varprojlim B_n \longrightarrow \hat{A} = \varprojlim \hat{A}_{K_n},\ (g_n)_n \longrightarrow (g_n \circ \hat{f}|_{K_n})_n$$

Then T is well-defined (use (1.4.3), note that $M(\hat{A}_{K_n}) = K_n$ by (4.3.3) and that $\hat{A}_{K_n}$ is a uB-algebra, i.e. complete with respect to the norm $\|\cdot\|_{K_n}$). Moreover T is an algebra homomorphism which maps B isomorphically onto a closed subalgebra of $\hat{A}$. Hence B is reflexive by remark (18.1.1)ii).

Note that for each $n \in \mathbb{N}$ B_n is rationally singly generated by the function

$$z\colon \hat{f}(K_n) \longrightarrow \mathbb{C},\ x \longrightarrow x.$$

Set $L_n = \hat{z}(M(B_n))$. Suppose there exists $y \in L_n \backslash \hat{f}(M(A))$. Let $\varphi \in M(B_n)$ such that $\hat{z}(\varphi)=y$. Then

$$1 = \varphi(1) = \varphi((z-y)/(z-y)) = 0,$$

and we get a contradiction. Hence L_n is a compact subset of $\hat{f}(M(A))$ which contains $\hat{f}(K_n)$ and B_n is topologically and algebraically isomorphic to $R(L_n)$. Moreover

$$\dots \longrightarrow R(L_{n+1}) \longrightarrow R(L_n) \longrightarrow \dots$$

constitutes a dense projective system with respect to the restriction mappings. B is topologically and algebraically isomorphic to $C = \varprojlim R(L_n)$ and $(L_n)_n$ is an admissible exhaustion of $M(C)=\hat{f}(M(A))$ with respect to the Gelfand topology by (3.3.4) and (1.2.4)iii).

Suppose there exists a point $y \in M(A)$ such that (*) is not valid, then the Gleason part of $R(L_n)$ which contains $\hat{f}(y)$ is trivial for all $n \in \mathbb{N}$ by (1.5.4). Since $M(B)$ is connected this implies that $M(B)=\hat{f}(y)$ by (18.2.3), i.e. $\hat{f}$ is constant and we get a contradiction. □

Remark. We are not able to decide whether f - in the situation above - is an open mapping. Note that this question is closely related to the question whether rationally singly generated reflexive uF-algebras are always of type Hol(G), $G \subset \mathbb{C}$ a domain (remark (18.2.9)i)).

(18.2.11) We call a uF-algebra (A,X) antisymmetric provided that each real-valued function $f \in A$ is constant.
It follows from the open mapping theorem (cf. (2.1.8) resp. remark (2.3.1)) that (Hol(X),X), X a connected reduced holomorphically separable space, is antisymmetric. We show that the same is true for reflexive uF-algebras.

Theorem. Let (A,X) be a reflexive uF-algebra on a connected space X. Then (A,X) is antisymmetric.

Proof. Let $f \in A$ be a real-valued function. We claim that $\hat{f}$ is real-valued, too. Since M(A) is connected by (6.1.7) this implies that $\hat{f}$ is constant by (18.2.10), hence f is constant.
Let $(K_n)_n$ be an admissible exhaustion of X, then $(j(K_n)\hat{}_A)_n$ is an admissible $\hat{A}$-convex exhaustion of M(A) by (4.3.4)ii). Assume there exists $\varphi \in j(K_n)\hat{}_A$ such that $\hat{f}(\varphi)$ is not contained in $\mathbb{R}$, i.e.

$$\hat{f}(\varphi) = a+ib, \text{ w.l.o.g. } b>0.$$

Since $f(K_n) \subset \mathbb{R}$ it is easy to see that we can choose b'>0 such that

(*) $\quad |a+ib-(a-ib')| > |x-(a-ib')|$ for all $x \in f(K_n)$.

Set

$$g = f - (a-ib'),$$

then $g \in A$ and

$$|\hat{g}(\varphi)| = |a+ib-(a-ib')| > \|g\|_{K_n}$$

by (*), and we get a contradiction to $\varphi \in j(K_n)^{\wedge}_{A}$. □

CHAPTER 19

UNIFORM FRECHET SCHWARTZ ALGEBRAS

As mentioned in remark (18.1.1)ii) in general quotients of reflexive Fréchet spaces are no longer reflexive. In this chapter we consider uF-algebras which are Schwartz spaces. These algebras will be called uniform Fréchet Schwartz algebras (uFS-algebras). We note that uFS-algebras are in particular reflexive uF-algebras and that they are stable with respect to the formation of Hausdorff quotients.

As proved in (2.3.8) Hol(X), X a hemicompact reduced complex space, is a uFS-algebra. Conversely we shall use the Schwartz property to characterize certain Stein algebras.

Vaguely spoken we use the following idea: Let A be a uFS-algebra with locally compact spectrum M(A), $F = (f_1,\ldots,f_n) \in A^n$, and let L be an affine complex line in $\mathbb{C}^n$. Set

$$I_L = \{f \in A: \hat{f}|_{\hat{F}^{-1}(L)} = 0\}.$$

Then I_L is a closed ideal in A and

$$M(A/I_L) = \hat{F}^{-1}(L) \quad \text{by (3.2.10).}$$

Suppose that A/I_L is a uF-algebra for each affine complex line L - in general we only know that A/I_L is an F-algebra - then we can show that $(\hat{A}_{\hat{F}^{-1}(L)},\hat{F}^{-1}(L))$ is isomorphic to A/I_L. So $(\hat{A}_{\hat{F}^{-1}(L)},\hat{F}^{-1}(L))$ is a uFS-algebra, since it is isomorphic to the quotient of a Schwartz space, in particular $(\hat{A}_{\hat{F}^{-1}(L)},\hat{F}^{-1}(L))$ is a reflexive uF-algebra. Hence $(\hat{A}_{\hat{F}^{-1}(L)},\hat{F}^{-1}(L))$ is a maximum modulus algebra and we can apply the results of chapter 15 and chapter 16.

It is Kramm's work [KRA 1],[KRA 2] we consider in this chapter.

(19.1) Fréchet Schwartz algebras

(19.1.1) The concept of Schwartz spaces was introduced for locally convex spaces. For the general theory of Schwartz spaces we refer the reader to the books of Jarchow [JAR] and Horvath [HOR].
Let A be an F-algebra with defining system of seminorms $(p_n)_n$, where $p_n \leq p_{n+1}$ for all $n \in \mathbb{N}$, cf. (3.1.7). As in (3.2.6) we denote by A_k the completion of the algebra

$$A/\{f \in A: p_k(f) = 0\}$$

with respect to the norm

$$p_k'(f+\{p_k=0\}) = p_k(f).$$

Definition. A is called Fréchet Schwartz algebra (FS-algebra) if for each $n \in \mathbb{N}$ there exists $m(n) \geq n$ such that the natural restriction mapping

$$\pi_{m(n),n}: A_{m(n)} \longrightarrow A_n, \text{ cf. (3.3.6)},$$

is a compact operator, i.e. the image $\pi_{m(n),n}(B)$ of each bounded set $B \subset A_{m(n)}$ is relatively compact in A_n.

Remarks. i) The definition is equivalent to the following statement: A - considered as a locally convex space - is a Schwartz space. We omit the proof, since we do not want to deal with the theory of locally convex spaces in this book. (It follows from [JAR], p. 201 that FS-algebras - according to our definition - are Schwartz spaces by considering the neighbourhoodbasis $Q = (U_n)_n$ of 0, where

$$U_n = \{f \in A: p_n(f) \leq 1/n\}.)$$

ii) As already mentioned Schwartz spaces have nice stability properties. Let E be a Schwartz space. Then every subspace F of E and every quotient E/F, F a closed subspace, is again a Schwartz space. Moreover cartesian products of arbitrarily many Schwartz spaces are Schwartz spaces ([JAR], p. 481).

iii) It is not hard to prove that Fréchet Schwartz spaces are Fréchet Montel spaces , cf. (2.1.14) for the definition, in particular Fréchet Schwartz spaces are reflexive, hence all results of chapter 18 remain valid if we replace "reflexive uF-algebra" by "uFS-algebra".

iv) A normed space E is a Schwartz space iff E is finite dimensional ([JAR], p. 203).

v) Finally we state without further use that every nuclear space, cf. [JAR] for the definition, is a Schwartz space.

(19.1.2) Examples. i) Let X be a hemicompact holomorphically separable reduced complex space. Then Hol(X) is a uFS-algebra by (2.3.8).

ii) Consider the algebra $A = C^{\infty}([0,1])$, cf. (3.3.5), with the defining sequence of seminorms

$$p_k(f) = 2^{k-1}\sup\{|f^{(l)}(x)| : x \in [0,1],\ l=0,1,\dots,k-1\},$$

where $f^{(l)}$ denotes the l-th derivative of f.

We claim that A is an FS-algebra.

Fix $k \in \mathbb{N}$, and let $(f_n)_n$ be a bounded sequence in A_{k+1}, i.e. there is $c>0$ so that $p_{k+1}(f_n) \leq c$ for all $n \in \mathbb{N}$. The elements of A_{k+1} are k-times differentiable and we have for $x,y \in [0,1]$

$$f_n^{(k-1)}(x) - f_n^{(k-1)}(y) = \int_0^x f_n^{(k)}(s)\,ds - \int_0^y f_n^{(k)}(s)\,ds,$$

hence

$$|f_n^{(k-1)}(x) - f_n^{(k-1)}(y)| \leq c|x-y|.$$

It follows from this that the sequence $(f_n^{(k-1)})_n \subset C([0,1])$ is equicontinuous. Since $(f_n^{(k-1)}(x))_n$ is relatively compact for each $x \in [0,1]$ we see that $(f_n^{(k-1)})_n$ is relatively compact in $C([0,1])$ (with respect to the norm $\|\cdot\|_{[0,1]}$) by the theorem of Arzela-Ascoli, hence a subsequence $(f_{n_i}^{(k-1)})_i$ converges to $g_{k-1} \in C([0,1])$. Repeating our ar-

guments we get a subsequence of $(f_{n_i}^{(k-2)})_i$, which converges to $g_{k-2} \in C([0,1])$. Hence g_{k-2} is differentiable and $g_{k-2}^{(1)} = g_{k-1}$. Continuing in this manner we get a subsequence of $(f_n)_n$ which converges in A_k and our claim is proved.
With regard to (19.2.6) we note that $C^\infty([0,1])$ is a semisimple FS-algebra with compact spectrum $[0,1]$, cf. (3.3.5).

(19.2) Strongly uniform Fréchet algebras

(19.2.1) Let I be a closed ideal in a uF-algebra A. A/I endowed with the quotient topology is an F-algebra and

$$M(A/I) = V(I) = \{\varphi \in M(A): \varphi(f) = 0 \text{ for all } f \in I\},$$

cf. (3.2.10). As already mentioned in the introduction we have to ensure that A/I is again a uF-algebra for certain ideals.

Definition. i) Let A be a uF-algebra, and let $M \subset M(A)$ be a subset. The ideal

$$k(M) = \{f \in A: \hat{f}|_M = 0\}$$

is called kernel (with respect to M). An ideal I is said to be a kernel ideal if $I = k(V(I))$.
ii) A is called strongly uniform (u^*F-algebra) if for each kernel ideal I, A/I endowed with the quotient topology is again a uF-algebra.

(19.2.2) Let A be a u^*F-algebra, I a kernel ideal in A, J a kernel ideal in A/I, and let $(p_n)_n$ be a defining sequence of seminorms for A, cf. (3.1.7).Since $M(A/I) = V(I)$ by (3.2.10), V(J) is a subset of V(I) and

$$(*) \qquad \tilde{J} = \{f \in A: \hat{f}|_{V(J)} = 0\} \supset I.$$

Note that

$$J = \{f+I: f \in \tilde{J}\}.$$

The map

$$T: (A/I)/J \longrightarrow A/\tilde{J},\ (f+I) + J \longrightarrow f+\tilde{J},$$

defines a bijective algebra homomorphism. Fix $n \in \mathbb{N}$, and let $f \in A$, then

$$\inf\{p_n(f+g): g \in \tilde{J}\} = \inf\{p_n(f+g+h): g \in \tilde{J},\ h \in I\}$$

by (*). It follows from this equality that T is a topological Isomorphism. Since $A/\tilde{J}$ is a uF-algebra by our hypothesis the same is true for $(A/I)/J$ and it follows that A/I is a u^*F-algebra.

<u>(19.2.3) Proposition.</u> Let A be a uF-algebra. Then A is a u^*F-algebra iff the restriction map

$$i: \hat{A} \longrightarrow \hat{A}_{V(I)},\ \hat{f} \longrightarrow \hat{f}|_{V(I)},$$

is surjective for each kernel ideal I.
(More precisely: Let $(K_n)_n$ be an admissible $\hat{A}$-convex exhaustion of M(A), set $K'_n = V(I)\cap K_n$. Then $(K'_n)_n$ is an admissible $\hat{A}$-convex exhaustion of the closed subset V(I), and we consider the map

$$i: \hat{A} \longrightarrow \hat{A}_{V(I)} = \varprojlim A_{K'_n},\ \hat{f} \longrightarrow (\hat{f}|_{K'_n})_n.)$$

<u>Proof.</u> i) Suppose the restriction map is surjective. The map

$$T: A/I \longrightarrow \hat{A}_{V(I)},\ f+I \longrightarrow (\hat{f}|_{K'_n})_n,$$

is an injective and continuous homomorphism. By our hypothesis T is even surjective, hence a homeomorphism by the open mapping theorem for Fréchet spaces. Our assertion follows, since $\hat{A}_{V(I)}$ is a uF-algebra.

ii) Suppose that A/I is a uF-algebra. By (3.2.10) M(A/I) = V(I). Let $(K_n)_n$ and $(K'_n)_n$ be as in i). Then the topology of $(A/I)^\wedge$, the algebra of the Gelfand transforms of A/I, and the topology of $\hat{A}_{V(I)}$ is defined by the same sequence of seminorms $(\|\cdot\|_{K'_n})_n$. Hence the map

$$\tilde{T}: (A/I)^\wedge \longrightarrow \hat{A}_{V(I)},\ (f+I)^\wedge \longrightarrow (\hat{f}|_{K'_n})_n,$$

defines a topological homomorphism onto its image.

Since $(A/I)^\wedge$ is a uF-algebra by our hypothesis and since $\tilde{T}$ has dense range we see that $(A/I)^\wedge$ is isomorphic to $\hat{A}_{V(I)}$.
Let $f \in \hat{A}_{V(I)}$ be an arbitrary element. Choose $g \in A$ such that $\tilde{T}((g+I)^\wedge) = f$. Then $i(g) = f$.

□

<u>(19.2.4) Remark.</u> Let A be a u*F-algebra. It follows from the proof of ii) that in this case A/I is topologically and algebraically isomorphic to $\hat{A}_{V(I)}$ for each kernel ideal $I \in A$. Hence $M(\hat{A}_{V(I)})$ can be identified as a topological space with V(I). (In the case that M(A) is locally compact this follows by (4.3.4)iii).)
If A is moreover a u*FS-algebra, then $\hat{A}_{V(I)}$ is a u*FS-algebra by remark (19.1.1)ii) and (19.2.2).

<u>(19.2.5) Examples.</u>i) Let X be a hemicompact k-space. Then (C(X),X) is a uF-algebra by (4.1.4)ii).
Let $Y \subset X$ be a closed subset, and let $(K_n)_n$ be an admissible exhaustion of X. Set $K'_n = Y \cap K_n$, then $(K'_n)_n$ is an admissible exhaustion of Y. Let $(f_n)_n \in C(X)_Y = \varprojlim C(K'_n)$ be an arbitrary element.
Using the extension theorem of Tietze (3.1.9) we find $F_1 \in C(K_1)$ so that $F_1|_{K_1} = f_1$. Set

$$g_2 : K'_2 \cup K_1 \longrightarrow \mathbb{C},\ x \longrightarrow \begin{cases} F_1(x) & \text{if } x \in K_1 \\ f_2(x) & \text{if } x \text{ is not contained in } K_1. \end{cases}$$

Then g_2 is continuous. Using again Tietze's theorem we find $F_2 \in C(K_2)$ such that $F_2|_{K'_2 \cup K_1} = g_2$. Continuing in this manner we get a complex-valued function F on X, such that $F|_{K'_n} = f_n$ for all $n \in \mathbb{N}$. F is continuous, since $F|_{K_n} = F_n$ is continuous for all $n \in \mathbb{N}$ and X is a k-space. Hence (C(X),X) is a u*F-algebra by (19.2.3).
ii) Let A be a Stein algebra, i.e. A is isomorphic to Hol(X), X a hemicompact reduced Stein space. Let $I \subset \text{Hol}(X)$ be a kernel ideal. Then

each element of Hol(V(I)) can be extended to an element of Hol(X) by (2.3.3), hence A is a u^*F-algebra by (19.2.3).
iii) In [KRA 5] Kramm gives an example of a uFS-algebra which is not strongly uniform.

(19.2.6) Finite sets are the only compact analytic subsets of a domain in $\mathbb{C}^n$, cf. (7.1.1) . We prove an analogue for u^*FS-algebras.

Definition. Let A be a uF-algebra, and let $S \subset A$ be an arbitrary set. The set

$$V(S) = \{\varphi \in M(A): \varphi(f) = 0 \text{ for all } f \in S\}$$

is called the hull of S.

Theorem [KRA 1]. Let A be a u^*FS-algebra, and let Y be a compact hull in M(A). Then Y is a finite set.

Proof. Let $I \in A$ be the kernel with respect to Y. Then A/I is a uFS-algebra. We have

$$M(A/I) = V(I) = Y.$$

A/I is isomorphic to $(A/I)^\wedge$, since it is a uF-algebra. The topology of $(A/I)^\wedge$ is defined by the norm $\|\cdot\|_Y$, i.e. $(A/I)^\wedge$ and hence A/I is a uB-algebra. By remark (19.1.1)iv) A/I is finite dimensional and our assertion follows. □

Remarks. i) We need the strong uniformity to conclude from the compactness of M(A/I)=Y that A/I is a uB-algebra, cf. (19.1.2)ii).
ii) Let A be a u^*FS-algebra, and let $Y \subset M(A)$ be a hull. Suppose that X is an open and closed subset of Y (with respect to the relative topology). Then $\hat{A}_Y$ is a u^*FS-algebra, $Y=M(\hat{A}_Y)$, by (19.2.4), and X is a hull in $M(\hat{A}_Y)$ by Shilov's idempotent theorem (6.1.5). Hence if X is moreover compact then X is a finite set by (19.2.6).

(19.3) Chevalley dimension for uF-algebras

(19.3.1) We introduce the notion of dimension for uF-algebras. It is motivated by the Chevalley dimension in complex analysis (cf. (7.1.1)) or commutative algebra.

Definition. i) Let A be a uF-algebra, and let $\varphi \in M(A)$. Define $d(\varphi)$ to be the minimum of all $n \in \mathbb{N}$ such that there exist $f_1,\dots,f_n \in \ker \varphi$ and a neighbourhood U of φ such that the map

$$\hat{F}: U \longrightarrow \mathbb{C}^n,\ \psi \longrightarrow (\hat{f}_1(\psi),\dots,\hat{f}_n(\psi)),$$

has finite fibers, i.e. the set $U\cap\hat{F}^{-1}(\hat{F}(\psi))$ is finite for each $\psi \in U$. If the minimum does not exist set $d(\varphi)=\infty$.

The Chevalley dimension of φ in M(A) is defined by

$$\dim_\varphi M(A) = \begin{cases} 0 & \text{if } \varphi \text{ is an isolated point in } M(A), \\ d(\varphi) & \text{otherwise}. \end{cases}$$

ii) Let A be a u^*F-algebra, $Y \subset M(A)$ a hull, and let $\varphi \in Y$. Then

$$\tilde{\varphi}: A/k(Y) \longrightarrow \mathbb{C},\ f+k(Y) \longrightarrow \varphi(f),$$

defines an element of $M(A/k(Y))$ by (3.2.10). We set

$$\dim_\varphi Y = \dim_{\tilde{\varphi}} M(A/k(Y))$$

and call it the dimension of φ with respect to the hull Y.

Remarks. i) We can identify $M(A/k(Y))$ with $V(k(Y))=Y$ as topological spaces by (3.2.10), hence we can interpret φ as an element of $M(A/k(Y))$ and shall use henceforth the more suggestive formulation

$$\dim_\varphi Y = \dim_\varphi M(A/k(Y)).$$

ii) If X is a hemicompact Stein space, i.e. $X = M(\mathrm{Hol}(X))$ by (2.3.2), the above defined dimension equals the usual notion of dimension, cf.

(7.1.1).

(19.3.2) Definition (19.3.1) immediately yields the following remarks.
Remarks. i) Let A be a u^*F-algebra, and let $Y \subset M(A)$ be a hull. Then the mapping

$$Y \longrightarrow \mathbb{N} \cup \{0,\infty\},\ \varphi \longrightarrow \dim_{\varphi} Y,$$

is upper semi-continuous, i.e. for every $\varphi \in Y$ there exists a neighbourhood $U \subset Y$ of φ such that

$$\dim_{\varphi} Y \geq \dim_{\psi} Y \text{ for all } \psi \in U.$$

If $\dim_{\varphi} Y < \infty$ for all $\varphi \in Y$, this mapping is bounded on any relatively compact subset of Y.
ii) The sets

$$\{\varphi \in Y: \dim_{\varphi} Y \leq n\}, \{\varphi \in Y: \dim_{\varphi} Y < n\},$$

are open, the sets

$$\{\varphi \in Y: \dim_{\varphi} Y > n\},\ \{\varphi \in Y: \dim_{\varphi} Y \geq n\},$$

are closed for all $n \in \mathbb{N}$.

(19.3.3) Next we prove a theorem which is an analogue to the semi-continuity of fiber dimensions of holomorphic mappings (cf. [G/R], p. 114). In complex analysis this theorem is proved by the classical Weierstraß theorems, which are not available in our setting.

Theorem [KRA 1]. Let A be a u^*FS-algebra with locally compact spectrum M(A), and let $F = (f_1,\dots,f_n) \in A^n$. Then the map

$$M(A) \longrightarrow \mathbb{N} \cup \{0,\infty\},\ \varphi \longrightarrow \dim_{\varphi} \hat{F}^{-1}(\hat{F}(\varphi)),$$

is upper semi-continuous, i.e. for every $\varphi \in M(A)$ there exists a neighbourhood U of φ such that

$$\dim_{\varphi} \hat{F}^{-1}(\hat{F}(\varphi)) \geq \dim_{\psi} \hat{F}^{-1}(\hat{F}(\psi)) \text{ for all } \psi \in U.$$

(Note that $\hat{F}^{-1}(\hat{F}(\varphi))$ is a hull in M(A) for all $\varphi \in M(A)$.)

We shall deduce theorem (19.3.3) from part i) of the next theorem. Part ii) will be needed later.

<u>(19.3.4) Theorem.</u> Let A be a u^*FS-algebra with locally compact spectrum M(A), and let $F = (f_1,\dots,f_n) \in A^n$.

i) Suppose that $\varphi \in M(A)$ is an isolated point of the fiber $\hat{F}^{-1}(\hat{F}(\varphi))$. Then there is a neighbourhood $W \subset M(A)$ of φ such that $\hat{F}$ has finite fibers on W, i.e. $\dim_\psi M(A) \le n$ for all $\psi \in W$.

ii) Let U be an open, hemicompact and $\hat{A}$-convex subset of M(A), and let $\varphi \in U$. (By (4.3.4)iii) we can identify $M(\hat{A}_U)$ as a topological space with U, hence we can interpret φ as an element of $M(\hat{A}_U)$.) Then

$$\dim_\varphi M(A) = \dim_\varphi M(\hat{A}_U) \text{ for all } \varphi \in U.$$

<u>Proof.</u> i) Denote by $[\varphi]$ the Gleason part which contains φ. By remark (19.1.1)iii),(18.2.3) $[\varphi]$ is an open subset of M(A). If $[\varphi]$ is compact then φ is an isolated point of M(A) and our assertion becomes trivial.

If $[\varphi]$ is not compact we can choose a compact neighbourhood K of φ with $\partial K \neq \emptyset$ and

$$(*) \quad K \cap \hat{F}^{-1}(\hat{F}(\varphi)) = \{\varphi\}.$$

W.l.o.g. we can assume that $\hat{F}(\varphi) = 0$, otherwise consider the functions $\hat{f}_1-\hat{f}_1(\varphi),\dots,\hat{f}_n-\hat{f}_n(\varphi)$. Set

$$S: M(A) \longrightarrow \mathbb{R},\ \psi \longrightarrow \max\{|\hat{f}_1(\psi)|,\dots,|\hat{f}_n(\psi)|\}.$$

Then S is continuous and we have

$$(**) \qquad \delta = \min\{S(\psi): \psi \in \partial K\} > 0$$

by the compactness of ∂K and by (*). Set

$$W = \{\psi \in K: S(\psi) < \delta/2\}.$$

Then W is an open neighbourhood of φ. For an arbitrary point $\psi_0 \in W$ the set

$$Y_{\psi_o} = \hat{F}^{-1}(\hat{F}(\psi_o))\cap K$$

is compact and relatively open in $\hat{F}^{-1}(\hat{F}(\psi_o))$, since

$$\hat{F}^{-1}(\hat{F}(\psi_o))\cap \partial K = \emptyset$$

by (**). Since $\hat{F}^{-1}(\hat{F}(\psi_o))$ is a hull in M(A), Y_{ψ_o} is a finite set by remark (19.2.6)ii).

ii) By our assumptions $(\hat{A}_U, U)$ is a uF-algebra and $M(\hat{A}_U) = U$ by (4.3.4)iii).

Clearly $\dim_\varphi M(A) \geq \dim_\varphi M(\hat{A}_U)$.

To show the inverse inequality choose an arbitrary point $\varphi \in U$ and set

$$r = \dim_\varphi M(\hat{A}_U).$$

The assertion is trivial if $r=\infty$ or $r=0$.

Let $0<r<\infty$. We choose a compact neighbourhood K of φ in U with $\partial K \neq \emptyset$, $G = (g_1,\dots,g_r) \in (\hat{A}_U)^r$ such that $\hat{G}(\varphi)=0$ and

$$K\cap\hat{G}^{-1}(\hat{G}(\varphi)) = \{\varphi\}.$$

Let S and δ be as in the proof of i) with respect to $\hat{g}_1,\dots,\hat{g}_r$. Choose $f_1,\dots,f_r \in A$ such that

$$\|\hat{f}_i-\hat{g}_i\|_K < \delta/4 \text{ for } i=1,\dots,r.$$

Then

$$Y_\varphi = \hat{F}^{-1}(\hat{F}(\varphi))\cap K$$

is compact and relatively open in $\hat{F}^{-1}(\hat{F}(\varphi))$, since $\hat{F}^{-1}(\hat{F}(\varphi))\cap\partial K=\emptyset$.

So Y_φ is a finite set by remark (19.2.6)ii). Hence we can choose a compact neighbourhood L of φ such that $\hat{F}^{-1}(\hat{F}(\varphi))\cap L = \{\varphi\}$, hence

$$\dim_\varphi M(A) \leq r \text{ by i)}.$$

□

<u>(19.3.5) Proof of (19.3.3).</u> Set

$$r = \dim_\varphi \hat{F}^{-1}(\hat{F}(\varphi)).$$

We can assume that $r < \infty$.

If r=0 choose a neighbourhood W of φ according to (19.3.4)i) such that $\hat{F}$ has finite fibers on W. Let $\psi \in W$, then ψ is an isolated point of $\hat{F}^{-1}(\hat{F}(\psi))$, hence

$$\dim_\psi \hat{F}^{-1}(\hat{F}(\psi)) = 0.$$

Let $0<r<\infty$. In the following we use frequently the identification

$$M(A/k(\hat{F}^{-1}(\hat{F}(\psi)))) = \hat{F}^{-1}(\hat{F}(\psi)), \text{ cf. (3.2.10).}$$

Choose $g_1',\ldots,g_r' \in A/k(\hat{F}^{-1}(\hat{F}(\varphi)))$ such that the mapping

$$\hat{F}^{-1}(\hat{F}(\varphi)) \longrightarrow \mathbb{C}^n,\ \psi \longrightarrow (\hat{g}_1'(\psi),\ldots,\hat{g}_r'(\psi))$$

has finite fibers in a neighbourhood W of φ in

$$\hat{F}^{-1}(\hat{F}(\varphi)) = M(A/k(\hat{F}^{-1}(\hat{F}(\varphi)))).$$

Choose $g_1,\ldots,g_r \in A$ such that $g_i+k(\hat{F}^{-1}(\hat{F}(\varphi)))=g_i'$, $i=1,\ldots,r$, and set $\hat{G}=(\hat{g}_1,\ldots,\hat{g}_r)$. Then we can choose a neighbourhood V of φ in M(A) such that

$$\{\psi \in V\colon \hat{G}(\psi) = \hat{G}(\varphi) \text{ and } \hat{F}(\psi) = \hat{F}(\varphi)\} = \{\varphi\}.$$

Theorem (19.3.4) implies that there is a neighbourhood U of φ in M(A) such that the map

$$U \longrightarrow \mathbb{C}^{n+r},\ \psi \longrightarrow (\hat{F}(\psi),\hat{G}(\psi)),$$

has finite fibers on U.

Let $\psi \in U$. Consider the elements

$$g_i'' = g_i + k(\hat{F}^{-1}(\hat{F}(\psi))) \in A/k(\hat{F}^{-1}(\hat{F}(\psi))),\ i=1,\ldots r.$$

The mapping

$$M(A/k(\hat{F}^{-1}(\hat{F}(\psi)))) \longrightarrow \mathbb{C}^n,\ x \longrightarrow (\hat{g}_1''(x),\ldots,\hat{g}_r''(x)),$$

has finite fibers in a neighbourhood $W \subset M(A/k(\hat{F}^{-1}(\hat{F}(\psi))))$ of ψ, thus

$$\dim_\psi \hat{F}^{-1}(\hat{F}(\psi)) \le r.$$

□

(19.4) A characterization of pure dimensional Stein algebras

(19.4.1) Theorem [KRA 2]. Let A be a u^*FS-algebra with locally compact spectrum M(A). Let $U \subset M(A)$ be an open, hemicompact and $\hat{A}$-convex set such that $\dim_\varphi M(A) = n < \infty$ for all $\varphi \in U$.
Then U can be given the structure of an n-dimensional Stein space such that $\hat{A}_U = \mathrm{Hol}(U)$.

Proof. i) The assertion becomes trivial if $n = 0$.
ii) Let $n \geq 2$. Let $F=(f_1,\dots,f_{n-1}) \in A^{n-1}$ be an arbitrary element, and let $\varphi \in U$ such that $\hat{F}(\varphi)=0$.
Then $\hat{F}^{-1}(0)$ is a hull in M(A) and $\hat{A}_{\hat{F}^{-1}(0)}$ is a u^*FS-algebra with spectrum $\hat{F}^{-1}(0)$ by remark (19.2.4).
Suppose there exists a compact set $L \subset \hat{F}^{-1}(0) \cap U$ such that L contains an isolated point ψ of $\hat{F}^{-1}(0)$. Then $\dim_\psi M(A) \leq n-1$ by (19.3.4)i), and we get a contradiction to our hypothesis by (19.3.4)ii).
By remark (19.1.1)iii) $\hat{A}_{\hat{F}^{-1}(0)}$ is a reflexive uF-algebra, hence $\|\hat{f}\|_L = \|\hat{f}\|_{\partial L}$ for each $\hat{f} \in \hat{A}$ and each compact set $L \subset \hat{F}^{-1}(0) \cap U$, where ∂L denotes the boundary of L relative to $\hat{F}^{-1}(0)$, by remark (18.2.7). Now it follows immediately from (16.1.5) that

$$(*) \qquad \gamma_{n-1}\hat{A}_K \subset \partial K,$$

for each $\hat{A}$-convex compact set $K \subset U$. (For the definition of the (n-1)-th Shilov boundary $\gamma_{n-1}\hat{A}_K$ cf. chapter 16.)

If $n = 1$, U contains no isolated points and we have

$$\|\hat{f}\|_K = \|\hat{f}\|_{\partial K}$$

for each compact subset $K \subset U$ and each $f \in A$ by remark (18.2.7), hence we obtain (*) in the case $n = 1$, too.
iii) Let $\varphi \in U$ be an arbitrary point. Choose an $\hat{A}$-convex compact neighbourhood K of φ and $F = (f_1,\dots,f_n) \in A^n$ such that $\hat{F}$ has finite fibers on K and

$$\hat{F}^{-1}(\hat{F}(\varphi)) \cap K = \{\varphi\}.$$

We can now follow the proof of theorem (17.1.7)γ) to conclude that there is an open, relatively compact, $\hat{A}$-convex and hemicompact neighbourhood V_φ of φ in U such that V_φ can be endowed with the structure of an n-dimensional Stein space such that

$$(**) \qquad \mathrm{Hol}(V_\varphi) = \hat{A}_{V_\varphi}.$$

Let $\varphi, \psi \in U$ so that $V_\varphi \cap V_\psi \neq \emptyset$. Then as in the proof of (11.1.4) part ii)⇒i) we can show that both structures coincide on $V_\varphi \cap V_\psi$. So U can be endowed with the structure of an n-dimensional Stein space such that $\hat{A}_U \subset \mathrm{Hol}(U,\mathcal{O})$, where $\mathrm{Hol}(U,\mathcal{O})$ denotes the algebra of all holomorphic functions with respect to this structure.

By (11.1.3) U can be given the structure of a Stein space such that $\hat{A}_U = \mathrm{Hol}(U,\mathcal{O}')$, hence $\mathrm{Hol}(U,\mathcal{O}') \subset \mathrm{Hol}(U,\mathcal{O})$.

Now let $f \in \mathrm{Hol}(U,\mathcal{O})$ be an arbitrary element,then for each $\varphi \in U$,

$$f|_{V_\varphi} \in \mathrm{Hol}(U,\mathcal{O})_{V_\varphi} = \hat{A}_{V_\varphi} = \mathrm{Hol}(U,\mathcal{O}')_{V_\varphi}.$$

For the first equality use (2.3.6) and (**). So $f|_{V_\varphi}$ is holomorphic with respect to the "new" structure , since the algebra of all holomorphic functions on a reduced complex space is complete with respect to the compact open topology. It follows that $\mathrm{Hol}(U,\mathcal{O}) = \mathrm{Hol}(U,\mathcal{O}')$, and we are done.

□

<u>(19.4.2) Corollary.</u> Let A be a u^*FS-algebra with locally compact spectrum M(A) such that $\dim_\varphi M(A) = n < \infty$ for all $\varphi \in M(A)$. Then M(A) can be given the structure of an n-dimensional Stein space so that $\hat{A} = \mathrm{Hol}(M(A))$.

Conversely, if X is an n-dimensional Stein space, then Hol(X) satisfies the above conditions.

<u>Proof.</u> The conversely part follows from (2.3.8), (19.2.5)ii), and remark (11.1.1).

□

(19.4.3) We want to show that the assumption "locally compact spectrum" is necessary for theorem (19.4.1).

Example [KRA 2]. We construct a doubly generated u^*FS-algebra A which is not a Stein algebra.
Beyond that there exists a one dimensional Stein space X such that A is a closed subalgebra of Hol(X).

i) For $n \in \mathbb{N}$ set

$$D_n = \{(z,w) \in \mathbb{C}^2 : |z| \le n, |w| \le n\},$$

$$K_n = \{(z,w) \in D_n : w \cdot \prod_{i=1}^{n} (z-(1/i)) = 0\},$$

and $A = \varprojlim P(K_n)$. A is a doubly generated uF-algebra with spectrum

$$M(A) = \cup K_n \text{ (as sets)},$$

since each K_n is polynomially convex, and $(K_n)_n$ is an admissible exhaustion of M(A) with respect to the Gelfand topology by (5.1.6). As usual interpret the elements of $\hat{A}$ as complex-valued functions on $\cup K_n$ which can be approximated uniformly on each K_n by polynomials.
Further set

$$X_o = \mathbb{C} \times \{0\},$$

$$X_i = (1/i) \times \mathbb{C} \quad \text{for } i \in \mathbb{N},$$

and

$$X = \bigcup_{i=0}^{\infty} X_i \setminus \{(0,0)\}.$$

Note that $\hat{f}|_{X_i}$ can be interpreted as an element of $\mathrm{Hol}(\mathbb{C})$ for each $i \in \mathbb{N}_o$.
X is an analytic subset of the domain of holomorphy $(\mathbb{C}\setminus\{0\}) \times \mathbb{C}$ and it is easily proved that X is holomorphically convex and holomorphically separable, hence Hol(X) is a Stein algebra, in particular a uFS-algebra by (2.3.8).
$M(A) = \bigcup_{i=0}^{\infty} X_i$ may be viewed in the following way:

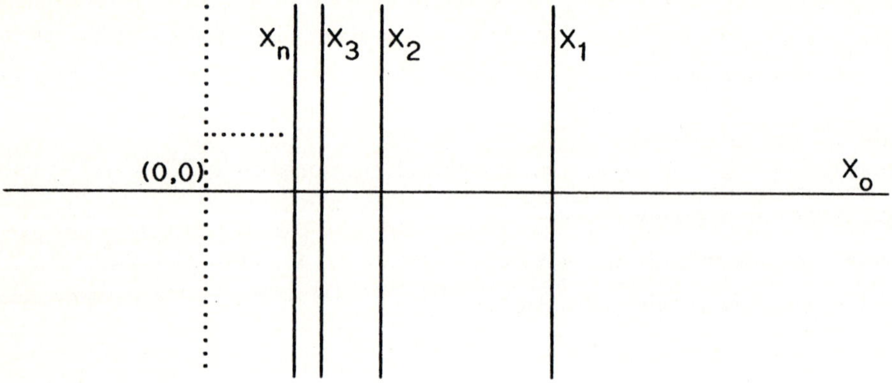

ii) Set

$$K'_n = K_n \setminus \{(z,0) \in \mathbb{C}^2 : |z| < 1/n\}.$$

Then K_n is the polynomially convex hull of K'_n, so $P(K_n)$ is isomorphic to $P(K'_n)$, thus $\varprojlim P(K_n)$ is isomorphic to $\varprojlim P(K'_n)$. Since $(K'_n)_n$ is an admissible exhaustion of X with respect to the induced euclidean topology and since Hol(X) is complete with respect to the compact open topology, we get

$$\varprojlim P(K'_n) \subset \mathrm{Hol}(X),$$

and the map

$$i: \hat{A} \longrightarrow \mathrm{Hol}(X),\ \hat{f} \longrightarrow \hat{f}|_{X'}$$

is a well-defined continuous and injective algebra homomorphism. Let $(\hat{f}_n)_n \subset \hat{A}$ be a sequence such that $i(\hat{f}_n)$ converges to $g \in \mathrm{Hol}(X)$, i.e. $(\hat{f}_n)_n$ converges uniformly on each K'_m to g. It follows from the maximum principle for holomorphic functions - and since $\hat{A}$ is complete - that $(\hat{f}_n)_n$ converges uniformly on each K_m to an element $\hat{h} \in \hat{A}$. Then $i(\hat{h}) = g$, so - by the open mapping theorem - A can be considered as a closed subalgebra of the Stein algebra Hol(X) via the map i, hence A is in particular a uFS-algebra by remark (19.1.1)ii).

iii) Let D be the open unit disc. We claim that

$$i(\hat{A}) = \{f \in \mathrm{Hol}(X): f|_{(D\setminus\{0\})\times\{0\}} \text{ is bounded}\}.$$

Let g be an arbitrary element of the right side, and let $n \in \mathbb{N}$. Then

$\bigcup_{i=0}^{n} X_i$ is an analytic subset of $\mathbb{C}^2$ and there is $\tilde{g}_n \in \mathrm{Hol}(\bigcup_{i=0}^{n} X_i)$ by the Riemann extension theorem such that

$$\tilde{g}_n\Big|_{\bigcup_{i=0}^{n} X_i\setminus\{(0,0)\}} = g\Big|_{\bigcup_{i=0}^{n} X_i\setminus\{(0,0)\}}.$$

In the same way as in remark (18.2.9)iii) we can show that every $h \in \mathrm{Hol}(\bigcup_{i=0}^{n} X_i)$ can be approximated uniformly on each compact subset of $\bigcup_{i=0}^{n} X_i$ by polynomials, thus $\tilde{g}_n|_{K_n} \in P(K_n)$. Note that $\tilde{g}_{n+1}|_{K_n} = \tilde{g}_n|_{K_n}$, hence $h = (\tilde{g}_n|_{K_n})_n \in \varprojlim P(K_n)$ and $i(\hat{h})=g$. The inverse inclusion is trivial.

iv) In the same way as in example (5.1.12) we can show that (0,0) has no compact neighbourhood (with respect to the Gelfand topology) in M(A). So M(A) is not locally compact, hence A is not a Stein algebra by remark (11.1.1). Furthermore

(*) there is no sequence $(\varphi_n)_n \subset M(A)\setminus X_o$ which converges to (0,0) with respect to the Gelfand topology.

Otherwise

$$K = \{(0,0)\}\cup\{\varphi_n : n \in \mathbb{N}\}$$

would be a compact subset of M(A), but there is no $n \in \mathbb{N}$ such that $K \subset K_n$.

v) A is a u^*F-algebra.

Let $I \subset A$ be a kernel ideal. We shall show that the restriction map $\hat{A} \longrightarrow \hat{A}_{V(I)}$ is surjective. This yields our assertion by (19.2.3).

We endow X_i always with the induced euclidean topology. If $V(I)\cap X_i$ has an accumulation point, then $X_i \subset V(I)$ by the identity theorem. So

$$V(I) = \bigcup_{i\in J} X_i \cup S,$$

where J is a subset of $\mathbb{N}\cup\{0\}$ (possibly empty) and S is a discrete subset of $M(A)\setminus \bigcup_{i\in J} X_i$ with respect to the Gelfand topology (possibly empty). This follows by (*) and since the induced Gelfand topology and the induced euclidean topology are equivalent on X and on every X_i.

Let $\hat{f} \in \hat{A}_{V(I)}$ be an arbitrary element.

Case I: J is a finite set.

Then $\bigcup_{i\in J} X_i$ is an analytic subset of $\mathbb{C}^2$, hence by (2.3.3) there exists $f' \in \mathrm{Hol}(\mathbb{C}^2)$ such that

$$f'|_{\bigcup_{i\in J} X_i} = f|_{\bigcup_{i\in J} X_i}.$$

We note that $\tilde{f} = f'|_{M(A)} \in \hat{A}$.

Case II. J is not finite.

It follows by the identity theorem that $0 \in J$. Set

$$\tilde{f}(x) = \begin{cases} f(x) & \text{if } x \in \bigcup_{i\in J} X_i, \\ f(1/i,0) & \text{if } x \in X_i,\ i \text{ not contained in } J. \end{cases}$$

Then $\tilde{f}|_X \in \mathrm{Hol}(X)$ and $\tilde{f}|_{(D\setminus\{0\})\times\{0\}}$ is bounded, hence $\tilde{f} \in \hat{A}$ by iii).

In both cases we have found an element $\tilde{f} \in \hat{A}$ such that

$$\tilde{f}|_{\bigcup_{i\in J} X_i} = f|_{\bigcup_{i\in J} X_i}.$$

If $S\cap X_i \neq \emptyset$, then $S\cap X_i$ consists of at most countable many points without an accumulation point in X_i. If $S\cap X_o \neq \emptyset$, then J is finite or empty and $S\cup\{(1/i,0): i\in J\}$ is a discrete subset of X_o. Choose $g_o \in \mathrm{Hol}(X_o)$ such that

$$g_o(y)=f(y)-\tilde{f}(y) \text{ for all } y \in S\cap X_o, \text{ and } g_o(1/i,0)=0 \text{ for } i \in J,$$

by the theorems of Weierstraß and Mittag-Leffler. The function

$$\tilde{g}_o(x) = \begin{cases} g_o(1/i,0) & \text{if } x \in X_i, \\ g_o(x) & \text{if } x \in X_o \end{cases}$$

defines an element of $\hat{A}$ by iii). If $S\cap X_o = \emptyset$ we set $\tilde{g}_o = 0$.

Set

$$J' = \{i \in \mathbb{N}: S\cap(X_i\setminus\{(1/i,0)\}) \neq \emptyset\}.$$

For i $\in$ J' we use again the theorems of Weierstraß and Mittag-Leffler to find $g_i \in \mathrm{Hol}(X_i)$ so that

$$g_i(y)=f(y)-\tilde{f}(y)-\tilde{g}_o(y) \text{ if } y \in S\cap(X_i\setminus\{((1/i),0)\}) \text{ and } g_i(1/i,0)=0.$$

By iii) the function

$$\tilde{g}_i(x) = \begin{cases} 0 & \text{if } x \text{ is not contained in } X_i \\ g_i(x) & \text{if } x \in X_i \end{cases}$$

lies in $\hat{A}$. Hence

$$\tilde{g} = \tilde{f} + \tilde{g}_o + \sum_{i\in J'} \tilde{g}_i \in \hat{A},$$

and $\tilde{g}|_{V(I)}=f$.

vi) M(A) is purely one dimensional.

The map

$$p: M(A) \longrightarrow \mathbb{C},\ (z,w) \longrightarrow z-w,$$

defines an element of $\hat{A}$. For x $\in$ X it is easy to find a (Gelfand open) neighbourhood U of x such that p has finite fibers on U.

Now set

$$f(z,w) = \begin{cases} n^2 w & \text{if } (z,w) \in X_n,\ n\geq 1, \\ 0 & \text{if } (z,w) \in X_o. \end{cases}$$

Then f $\in$ $\hat{A}$ by iii). Set

$$U = \{(z,w) \in M(A): |f(z,w)| < 1\}$$

Then U is a (Gelfand open) neighbourhood of (0,0) by (3.2.2)i).

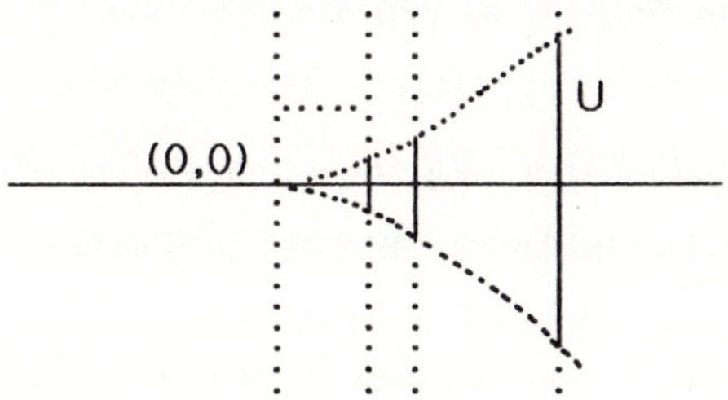

Choose an arbitrary point $z_o \in \mathbb{C}\setminus\{0\}$. For $n\geq 1$ set

$$S_n = p^{-1}(z_o)\cap U\cap X_n = \{(1/n,w): (1/n)-w=z_o, |w|<1/n^2\}.$$

Hence $S_n = \emptyset$ for all $n \in \mathbb{N}$ such that $(1/n)+(1/n^2)<|z_o|$, and it follows that p has finite fibers on U, since $p^{-1}(0)\cap U = \{(0,0)\}$.

vii) Finally we note that ker φ is (algebraically) doubly generated for

$$\varphi \in S = \{(1/n,0): n \in \mathbb{N}\},$$

and (algebraically) singly generated for $\varphi \in M(A)\setminus S$.

The assertion is clear for $\varphi \neq (0,0)$. If $f \in \ker (0,0)$, then $g=f/(z_1|_X) \in Hol(X)$, and g is bounded on $(D\setminus\{0\})\times\{0\}$, hence $g \in i(\hat{A})$ by iii), i.e. ker (0,0) is singly generated by the coordinate function z_1.

<u>(19.4.4)</u> We wish to characterize Stein algebras, which are isomorphic to Hol(X), X a Stein manifold.

Let X be an analytic subset of a domain $G \subset \mathbb{C}^n$, and let $a \in X$. Denote by ${}_X\mathcal{O}_a$ the algebra of germs of holomorphic functions on X at a (analogous definition as in (17.1.1)). Set

$$m_a = \{f \in {}_X\mathcal{O}_a : f(a) = 0\},$$

and

$$m_a^2 = \{f \in {}_X\mathcal{O}_a : \text{There is } r \in \mathbb{N}, f_1,\dots,f_r,g_1,\dots,g_r \in m_a \text{ so that } f = \sum_{i=1}^{r} f_i g_i\}.$$

Then

$$\mathrm{emb}_a X = \dim_{\mathbb{C}}(m_a/m_a^2)$$

is called the embedding dimension of X at a. We always have

$$\dim_a X \leq \mathrm{emb}_a X.$$

Equality holds iff a is a regular point of X, i.e. if there exists a neighbourhood U of a in X such that U is biholomorphically equivalent to an open subset of $\mathbb{C}^m$, $m = \dim_a X$, cf. [K/K], p. 190.

Definition. i) An ideal I in a uF-algebra is said to be topologically n-generated, if there are $f_1,\dots,f_n \in I$ such that the ideal which is algebraically generated by $f_1,\dots,f_n$ lies dense in I.
ii) Let $\varphi \in M(A)$. Then ker φ is called locally topologically n-generated if there exists a compact neighbourhood K of φ such that

$$(\ker \varphi)_K = \{f \in \hat{A}_K : \varphi(f) = 0\}$$

is topologically n-generated in $\hat{A}_K$.

Notation. In the following theorem "n-generated" refers to the minimal number of generators.

Theorem [KRA 2]. Let $A \neq C(\{p\})$ be a u^*FS-algebra with locally compact and connected spectrum M(A). Then the following assertions are equivalent:

i) A is topologically and algebraically isomorphic to Hol(X), X a Stein manifold;

ii) there is $n \in \mathbb{N}$ such that ker φ is locally topologically n-generated for all $\varphi \in M(A)$.

Proof. i) ⇒ ii) Let X be an n-dimensional Stein manifold, and let $x \in X$ be an arbitrary point. The ideal $I_x = \{f \in \mathrm{Hol}(X): f(x)=0\}$ is algebraically n-generated and cannot be algebraically m-generated with an $m<n$ by [FOR 1], Satz 5.4. Hence I_x is locally topologically m-generated with an $m \leq n$.
Suppose that $m<n$. Let K be a compact neighbourhood of x such that $\{f \in \mathrm{Hol}(X)_K: f(x)=0\}$ is topologically m-generated by $f_1,\dots,f_m \in \mathrm{Hol}(X)_K$. Choose an open Hol(X)-convex neighbourhood U of x, $U \subset K$. Then U is an n-dimensional Stein manifold. Since finitely generated ideals in Stein algebras are closed by (2.3.4) we see that $f_1|_U,\dots,f_m|_U$ generate algebraically the ideal $\{f \in \mathrm{Hol}(U): f(x)=0\}$ and we get a contradiction.

ii) $\Rightarrow$ i) We first show that M(A) is purely n-dimensional. Let $\varphi \in M(A)$ be an arbitrary point, and let K be a compact neighbourhood of φ such that $(\ker\varphi)_K$ is topologically n-generated by $f'_1,\dots,f'_n \in \hat{A}_K$. Note that

$$\{\psi \in K: f'_i(\psi) = 0, i=1,\dots,n\} = \{\varphi\}.$$

Hence we find $F = (f_1,\dots,f_n) \in A^n$ such that $\hat{F}^{-1}(\hat{F}(\varphi))\cap K$ is a compact and relatively open subset of $\hat{F}^{-1}(\hat{F}(\varphi))$. So $\hat{F}^{-1}(\hat{F}(\varphi))\cap K$ is a finite subset by remark (19.2.6)ii), thus $\dim_\varphi M(A) \le n$ by (19.3.4)i). Set

$$m = \min\{\dim_\varphi M(A): \varphi \in M(A)\}.$$

Suppose $m<n$.
If $m=0$ then M(A) consists of a singelton, since M(A) is connected, and we get a contradiction to $A \neq C(\{p\})$.
So $m>0$. By (19.3.2)ii)

$$X = \{\psi \in M(A): \dim_\psi M(A) = m\} = \{\psi \in M(A): \dim_\psi M(A) \le m\}$$

is an open purely m-dimensional subset of M(A). Choose an open, hemicompact and $\hat{A}$-convex set $U \subset X$ by (4.3.7), then U can be given the structure of an m-dimensional Stein space such that $\hat{A}_U = \mathrm{Hol}(U)$ by (19.4.1).
The set of regular points of U is open and dense in U by remark (2.3.1). But for these points $\psi \in U$ the ideals $\ker\psi$ are locally topologically m-generated, a contradiction to the minimal choice of n.

By corollary (19.4.2) M(A) can be endowed with the structure of an n-dimensional Stein space such that $\hat{A} = \mathrm{Hol}(M(A))$ with respect to this structure. We claim that M(A) is in fact a Stein manifold. Let $\varphi \in M(A)$ be an arbitrary point. Choose an open, relatively compact, hemicompact and $\hat{A}$-convex neighbourhood U of φ such that

$$(\ker\varphi)_{\overline{U}} = \{f \in \hat{A}_{\overline{U}}: \varphi(f) = 0\}$$

can be topologically generated by $f_1,\dots,f_n \in \hat{A}_{\overline{U}}$, where $\overline{U}$ denotes the the closure of U. Since finitely generated ideals in Stein algebras are closed by (2.3.4) $f_1|_U,\dots,f_n|_U$ generate $(\ker\varphi)_U = \{f \in \hat{A}_U: \varphi(f) = 0\}$

algebraically.
Choose an open neighbourhoodbasis $(U_i)_i$ of φ, $U_i \subset U$ for all $i \in \mathbb{N}$, so that each U_i is a Stein space. Then $f_k|_{U_i}$, $k=1,\dots,n$, generate $(\ker \varphi)_{U_i}$ and we conclude that $\tilde{f}_1,\dots,\tilde{f}_n$, the germs of $f_1,\dots,f_n$ at φ, generate m_φ.
Now let $\tilde{g}+m_\varphi^2$ be an arbitrary element of m_φ/m_φ^2, i.e. there are $\tilde{h}_i \in {}_{M(A)}\mathcal{O}_\varphi$, $i=1,\dots,n$, such that

$$\tilde{g} = \sum_{i=1}^{n} \tilde{h}_i \tilde{f}_i .$$

Set $\lambda_i = \tilde{h}_i(\varphi)$, $i=1,\dots,n$, then

$$\sum_{i=1}^{n} \lambda_i \tilde{f}_i - \tilde{g} \in m_\varphi^2,$$

and it follows that $\dim_{\mathbb{C}}(m_\varphi/m_\varphi^2) \leq n$, i.e. $\mathrm{emb}_\varphi M(A) = n$, since $\dim_\varphi M(A) = n$, hence φ is a regular point. Consequently $M(A)$ carries the structure of an n-dimensional Stein manifold. □

(19.5) Riemann algebras

(19.5.1) We call a uF-algebra A Riemann algebra if A is topologically and algebraically isomorphic to Hol(X), X a Riemann surface.
We use Gleason's theorem (1.5.5) to characterize Riemann algebras. In particular we strengthen theorem (19.4.4) considerably in the case n=1. The theorem stated below can be found in [KRA 4] and is based on ideas of Carpenter [CAR 3].

Lemma. Let B be a uB-algebra, and let $f \in B$. If $\hat{f}(\varphi) \neq 0$ for all $\varphi \in \gamma B$, then the principal ideal (f) is closed in B.

Proof. Set $V(\hat{f}) = \{\psi \in M(B) : \hat{f}(\psi) = 0\}$.
Let $h_n = f \cdot g_n$, $n \in \mathbb{N}$, be a Cauchy sequence in (f). For $n,m \in \mathbb{N}$ we

find $\varphi_o \in \gamma B$ such that

$$\|g_n - g_m\| = |\hat{g}_n(\varphi_o) - \hat{g}_m(\varphi_o)| = |\hat{f}(\varphi_o)|^{-1} \cdot |\hat{h}_n(\varphi_o) - \hat{h}_m(\varphi_o)|$$

$$\leq \|h_n - h_m\| (\inf\{|\hat{f}(\varphi)| : \varphi \in \gamma B\})^{-1}.$$

The second equality resp. the inequality are valid, since $V(\hat{f}) \cap \gamma B = \emptyset$. It follows that $(g_n)_n$ is a Cauchy sequence in B and we are done. □

(19.5.2) Lemma. Let A be a uF-algebra, and let $\varphi \in M(A)$ be a point, which is not isolated in M(A). Suppose that φ has a compact neighbourhood and that ker φ is a principal ideal. Then there exists an admissible exhaustion $(\tilde{K}_n)_n$ of M(A) such that

$$\varphi \in \tilde{K}_n \setminus \gamma \hat{A}_{\tilde{K}_n} \quad \text{for all } n \in \mathbb{N}.$$

Proof. Let

(*) $\qquad (f) = \ker \varphi.$

Note that ker φ - as a closed subspace of A - is a Fréchet space, and that

(**) $\qquad V(\hat{f}) = \{\varphi\}$

by (*). Consider the map

$$i: A \longrightarrow \ker \varphi, \; g \longrightarrow fg.$$

Clearly i is a continuous and surjective linear mapping. If

$$0 = i(g) = fg,$$

then

$$\hat{g}|_{M(A)\setminus\{\varphi\}} = 0$$

by (**), hence $g \equiv 0$, since $\{\varphi\}$ is not an isolated point of M(A). This proves that i is injective, hence a homeomorphism by the open mapping theorem.

By the proof of (4.4.2) there exists a countable open neighbourhood-basis $\ldots \supset U_n \supset U_{n+1} \supset \ldots$ of φ. Let $(K_n)_n$ be an admissible exhaustion of M(A), w.l.o.g. we can assume that $U_1 \subset K_1$.

Suppose that $\varphi \in \gamma\hat{A}_{K_n}$ for all $n \in \mathbb{N}$. Since $\hat{A}|_{K_n}$ is dense in $\hat{A}_{K_n}$, and since the set of the strong boundary points of $\hat{A}_{K_n}$ lies dense in $\gamma\hat{A}_{K_n}$ by (1.3.12) we find for each $n \in \mathbb{N}$ an element $g_n \in A$ such that

$$1 = \sup\{|\hat{g}_n(x)|: x \in U_n\} > 1/n \geq \|\hat{g}_n\|_{K_n \setminus U_n}.$$

Hence $i(g_n)=fg_n$ tends to zero for $n \longrightarrow \infty$. Since i is a linear homeomorphism this implies that g_n tends to zero for $n \longrightarrow \infty$, a contradiction to $\|\hat{g}_n\|_{K_n} = 1$ for all $n \in \mathbb{N}$. Hence there is $m \in \mathbb{N}$ such that

$$\varphi \in K_m \setminus \gamma\hat{A}_{K_m}.$$

Set $\tilde{K}_n = K_{n+m}$ for all $n \in \mathbb{N}$. □

<u>(19.5.3) Theorem.</u> Let A be a uF-algebra, and let $\varphi \in M(A)$ be a point which is not isolated in $M(A)$. Suppose that φ has a compact neighbourhood in $M(A)$, and that $\ker \varphi$ is a principal ideal, generated by $f \in A$. Then there exist an open set $V \subset \mathbb{C}$ and an open neighbourhood U of φ in $M(A)$ such that $\hat{f}: U \longrightarrow V$ is a homeomorphism and the map

$$\mathrm{Hol}(V) \longrightarrow \hat{A}_U, \; h \longrightarrow h \circ \hat{f},$$

defines a topological algebra homomorphism.

<u>Proof.</u> According to (19.5.2) there is a compact neighbourhood K of φ in $M(A)$ - w.l.o.g. we can assume that K is $\hat{A}$-convex - such that $\varphi \in K \setminus \gamma\hat{A}_K$.

This implies

(*) $\qquad V(\hat{f}) \cap \gamma\hat{A}_K = \emptyset.$

Clearly the ideal $\hat{f}|_K \cdot \hat{A}_K$ lies dense in

$$(\ker \varphi)_K = \{g \in \hat{A}_K: \varphi(g)=0\}.$$

By (*) and lemma (19.5.1) we have in fact

$$\hat{f}|_K \cdot \hat{A}_K = (\ker \varphi)_K.$$

So we can apply Gleason's theorem (1.5.5) to the uB-algebra $\hat{A}_K$, φ and $\hat{f}|_K$. It follows that there are an open set $V \subset \mathbb{C}$ and an open neighbourhood U of φ in $M(\hat{A}_K)=K$ such that

i) $\hat{f}: U \longrightarrow V$ is a homeomorphism

and

ii) $\hat{g} \circ \hat{f}^{-1} \in Hol(V)$ for all $g \in \hat{A}_K$.

By (4.3.7) we can w.l.o.g. assume that U is a hemicompact and $\hat{A}$-convex set, i.e. $M(\hat{A}_U) = U$ by (4.3.4)iii). It follows that the map

$$\hat{A}_U \longrightarrow Hol(V),\ g \longrightarrow \hat{g} \circ \hat{f}^{-1},$$

is a continuous and injective algebra homomorphism. By (6.1.2) this map is even surjective and our assertion follows from the open mapping theorem.

□

<u>Remark.</u> Note that in general (19.5.3) becomes false if φ has no compact neighbourhood. In example (19.4.3) $\ker \varphi = (z_1)$ for $\varphi=(0,0)$ and the first coordinate function, cf. (19.4.3)vii) , but there is no neighbourhood U of φ such that z_1 is injective on U or such that $\hat{A}_U$ is a Stein algebra.

<u>(19.5.4) Theorem.</u> Let $A \neq C(\{p\})$ be a uF-algebra with locally compact and connected spectrum M(A). Then the following statements are equivalent:

i) A is algebraically and topologically isomorphic to Hol(X), X a Riemann surface;

ii) $\ker \varphi$ is a principal ideal for all $\varphi \in M(A)$.

<u>Proof.</u> ii) $\Rightarrow$ i) Let $\varphi \in M(A)$ be a point with $\ker \varphi = (f_\varphi)$. Choose an open neighbourhood U_φ of φ in M(A) and an open set $V_\varphi \subset \mathbb{C}$ according to (19.5.3). Now we conclude as in the proof of (17.1.6) that the

family $(U_\varphi, f_\varphi, V_\varphi)_{\varphi\in M(A)}$ defines an atlas on M(A) such that M(A) becomes a Riemann surface and such that $\widehat{A} = Hol(M(A))$.

ii) ⇒ i) This follows by [FOR 1], Satz 5.4. □

APPENDIX A

SUBHARMONIC FUNCTIONS, POISSON INTEGRAL

We collect some results which are needed in the text. For the proofs and further results cf. for example the books of Narasimhan [NAR 1] and Garnett [GAR].

A.1 Definition. Let X be a metric space, and let $u: X \longrightarrow \mathbb{R} \cup \{-\infty\}$ be a map. Then u is called upper semi-continuous (usc.) if, for any $\lambda \in \mathbb{R}$, the set

$$\{x \in X: u(x) < \lambda\}$$

is open in X.

A.2 Let X be a metric space, and let u be an usc. function on X. Suppose that u is bounded above. Then there exists a sequence $(u_n)_n$ of continuous functions on X such that

$$u_{n+1}(x) \leq u_n(x) \text{ for all } n \geq 1,\ x \in X$$

and

$$u_n(x) \longrightarrow u(x) \text{ as } n \longrightarrow \infty.$$

A.3 Let $G \subset \mathbb{C}$ be an open set. Denote by $C^2(G)$ the set of all twice continuously differentiable complex-valued functions on G.

Definition. We say that $u \in C^2(G)$ is harmonic if

$$\frac{\partial^2 u}{\partial x^2} + \frac{\partial^2 u}{\partial y^2} \equiv 0 \text{ on } G.$$

We remark that if $f \in \mathrm{Hol}(G)$, then real- and imaginary part of f are harmonic functions on G. Vice versa if f is a harmonic function on an open disc $D \subset \mathbb{C}$, then there exists $g \in Hol(D)$ such that Re g = f on D, where Re g denotes the real part of g.

A.4 Harmonic functions satisfy the mean value property: If f is harmonic on the open set G and the disc $\{z: |a-z|\leq R\}$, R>0, is contained in G, then

$$f(a) = 1/2\pi \int_0^{2\pi} f(a+R\exp(it))\, dt.$$

A.5 If $\overline{D}$ is a closed disc in $\mathbb{C}$ and f is a continuous function on $\partial\overline{D}$, the boundary of $\overline{D}$, then there exists a continuous function h on $\overline{D}$, which is harmonic on the open disc D, such that $h|_{\partial\overline{D}} = f$.

Combining this result with A.3 it is easy to show that for each real-valued function $f \in C(\partial\overline{D})$ and each $k \in \mathbb{N}$ there exists a polynomial p_k such that

$$\|f-\text{Re } p_k\|_{\partial\overline{D}} < 1/k.$$

<u>A.6 Definition.</u> Let u be an usc. function on an open set $G \subset \mathbb{C}$. Then u is called subharmonic on G provided that the following condition is satisfied:

Let U be a relatively compact open subset of G, and let h be a continuous real-valued function on $\overline{U}$, which is harmonic on U. Then if $u(z)\leq h(z)$ for all $z \in \partial U$ we have $u(z)\leq h(z)$ for all $z \in U$.

A.7 A theorem of Littlewood states that if u is an usc. function on an open set $G \subset \mathbb{C}$ then u is subharmonic, if for every $a \in G$, there exists a sequence $(r_n)_n$ of positive numbers such that $r_n \to 0$ as $n \to \infty$ and

$$u(a) \leq 1/2\pi \int_0^{2\pi} u(a+r_n\exp(it))dt \quad \text{for all } n\geq 1.$$

A.8 The next theorem is due to Hartogs, for a proof see [AUP 2],p. 174:

Let $D \subset \mathbb{C}$ be an open disc, and let $f: D \longrightarrow \mathbb{C}$ be a bounded function such that $\log|f-a|$ is subharmonic on D for every $a \in \mathbb{C}$. Then either f or the conjugate of f is holomorphic on D.

A.9 We now consider the Poisson kernel for the open unit disc D.

<u>Definition.</u> We define the Poisson kernel for D to be the function

$$D \longrightarrow \mathbb{R},\ r\exp(it) \longrightarrow (1-r^2)/(1+r^2-2r\cos(t)) =$$

$$= (1-r^2)/|\exp(it)-r|^2,\ 0\le r\le 1,\ 0\le t<2\pi.$$

A.10 We collect some properties of P. First note that $P(z)>0$ for all $z \in D$. Furthermore we have

i) $1/2\pi \int_0^{2\pi} P(r\exp(it))\, dt = 1,\ 0\le r<1,$

ii) Let $0<\delta<\pi/2$. Then, for $\delta\le t\le 2\pi-\delta$, we have

$0 < P(r\exp(it)) \le (1-r^2)/(1-(\cos(\delta))^2)$ for $0 \le r \le 1$.

In particular, $P(r\exp(it)) \to 0$ as $r \to 1$, uniformly for $\delta\le t\le 2\pi-\delta$.

A.11 Let h be a (Lebesgue) integrable complex-valued function on $[0,2\pi]$ such that $h(0)=h(2\pi)$. Set

$$H(r\exp(i\vartheta)) = 1/2\pi \int_0^{2\pi} P(r\exp(i(\vartheta-t)))h(t)\, dt.$$

Then H is harmonic on D. Furthermore, if h is continuous at $t_o \in [0,2\pi]$, then $H(z) \to h(t_o)$ as $z \to \exp(it_o)$.

A.12 Let $f \in \mathrm{Hol}(D)$ be a bounded function. Then for almost all $t \in [0,2\pi]$ the non-tangential limit

$$f^*(\exp(it)) = \lim_{z \to \exp(it)} f(z)$$

exists, i.e. the limit exists if z approaches exp(it) through the angle between two rays emanating from exp(it) which are not tangential to the unit circle and which approach exp(it) through D.

A.13 Let f, f^* be as in A.12, then

$$f(r\exp(i\vartheta)) = 1/2\pi \int_0^{2\pi} P(r\exp(i(\vartheta-t))f^*(\exp(it))\,dt.$$

A.14 Definition. i) Let $G \subset \mathbb{C}^n$ be an open set, and let $f:G \longrightarrow \mathbb{R}\cup\{-\infty\}$ be an usc. function. We say that f is plurisubharmonic if for each complex line $L = \{az+b\} \subset \mathbb{C}^n$, the function

$$\{z \in \mathbb{C}: az+b \in G\} \longrightarrow \mathbb{R}\cup\{-\infty\},\ z \longrightarrow f(az+b),$$

is subharmonic.

ii) A set $E \subset G$ is called pluripolar if there exists a plurisubharmonic function f on G, $f \not\equiv -\infty$, such that $f(x) = -\infty$ for all $x \in E$.

Remark. If E is a pluripolar set, then $\lambda_{2n}(E) = 0$, where λ_{2n} denotes the Lebesgue maesure in $\mathbb{R}^{2n}$, cf. [G/L], p. 234.

APPENDIX B

FUNCTIONAL ANALYSIS

We state three theorems which we often used in this book.

B.1 Open mapping theorem. If E, F are Fréchet spaces and $T: E \longrightarrow F$ is a continuous, linear and surjective mapping, then T is an open mapping, i.e. T(U) is open in F if U is open in E. In particular T is a homeomorphism if T is bijective.

B.2 Closed graph theorem. If E, F are Fréchet spaces and $T: E \longrightarrow F$ is a linear mapping, then T is continuous if its graph $\{(x,T(x)): x \in E\}$ is closed in ExF.

B.3 A subset E of a topological space X is called nowhere dense if interior $\overline{E} = \emptyset$, where $\overline{E}$ denotes the closure of E.

Baire category theorem. Let X be a complete metrizable space. Then the countable union of nowhere dense subsets of X contains no interior point.

Equivalently: Let $(U_i)_i$ be a sequence of dense open subsets of X. Then $\bigcap U_i$ is a dense subset of X.

LIST OF SYMBOLS

$\mathbb{C}$, $\mathbb{R}$, $\mathbb{N}$, complex, real, natural numbers

$\mathbb{N}_o = \mathbb{N} \cup \{0\}$

Re z, Im z denote the real resp. the imaginary part of a point $z \in \mathbb{C}$

∂K denotes the boundary, $\overline{K}$ denotes the closure of a set,

REFERENCES

[A/F] Aron, R., and Fricke, G., Homomorphisms on C(ℝ), Amer. Math. Monthly 93 (1986), 555.

[ARE 1] Arens, R., Dense inverse limit rings, Michigan Math. J. 5 (1958), 169-182.

[ARE 2] —— , The analytic-functional calculus in commutative topological algebras, Pacific J. Math. 11 (1961), 405-429.

[ARE 3] —— , The group of invertible elements of a commutative Banach algebra, Studia Math. Zeszyt (Ser. Specjalna) 1 (1963), 21-23.

[ARE 4] —— , Weierstrass products for inverse limits of Banach algebras, Math. Ann. 163 (1966), 115-160.

[ARE 5] —— , The limit of a sequence of squares in an algebra need not be a square, Pacific J. Math. 137 (1989), 15-18.

[ARE 6] —— , A generalization of normed rings, Pacific J. Math. 2 (1952), 455-471.

[AUP 1] Aupetit, B., Analytic multivalued functions in Banach algebras and uniform algebras, Adv. in Math. 44 (1982), 18-60.

[AUP 2] —— , Propriétés spectrales des algèbres de Banach (Lecture Notes in Math. 735, Springer, Heidelberg, 1979).

[BAS] Basener, R., A generalized Shilov boundary and analytic structure, Proc. Amer. Math. Soc. 47 (1975), 98-104.

[BEK] Bekken, O. B., et al., eds.,Spaces of analytic functions, Kristiansand, Norway 1975 (Lecture Notes in Math. 512, Springer, Heidelberg, 1976)

[B/F] Bierstedt, K.-D., and Fuchssteiner, B., eds., Functional analysis: Surveys and recent results III (Notas Mat. 94, North-Holland, Amsterdam, 1984).

[B/H] Bear, H. S., and Hile, G. N., Analytic structure in function algebras, Houston J. Math. 5 (1979), 21-28.

[BIR 1] Birtel , F. T., Singly-generated Liouville F-algebras, Michigan Math. J. 11 (1964), 89-94.

[BIR 2] —— , ed., Function algebras (Scott-Foresman, Chicago, 1965).

[BIR 3] —— , Some aspects of function algebras, Nieuw Arch. Wisk. (3) 42 (1969), 171-188.

[B/L] —— , and Lindberg Jr., J. A., A Liouville algebra of non-entire functions, Studia math. 25 (1964), 27-37.

[BIS 1] Bishop, E., Holomorphic completions, analytic continuations and the interpolation of seminorms, Ann. of Math. 78 (1963), 468-500.

[BNS] Beckenstein, E., Narici, L., Suffel C., Topological algebras (Notas Mat. 60, North-Holland, Amsterdam, 1977).

[BRI] Briggs J. M., Finitely generated ideals in regular F-algebras, Pacific J. Math. 61 (1975), 339-350.

[BRO 1] Brooks, R. M., On Arens' "Weierstrass products" paper, Math. Ann. 190 (1971), 329-337.

[BRO 2] —— ,Partitions of unity in F-algebras, Math. Ann. 177 (1968), 265-272.

[BRO 3] —— , On the spectrum of finitely-generated locally m-convex algebras, Studia Math. 29 (1969), 143-150.

[BRO 4] Brooks, R.M., Boundaries for natural systems, Indiana Univ. Math. J. 20 (1971), 865-875.

[BRO 5] — , Analytic structure in the spectra of certain uF-algebras, Math. Ann. 240 (1979), 27-33.

[BRO 6] — , Analytic structure in the spectrum of a natural system, Pacific J. Math. 49 (1973), 315-334.

[BRO 7] — , On singly-generated locally m-convex algebras, Duke Math. J. 37 (1970), 529-536.

[BRO 8] — , Boundaries for locally m-convex algebras, Duke Math. J. 34 (1967), 103-116.

[BRO 9] — , Boundaries for natural systems II, Math. Scand. 30 (1972), 281-289.

[BRO 10] — , ed., Symposium on several complex variables, Park City, Utah, 1970 (Lecture Notes in Math. 184, Springer, Heidelberg, 1971).

[BRO 11] — , On representing F^*-algebras, Pacific J. Math. 31 (1971), 51-69.

[BRO 12] — , On the spectrum of an inverse limit of holomorphic function algebras, Adv. in Math. 19 (1976), 238-243.

[BRW] Browder, A., Introduction to function algebras (Benjamin, New York and Amsterdam, 1969).

[B/S] Behnke, H., and Sommer, F., Theorie der analytischen Funktionen einer komplexen Veränderlichen (Grundlehren Math. Wiss. 77, Springer, Heidelberg, 1972).

[CAR 1] Carpenter, R. L., Uniqueness of topology for commutative semisimple F-algebras, Proc. Amer. Math. Soc. 29 (1971), 113-117.

[CAR 2] — , Singly generated homogeneous F-algebras, Trans. Amer. Math. Soc. 150 (1970), 457-469.

[CAR 3] Carpenter, R.L., Principal ideals in F-algebras, Pacific J. J. Math. 35 (1970), 559-563.

[CAR 4] —— , Accessible points in the spectrum of a Fréchet algebra, Proc. Amer. Math. Soc. 63 (1977), 101-106.

[CAR 5] —— , Continuity of derivations on F-algebras, Amer. J. Math. 93 (1971), 500-502.

[CAR 6] —— , Characterizing hol(Ω), Proc. Amer. Math. Soc. 53 (1975), 57-59.

[CAR 7] —— , Continuity of systems of derivations on F-algebras, Proc. Amer. Math. Soc. 30 (1971), 141-146.

[CAR 8] —— , Principal ideals in F-algebras, Pacific J. Math. 35 (1970), 559-563.

[CLA 1] Clayton, D., A reduction of the continuous homomorphism problem for F-algebras, Rocky Mountain J. Math. 5 (1975), 337-344.

[CLA 2] —— , Local analytic structure in Banach algebras, Indiana Univ. Math. J. 20 (1971), 507-513.

[CRA 1] Craw, I.G., Factorization in Fréchet algebras, J. London Math. Soc. 44 (1969), 607-611.

[CRA 2] —— , A condition equivalent to the continuity of characters on a Fréchet algebra, Proc. London Math. Soc. 22 (1971), 452-464.

[C/S] Corach, G., and Suarez, D., Extension of characters in commutative Banach algebras, Studia math. 85 (1987), 199-202.

[DAL] Daly, J.T., On Liouville F-algebras, Studia Math. 40 (1971), 1-16.

[D/E] Dixon, P.G., and Esterle, J., Michael's problem and the

Poincaré-Fatou-Bieberbach phenomenon, Bull. Amer. Math. Soc. (NS) 15 (1986), 127-187.

[D/S] Diederich, K., Sibony N.: Strange complex structures on euclidean space, J. Reine Angew. Math. 311/312 (1979), 397-407.

[DOR] Dors, A.G., A Fréchet algebra example, Proc. Amer. Math. Soc. 64 (1977), 62-64.

[DUG] Dugundji, J., Topology (Allyn and Bacon, Inc., Boston, 1966).

[EPH] Ephraim, R., Multiplicative linear functionals on Stein algebras, Pacific J. Math. 78 (1978), 89-93.

[EST] Esterle, J., Mittag-Leffler methods in the theory of Banach algebras and a new approach to Michael's problem, in [G/G], 107-129.

[FOR 1] Forster, O., Zur Theorie der Steinschen Algebren und Moduln, Math. Z. 97 (1967), 376-405.

[FOR 2] —— , Holomorphiegebiete, in Behnke, H., and Thullen, P., Theorie der Funktionen mehrerer komplexer Veränderlichen, (Ergeb. Math. Grenzgeb. 51, Springer, Heidelberg, 1970), 134-170.

[FOR 3] —— , Primärzerlegung in Steinschen Algebren, Math. Ann. 154 (1964), 307-329.

[FOR 4] —— , Uniqueness of topology in Stein algebras, in [BIR 2], 157-163.

[GAM] Gamelin, Th. W., Uniform algebras (Prentice Hall, Englewood Cliffs, 1969).

[GAR] Garnett, J.B., Bounded analytic functions (Academic Press, New York, 1981).

[G/G] Greenleaf, F., and Gulick, D., eds., Proceedings of the conference on Banach algebras and several complex variables, Contemp. Math. 32 (Amer. Math. Soc.),1984.

[G/L] Gruman, L., and Lelong, P., Entire functions of several complex variables (Grundlehren Math. Wiss. 282, Springer, Heidelberg, 1986).

[GOL 1] Goldmann, H., A uniform F-algebra with locally compact spectrum and with non-global peak points, Studia Math. 86 (1987), 273-275.

[GOL 2] —— , Finitely generated maximum modulus algebras, Ann. Polon. Math. 47 (1986), 149-153.

[GOL 3] —— , A remark on functional continuity of certain Fréchet algebras, to appear in Studia Math. 94.

[GOL 4] —— , An analytic algebra without analytic structure in the spectrum, Ark. Math. 27 (1989), 89-95.

[GOL 5] —— , Antisymmetry and analytic structure in the spectrum of a uniform Fréchet algebra, Manuscripta Math. 59 (1987), 229-244.

[GKV] —— , Kramm, B., and Vogt, D., Reflexive function algebras, in [B/F], 253-264.

[G/R] Gunning, R.C., and Rossi, H., Analytic functions of several complex variables (Prentice Hall, Englewood Cliffs, 1965).

[GR/R 1] Grauert, H., and Remmert, R., Komplexe Räume, Math. Ann. 136 (1958), 245-318.

[GR/R 2] Grauert, H., and Remmert, R., Theory of Stein spaces (Grundl. 236, Springer, Heidelberg, 1979).

[GUI] Guichardet, A., Special topics in topological algebras (Gordon and Breach, New York, 1968).

[HAY 1] Hayes, S., Maximal simultaneous continuations in complex analysis, Exposition. Math. 3 (1983), 261-273.

[Hay 2] —— , The weak Nullstellensatz for finite dimensional complex spaces, Pacific J. Math. 99 (1982), 45-56.

[HAY 3] —— , A holomorphically separable complex space without the Gelfand topology, Enseignement Math. (2) 31 (1985), 147-149.

[H/V 1] Hayes, S., and Vigué, J.-P., A complex space whose spectrum is not locally compact anywhere, Proc. Amer. Math. Soc. 98 (1986), 271-275.

[H/V 2] —— , Remark on the k-space problem for complex spaces, Resultate Math. 11 (1987), 229-238.

[HÖR] Hörmander, L., An introduction to complex analysis in several variables (North-Holland, Amsterdam, 1973).

[HOR] Horvath, J., Topological vectorspaces and distributions (I. Reading, Mass, 1966).

[HUS] Husain T., Multiplicative functionals on topological algebras (Research Notes in Math. 85, Pitmann Publishing, Boston, 1983).

[H/W] Heal, E.R., and Windham, M.P., Finitely generated F-algebras with applications to Stein manifolds, Pacific J. Math. 51 (1974), 459-465.

[JÄR] Järvi, P., A note on maximum modulus algebras, Ann. Acad. Sci. Fenn., Ser. A 1 Math. 6 (1981), 173-178.

[JAR] Jarchow, H., Locally convex spaces (Teubner, Stuttgart, 1981).

[JOH] Johnson, B.E., Continuity of derivations on commutative algebras, Amer. J. Math. 91 (1969), 1-10.

[KEL] Kelley, J.L., General topology (Graduate Texts in Math. 27, Springer, Heidelberg)

[KEO] Keown, E.R., Reflexive Banach algebras, Proc. Amer. Math. Soc. 6 (1955), 252-259.

[K/K] Kaup, B., and Kaup, L., Holomorphic functions of several variables (Walter de Gruyter, Berlin, 1983).

[KÖT] Köthe, G., Topological vector spaces I (Grundlehren Math. Wiss. 159, Springer, Heidelberg, 1969).

[KRA 1] Kramm, B., Complex analytic properties of certain uniform Fréchet-Schwartz algebras, Stud. Math. 66 (1978), 155-167.

[KRA 2] —— , A functionalanalytic characterization of pure-dimensional and regular Stein algebras, Adv. in Math. 42 (1981), 196-216.

[KRA 3] —— , Eine funktionalanalytische Charakterisierung Steinscher Algebren, Habilitationsschrift Frankfurt (1977).

[KRA 4] —— , A characterization of Riemann algebras, Pacific J. Math. 65 (1976), 393-397.

[KRA 5] —— , Analytische Struktur in Spektren - ein Zugang über die ∞-dimensionale Holomorphie, J. Func. Anal. 37 (1980),

249-270.

[KRA 6] Kramm, B., Nuclearity (resp. Schwarzity) helps to embed holomorphic structure into spectra - a survey, in [G/G], 143-162.

[KRA 7] —— , Gleason parts and boundary value integrals for certain Fréchet-nuclear algebras, Michigan Math. J. 24 (1977), 129-139.

[KRA 8] —— , Nuclearity and function algebras - a survey, in [B/F], 233-252.

[KRA 9] —— , Ein Approximationssatz für holomorphe Funktionen auf nicht-kompakten Riemannschen Flächen, Manuscripta Math. 26 (1978), 293-313.

[KRN] Krantz, S.G., Function theory of several complex variables (Wiley,1982).

[KUM 1] Kumagai, D., Maximum modulus algebras and multi-dimensional analytic structure, in [G/G], 163-168.

[KUM 2] —— , Plurisubharmonic functions associated with uniform algebras, Proc. Amer. Math. Soc. 87 (1983), 303-308.

[LAR] Larsen, R., Banach algebras, an introduction (Dekker, New York, 1973).

[LEV] Levenberg, N., On an example of Wermer, Ark. Math. 26 (1988), 155- 163.

[LOY] Loy, R.J., Local analytic structure in certain commutative topological algebras, Bull. Austral. Math. Soc. 6 (1972), 161-167.

[MAL] Mallios, A., Topological algebras, selected topics (Notas

Mat. 109, North-Holland, Amsterdam, 1986).

[MEY 1] Meyers, W.E., A uniform algebra with non-global peak points, Studia Math. 33 (1969), 207-211.

[MEY 2] —— , Montel algebras on the plane, Canad. J. Math. 22 (1970), 116-122.

[MIC] Michael, E.A., Locally m-convex topological algebras, Mem. Amer. Math. Soc. 11 (1952).

[NAG] Nagata, M., Local rings (Krieger Publishing Company, Huntington, 1975).

[NAR 1] Narasimhan, R., Complex analysis in one variable (Birkhäuser, Basel, 1985).

[NAR 2] —— , Introduction to the theory of analytic spaces (Lecture Notes in Math. 25, Springer, Heidelberg, 1966).

[REM] Remmert, R., Isomorphiesätze von Frobenius, Hopf und Gelfand-Mazur, in Ebbinghaus H.-D., et al., Zahlen (Grundwissen Mathematik 1, 2. Auflage, Springer, Heidelberg, 1988) 197-202.

[RIC 1] Rickart, C.E., Holomorphic convexity for general function algebras, Canad. J. Math. 20 (1968), 272-290.

[RIC 2] —— , Natural function algebras (Springer, Heidelberg, 1979).

[RIC 3] —— , General theory of Banach algebras (Krieger Publishing Company, Huntington, 1960).

[ROI] Rossi, H., Analytic spaces with compact subvarieties, Math. Ann. 146 (1962), 129-145.

[ROS] Rosenfeld, M., Commutative F-algebras, Pacific J. Math. 16 (1966), 159-166.

[RUS] Rusek, K., Analytic structure on locally compact spaces determined by algebras of continuous functions, Ann. Polon Math. 42 (1983), 301-307.

[SCH] Schottenloher, M., Michael's problem and algebras of holomorphic functions, Arch. Mat. 37 (1981), 241-247.

[SIB] Sibony, N., Multidimensional analytic structure in the spectrum of a uniform algebra, in [BEK], 139-165.

[SEN 1] Senichkin, V., The maximality of the algebra of functions that are continuous on the extended plane and analytic in the complement of a totally disconnected compactum, English translation in J. Soviet Math. 2 (1974), 166-173.

[SEN 2] —— , Subharmonic functions and the analytic structure in the maximal ideal space of a uniform algebra, Math. USSR-Sb. 36 (1980), 111-126.

[STO] Stout, E.L., The theory of uniform algebras (Bogden & Quigley, New York, 1971).

[S/W] Stewart J., Watson S., Topological algebras with finitely-generated bases, Math. Ann. 271 (1985), 315-318.

[WAE] Waelbroeck, L., Le calcule symbolique dans les algebres commutatives, J. Math. Pures Appl. 33 (1954), 147-186.

[WAR] Warner, S., The topology of compact convergence on continuous function spaces, Duke Math. J. 25 (1958), 265-282.

[WER 1] Wermer J., Polynomially convex hulls and analyticity, Ark. Math. 20 (1982), 129-135.

[WER 2] —— , Subharmonicity and hulls, Pacific J. Math. 58 (1975), 283-290.

[WER 3] —— , Potential theory and function algebras, Texas Tech. U. Math. Series 14 (1981), 113-125.

[WER 4] —— , Banach algebras and several complex variables (Graduate Texts in Math. 35, Springer, Heidelberg, 1976).

[ZAM] Zame, W., Existence, uniqueness and continuity of functional calculus homomorphisms, Proc. London Math. Soc. (3) 39 (1979), 73-92.

[ZEL 1] Zelazko, W., Selected topics in topological algebras (Aarhus Univ. Lecture Notes Series 31, 1971).

[ZEL 2] —— , Functional continuity of commutative B_0-algebras with countable maximal ideal spaces, Colloq. Math. 51 (1987), 395-399.

[ZEL 3] —— , Banach algebras (Polish Scientific Publishers, Warszawa, 1973).

INDEX